T0358635

Total Quality Management and Lean Thinking 5.0

In the era of Industry 4.0, the quality management paradigm is undergoing a dramatic transformation. The manufacturing and service industries are rapidly evolving, and businesses need to be agile and adaptive to stay competitive. *Total Quality Management and Lean Thinking 5.0: Theories and Methods* offers an integrated approach to quality management that combines the principles of Total Quality Management (TQM) and Lean Thinking.

Covering vital topics including Lean 4.0, Lean Six Sigma, problem solving, statistical tools, managerial tools, Quality Function Deployment (QFD), risk management and customer analysis, the authors also offer insight into possible and probable future directions. A dedicated chapter of case studies centred on TQM issues furnished the reader with rich in-depth examples with which to advance and inform their understanding of TQM.

Total Quality Management and Lean Thinking 5.0: Theories and Methods is an ideal textbook for quality management courses at the undergraduate or graduate level, and can also be used as a reference by managers, quality professionals, engineers, process improvement specialists, Six Sigma practitioners, engineers, data analysts, students studying quality management or related fields and anyone interested in learning about the latest concepts and tools of quality management.

Giuseppe Ioppolo is Full Professor of Commodity Science (SECS-P/13). He has been teaching Lean Management and TQM for 15 years and has several articles and research papers on Quality Management, Lean Management, Circular Economy and Sustainability. He has extensive experience in project management too. He also serves as Editor-in Chief of MDPI journal, *Sustainability*.

Cristina Ciliberto received a Ph.D. in Economics, Management and Statistics and has experience working with organizations to improve their manufacturing processes. She had published articles and research papers on the topics of Quality Management, Industry 4.0 and Lean Production. She is black belt of Lean Six Sigma.

Katarzyna Szopik-Depczyńska, PhD., DSc., is Associate Professor in Management and Quality Sciences at University of Szczecin, Institute of Management (Poland) and has several papers on TQM, Business Innovation and Lean Production.

Total Quality Management and Lean Thinking 5.0

Theories and Methods

Giuseppe Ioppolo, Cristina Ciliberto and Katarzyna Szopik-Depczyńska

LONDON AND NEW YORK

First published 2025
by Routledge
4 Park Square, Milton Park, Abingdon, Oxon OX14 4RN

and by Routledge
605 Third Avenue, New York, NY 10158

Routledge is an imprint of the Taylor & Francis Group, an informa business

British Library Cataloguing-in-Publication Data
A catalogue record for this book is available from the British Library

ISBN: 978-1-032-72675-5 (hbk)
ISBN: 978-1-032-72673-1 (pbk)
ISBN: 978-1-032-72674-8 (ebk)

DOI: 10.4324/9781032726748

Typeset in Times New Roman
by Apex CoVantage, LLC

Contents

Figures and Tables

Figures

Tables

Introduction

The concept of quality is a transversal concept that embraces the entire production organization. The organization must investigate to understand the market and identify the characteristics that its goods and services must possess to meet the needs of potential consumers. These characteristics must be translated into specifications, and then products conforming to them must be obtained. The end consumer is not just a stakeholder but has become the main recipient of quality over the years. Today, quality is subject to increasingly stringent evaluations by final consumers, given the development of emerging technologies and the ecological transition. This implies that an organization must not only comply with international standards but also develop competitive business models on the market that keep pace with the ecological and technological transition if it wants to survive in the market.

In particular, this volume will examine Total Quality Management (TQM) and Lean thinking, which are two managerial approaches aimed at optimization, which is why many tend to consider them the same thing. TQM is nothing more than the conscious management of Quality at every level of activity, from planning and designing the product or service to monitoring work to find ways to improve it. It focuses mainly on process control and customer satisfaction through activities such as statistical control, supply control, and quality design. It is a concept that originated in academic environments (Feigenbaum, Juran, and Deming) and had among its greatest expert scholars.

Lean thinking represents a philosophy that is based on the idea that all resources made available should be spent on providing value to the customer. Everything that does not contribute to adding value should ideally be eliminated because it is considered a waste. More value with fewer resources, therefore, an idea that began to take hold in a reality like Toyota, which applied it in the field and had among its greatest disseminators Womack with his famous book *The Machine that Changed the World.*

TQM has, among its main tools, Kaizen, as we can see well visualized in the Deming cycle. But also process mapping, definition of work methodologies, and statistical control are supporting tools of this methodology. TQM also emphasizes the importance of good people management, leadership, teamwork, training, and worker involvement because technical and human aspects cannot be separated if we truly seek excellence.

Total Quality Management is a managerial standpoint that emphasizes:

- continuous improvement;
- respect for customer requirements;
- reduction of rework;
- involvement of people in the work they do;

- attention to collaboration;
- redesign of processes in an improvement perspective;
- growth through comparison (for example with the benchmarking tool);
- problem-solving;
- constant measurement of results achieved;
- close relationships with suppliers.

It can be summarized with:

- attention to the customer and their satisfaction because it is the customer who "drives" our work;
- continuous improvement instead of criticism towards those who have worked poorly because it is necessary to understand why the process has allowed these people to work poorly;
- teamwork;
- deciding based on the collected data.

So, what differs from Lean thinking, which, as we know, uses the cited tools and focuses on very similar points? The first big difference is that Lean thinking focuses on the entire value stream while TQM and many other improvement methodologies tend to focus on one process at a time. The second major difference is that TQM focuses on improving productivity and process efficiency while Lean thinking seeks to eliminate non-value-added activities.

Lean thinking consists of constant control of resource optimization in the perspective of customer satisfaction and waste reduction. It can be summarized as:

- its starting point is the customer-value-creation;
- it seeks to eliminate waste;
- it focuses on the continuous improvement of processes.

Table 0.1 Similarities and differences between the two methodologies.

Topic	*TQM*	*Lean Management*
Approach	Quality; attention to the customer	Customer value; waste elimination
Processes	Process improvement	Flows improvement
Improvement	Continuous improvement	Continuous improvement
Main concepts	Data-based decisions; employee engagement; customer-centric approach	Continuous improvement of value delivered to customers; participation from everyone
	All, suppliers included	All, suppliers included
Methodologies	PDCA cycle; statistical tools	Analytical tools
Primary effects	Increase in customer satisfaction	Reduction of lead time
Secondary effects	Customer loyalty	Inventory reduction; increase in productivity; increase in customer satisfaction
Change	Slow and incremental	Fast or slow, depending on how you choose to intervene
Time needed for implementation	Long because there is a lot to learn	Long because there is a lot to learn
Critiques	Subtle improvements; resource requirements	It can cause congestion within the supply chain

Source: authors' own elaboration

Therefore, in this volume, both managerial approaches affected by the emerging technologies will be examined. Indeed, the advent of technological innovation, digitalization, and the Industry 4.0 revolution have completely transformed the way businesses are managed and have played a vital role in achieving sustainability objectives (Piccarozzi et al., 2023).

Organizations that have adopted Quality 4.0 are equipped to make data-driven decisions and are more flexible in their operations. While Industry 4.0 represents the current technological paradigm of efficient and intelligent machines, Industry 5.0 embodies the forthcoming technologies that emphasize collaboration between humans and machines to achieve progress. Unlike Industry 4.0, which places a heavy emphasis on technology, Industry 5.0 is revolutionizing the way we think by acknowledging that progress depends on the collaboration, between humans and machines. By utilizing personalized products, the industrial revolution has led to improved customer satisfaction. In today's business environment, Industry 5.0 is necessary to gain a competitive advantage and promote economic growth for smart factories (Adel, 2022). Nahavandi (2019) provides a definition of Industry 5.0 which involves close collaboration between human workers and machines within a factory to enhance the efficiency of processes by harnessing the thinking and cognitive abilities of humans while integrating workflows, with systems. Similarly, according to Friedman and Hendry (2019) Industry 5.0 will require professionals, in business, information technology and philosophy to prioritize aspects when introducing technologies in industrial systems. In essence Industry 5.0 can be seen as an era where factories become socially intelligent with robots collaborating with humans through conversations and enterprise social networks facilitating communication between workers and components of cyber physical production systems (CPPS).

As opposed to Industry 4.0 which is characterized by the digitalization and automation of processes the emerging Industry 5.0 places an emphasis on centricity (Golovianko et al., 2023). This shift can be attributed to the challenges faced by societies today including issues like climate change on a scale pandemics, hybrid warfare alongside conventional conflicts and refugee crises. To establish resilient processes in response to these challenges it becomes crucial for humans to play an active role, in organizational decision making.

Henceforth, it is imperative that we duly consider the concepts of Quality 4.0 and Lean 4.0, as well as their potential progression towards Quality 5.0 and Lean 5.0. Quality 4.0 signifies the advancement of the antiquated quality approach, wherein the scope of its processes is expanded, necessitating new capabilities from companies and their operators. While Connectivity, Data, and Analytics undoubtedly serve as indispensable prerequisites for the notion of Quality 4.0, it is important to note that they alone are not sufficient conditions. In order to fully embrace a truly modern approach, it is imperative to incorporate additional elements alongside the aforementioned prerequisites. **Collaboration**, for instance, plays a pivotal role in the Quality 4.0 approach, which is inherently cross-functional and global in nature. It is worth noting that relying on outdated communication methods such as email and paper-based systems hinders progress, whereas utilizing communication and collaboration methods that enable real-time operation between machines and humans is far more effective.

Furthermore, **scalability** is a key characteristic of Quality 4.0 systems. These systems possess an exceptional ability to efficiently handle substantial quantities of data, even when faced with unforeseen surges in growth. It is important to acknowledge that the fragmentation of information systems currently poses a significant challenge to the integration between machines. It becomes evident that integration leads to an increase in the volume of data that needs to be processed.

Quality Management Systems – Despite the considerable progress that markets have made in recent years in the adoption of EQMS (Electronic Quality Management Systems), it is worth noting that many companies, not only those in Italy, still lag behind. The existence of such systems is of utmost importance as they enable the collection of data from interconnected systems, which can then be harmonized with the control system and subsequently interpreted. It is crucial to recognize that Quality 4.0, an approach primarily centred around Data and Analytics, as well as Connection, relies heavily on the presence of these management systems. Without them, the fundamental principles of Quality 4.0 lose their significance and consequently devalue the entire approach.

Compliance – In an increasingly regulated world, Compliance stands as one of the primary objectives for Quality Managers and CEOs alike. Its purpose is to ensure the utmost reliability of products and mitigate the risks associated with damaging the reputation of the company. The implementation of the Quality 4.0 approach allows for the automation of various Compliance-related checks, thereby reducing the costs associated with these checks and minimizing the likelihood of regulatory violations. By embracing this approach, companies can effectively address the challenges posed by regulatory requirements while maintaining their commitment to quality and compliance.

Quality 4.0 Culture – a true Quality 4.0 approach can only be achieved through a deep Quality culture that is inherent in the company's DNA and in the professional profiles of all Managers and employees. This is especially due to the cross-functional nature typical of the Quality 4.0 approach and also because when working on a Quality 4.0 plan, Quality processes become more visible, connected, and relevant. The Culture theme is undeniably a challenging aspect that numerous companies encounter hurdles with, particularly when it comes to fostering a shared understanding among cross-functional teams regarding the profound influence that Quality has on the overall success of the strategy.

Effective leadership is crucial for the success of corporate initiatives, especially within Quality 4.0. This modern approach surpasses the traditional methods by transitioning from a functional system to a cross-functional system, ultimately culminating in a leadership system. This advancement stands out for its exceptional capacity to harmoniously blend diverse functions and foster a leadership-driven culture.

In the context of **Competence 4.0**, it is worth noting that less mature organizations heavily rely on the knowledge and experiences of individual employees, which are shared on a voluntary basis. However, in 4.0 systems, the exchange and transmission of skills are deliberately structured and supported by management systems. In fact, a staggering 50% of companies consider training to be the most critical process for their success. This highlights the significance of a well-structured and enforced approach to skill development in achieving organizational goals.

It is of utmost importance to duly consider that **Lean 4.0** signifies the natural progression of the Toyota Production System Lean methodology. This evolution is characterized by its seamless alignment with the principles of One to One marketing logic, which, in turn, is facilitated by the utilization of enabling technologies. Furthermore, it is worth noting that the implementation of Lean 4.0 necessitates the acquisition of new skills by the staff, as a direct result of the ongoing process of Digital transformation. In this context, the **Lean Six Sigma 4.0** methodology is capable of improving the productivity of people and facilities, the product quality, and reducing delivery times and transformation costs. Its success relies not only on technical tools but also significantly on the top management's commitment to supporting an improvement program and engaging employees. Future processes will, therefore, incorporate more technology thanks to emerging technologies and become smarter, but they will remain processes nonetheless. This observation, which may seem so trivial, is a way to highlight that the Lean Six Sigma methodology will continue to exist, despite massive automation: process capability and deviation will continue to be calculated.

Resilient and responsive production systems are the answer to our questions. They can be obtained by combining different methodologies and approaches that will be studied in this volume, while still considering the differences that characterize them.

Conversely, "Quality 5.0" typically refers to the highest level of quality achievable in a product or service. It suggests that the product or service meets or exceeds all expectations, standards, and requirements, ensuring superior performance, reliability, and customer satisfaction. Achieving a "Quality 5.0" rating often involves continuous improvement, rigorous testing, adherence to industry standards, and a commitment to excellence throughout the entire production or service delivery process. This term is commonly used in manufacturing, software development, and various other industries where quality is paramount. The difference between "Quality 5.0" and "Quality 4.0" lies in the degree of excellence and advancement achieved in the respective levels of quality.

Quality 5.0 implies the pinnacle of quality standards, reflects an even higher level of excellence compared to Quality 4.0, typically involves cutting-edge technology, state-of-the-art processes, and meticulous attention to detail. It often incorporates advanced quality management systems and methodologies to ensure superior performance and customer satisfaction and represents continuous improvement and innovation to push the boundaries of what is achievable in terms of quality.

Quality 4.0 signifies a high level of quality attainment, represents a significant advancement from previous quality standards, but may not reach the absolute peak. It involves efficient processes, effective quality control measures, and consistent adherence to standards and may incorporate elements of automation, digitization, and data-driven decision-making to enhance quality. It reflects a commitment to excellence, but there may still be room for further improvement and optimization.

In essence, while both Quality 4.0 and Quality 5.0 denote high-quality standards, Quality 5.0 represents a further evolution and refinement towards perfection, often incorporating the latest advancements and innovations in quality management and production processes.

1 Total Quality Management

Learning Objectives

- Trace the evolution of quality management and its historical development.
- Define quality and explore different perspectives on what constitutes quality.
- Familiarize yourself with the core principles of quality management.
- Comprehend the 9M Framework and its relevance in quality management.
- Explore the journey from inspection to Total Quality Management (TQM).
- Understand the ISO 9001:2015 standard and its evolution.
- Examine the eight principles of ISO 9000:2008.
- Explore ISO 14001, an international standard for environmental management, and understand its importance in promoting sustainable practices and reducing environmental impact.
- Understand ISO/IEC 27001, a standard for information security management systems, and recognize its significance in safeguarding sensitive information and ensuring data protection.
- Gain insights into the concept of Quality 5.0, which represents the future of quality management, and understand its potential impact on organizations.

The Concept of Quality

The definition of quality includes the known world of customer needs, customer expectations, the expectations that are intended to be met or "production aims", and their specifications.

Quality, as defined by UNI EN ISO 8402:1995, encompasses the qualities and characteristics of a product or service that enable it to meet expressed or implicit needs. It is the extent to which a collection of inherent attributes (distinctive elements) fulfils requirements (needs or expectations that can be expressed, typically implicit or obligatory). It is important to note that the term "quality" can be modified by adjectives such as poor, good, or excellent. Additionally, the adjective "intrinsic", in contrast to assigned, denotes that it is an enduring characteristic present within something.

The concept of quality should not only be referred to a product/service, but it can also be extended to the processes and systems that organizations use to generate a value proposition capable of satisfying the expectations of customers and stakeholders. However, it should be remembered that the concept of quality has a dynamic, rather than static nature, which follows

DOI: 10.4324/9781032726748-1

the evolution of societies and, consequently, the needs of the individuals who make them up. For this reason, quality has evolved over the centuries.

Evolution of the Concept of Quality

With the industrial revolutions, there was a so-called paradigm shift with which the economies of the most developed countries moved from being based on artisanal production to being centred on mass production of an industrial type.

Starting from 1760, the year of birth of the first industrial revolution, the first industries were born, which, thanks to important technological innovations, such as the steam engine, were able to increase their production capacity in a dizzying way.

In this type of production, qualitative results depended less and less on the abilities of individual operators and more and more on the design and formalization of production processes.

During the latter half of the nineteenth century, a significant transformation occurred as the second industrial revolution emerged. This momentous event brought about substantial advancements in industrial economies, which had a profound impact on various sectors. It is imperative to acknowledge the magnitude of this development and its far-reaching consequences.

The first industrial revolution was primarily marked by advancements in metallurgy and manufacturing, particularly due to the introduction of the steam engine. However, the second industrial revolution witnessed a broader expansion of industrialization into various sectors. Notably, the steel processing industry experienced significant growth during this period. Additionally, innovations in the energy sector, such as the emergence of oil and electricity, played a crucial role in accelerating the development of production. These advancements propelled the overall progress of industrialization, leading to substantial economic growth and societal transformation.

However, it was thanks to management innovations brought about by engineer Frederick Taylor that companies were able to maximize their production; Taylorism, which is based on the decomposition and parcelization of production processes and the assignment of highly specialized and standardized tasks to each worker, was adopted by numerous companies in those years, particularly by Henry Ford's company, which, by applying Taylor's concepts in the monograph *The Principles of Scientific Management* to his business and devising his production management model based on the assembly line, which became known as Fordism, represented the emblem of entrepreneurial philosophy of that period: maximize production.

Although in the context just described, quality did not have a primary role, and it was only applied in the testing phase, from the 1920s onwards, philosophies that gave greater importance to quality began to spread in the business world.

In particular, in this period, there was a transition from the pre-war cycle, where the concept of quantity reigned supreme, to the post-war cycle, where quality became the primary objective of companies, especially in Japanese organizations.

In 1945, Japan, which was in a devastating economic crisis, saw the concept of quality as the element to heal the nation's economic and productive system. For this reason, a new production philosophy began to spread in Japanese companies, which gave quality the role of a guiding function in order to generate better products at lower production costs.

With this new philosophy, the concept of quality was framed from a different perspective, moving from a passive approach to non-conformities to a proactive one, through which Japanese companies, through the design of a Quality System, also began to focus on incident prevention.

In the same year, Armand Feigenbaum published an article in which, describing his experience at General Electric, he first spoke of Total Quality Control, introducing a concept that would later be developed in the following decades in the Western world.

Definitions of Quality

Quality is defined in various ways. According to Robert A. Broh, it refers to achieving excellence at a reasonable cost while also managing variability within acceptable expenses.

The American Society for Quality (ASQ) defines quality as a subjective term, with each individual having their own interpretation. From a technical standpoint, quality can be understood in two ways: firstly, as the characteristics of a product or service that are capable of meeting stated or implicit needs, and secondly, as a product or service that is free from defects.

A process or product that possesses the necessary attributes to fulfil its intended purpose is commonly referred to as being of quality. This concept of quality can be applied to various processes, products, or services. However, it should be noted that quantifying quality in this manner can present challenges in terms of measurement. In the realm of production, quality is defined as a measure of excellence or the state of being free from defects and deficiencies.

In a seminal paper, Bob Garvin (1984) meticulously classified the multifaceted concept of quality into five distinct categories. These categories encompass the product-based, user-based, and value-based aspects of quality. Furthermore, Garvin astutely identified eight attributes that serve to define quality in a comprehensive manner. These attributes include performance, features, reliability, conformity, durability, ease of maintenance, aesthetics, and perceived quality.

Additionally, Joseph Juran contributed to the discourse on quality by offering a succinct definition. Juran's definition emphasizes the importance of suitability for use or purpose, which evaluates the degree to which a product aligns with its intended function. This criterion provides a valuable perspective on the concept of quality, further enriching our understanding of this complex subject matter.

In his work, Philip Crosby (1979) provides a concise definition of quality as "conformance to requirements/specifications" and emphasizes the importance of achieving "zero defects". This definition is particularly useful for quality control teams as it enables them to assess the quality of processes, systems, services, and products by evaluating their adherence to established criteria. By doing so, they can easily identify any instances of non-conformity. However, it is worth noting that this definition may present certain limitations, as it offers a subjective perspective on quality. In many cases, product requirements are determined by business stakeholders, which means that there is no objective means of validating whether these requirements truly result in quality outcomes.

Dr Genichi Taguchi, on the other hand, proposes an alternative perspective on quality. According to his viewpoint, quality can be measured by the loss incurred by society due to the variation in a product's function and any harmful effects it may have, excluding any losses resulting from its intrinsic functions.

Deming has provided a definition of quality as "continuous improvement" and "meeting or exceeding customer expectations". In the mathematical approach, quality is determined by the ratio of performance to expectation. This can be expressed as Quality (Q) = Performance (P) ÷ Expectation (A). There are three possible cases to consider:

1. If the performance (P) is greater than the expectation (A), it indicates that the quality is superior.
2. If the performance (P) is equal to the expectation (A), it indicates that the quality is satisfactory.
3. If the performance (P) is less than the expectation (A), it indicates that the quality is inferior.

Quality is often associated with several key factors. These factors include the quest for excellence, understanding customer needs, taking action to gain customer appreciation, demonstrating leadership, involving all people, fostering team spirit for a common goal, and using a yardstick to measure progress.

Traditionally, the cost of materials has been used to determine product quality. For example, a gold watch is generally considered to be of higher quality than a plastic watch.

Quality significantly influences the price customers are willing to pay for a product or service. In fact, quality is an essential component of many business models. Economists have proposed various definitions of quality, with some suggesting that it is directly linked to product cost. In other words, the higher the price of a product, the higher its perceived quality.

The manufacturing industry has been at the forefront of prioritizing quality. Manufacturers are increasingly expressing concerns about the quality of their products, which can be attributed to the quality of the manufacturing process.

If they produce a million cars a month, they cannot afford to produce lower quality products that will be returned by their customers. They cannot bear responsibility for products resulting from non-conforming products, nor can they afford inefficient processes.

Quality is a satisfying experience. In this sense, marketing professionals have developed definitions of quality that explain why customers buy services.

ASQ's 2015 *Future of Quality* report covers topics indicating that quality and performance improvement will be propelled by connectivity, intelligence, and automation. This will be made possible through the utilization of Industry 4.0 technologies, including the Internet of Things and Machine Learning. By establishing ecosystems based on these technologies, individuals and machines will collaborate to attain common objectives and utilize data to promptly generate valuable insights.

Quality Management Principles

The objective of quality management is to ensure that all activities involved in designing, developing, and implementing a product or service are efficient and effective in terms of the system and its performance. Quality management comprises three fundamental components: quality control, quality assurance, and quality enhancement. It's worth noting that quality management not only prioritizes product excellence but also the methodologies employed to attain it. In striving for more uniform quality, quality management incorporates process control, product control, and quality assurance. This comprehensive approach ensures that quality management addresses both the procedural and product aspects.

Quality improvement involves purposeful adjustments to processes to enhance the probability of achieving desired objectives. Conversely, quality control focuses on the continual preservation of consistency and dependability within a process to secure the desired outcomes.

Quality assurance, on the other hand, involves the implementation of planned or systematic actions to instil sufficient confidence that a product or service will meet the specified quality requirements.

Quality Control versus Quality Assurance

Quality control is vital in product development, aiming to identify and rectify defects while meeting all requirements. This process encompasses various activities that are carried out after the product has been developed. These activities are designed to verify that the deliverables are of satisfactory quality, fully complete, and accurate. Some of the key quality control activities include inspection, peer reviews of deliverables, and rigorous testing processes.

Table 1.1 Differences between quality control and quality assurance.

Quality Control	*Quality Assurance*
• It is a set of activities for ensuring quality in products. The activities focus on identifying defects in the actual products produced.	• It is a set of activities for ensuring quality in the processes by which products are developed.
• It aims to identify and co1Tect defects in the finished product and is a reactive process.	• It aims to prevent defects with a focus on the process used to make the product. It is a proactive quality process.
• The goal is to identify defects after a product is developed and before it is released.	• The goal is to improve development and test processes so that defects do not arise when the product is being developed.
• Finding and eliminating sources of quality problems through tools and equipment so that customer's requirements a.re continually met.	• It establishes a good quality management system and conducts assessment of its adequacy and periodic conformance audits of the operations of the system.
• The activities or techniques used to achieve and maintain the product quality, process, and service.	• Prevention of quality problems through planned and systematic activities including documentation is done.
• It is usually the responsibility of a specific team that tests the product for defects.	• All team members involved in developing the product are responsible for quality assurance.
• It is a corrective tool.	• It is a managerial tool.
• Statistical quality control (SQC) is a pm1 of quality control.	• Statistical process control is a part of quality assurance.
• Validation/Software testing is an example of quality control.	• Verification is an example of quality assurance

Quality assurance is a process-oriented approach that aims to prevent defects. It is not tailored to the specific requirements of the product being developed, but rather focuses on the process used to produce the deliverables. Quality assurance activities are determined prior to the start of production work and are carried out during the product development process. Examples of quality assurance activities include creating process checklists, conducting project audits and reviews, and establishing standards. Table 1.1 provides a concise overview of the distinctions between quality control and quality assurance.

The Five Quality Paradigms

Quality paradigms are the result of changes that have occurred over the years in the field of technology, society, and customer demands.

Artisanal Paradigm

The concept of quality and the methodologies to achieve it evolved starting from the early 1900s when production was exclusively artisanal. In this phase, the product or service is created exactly as desired by the customer. Therefore, attention to the product/service and its performance is focused on demand. Quality is the main goal of the producer/artisan who personally controls the finished product.

Mass Production Paradigm

This phase developed after the Industrial Revolution. Attention is primarily focused on the production rate. During this period, the concept of testing is introduced, performed by workshop supervisors, consisting of an inspection of the product (i.e. detecting defects and/or more or less standardized objective quality). The reworking of products and production of waste is very high, while the delivery time is generally low.

Statistical Quality Control Paradigm after the First World War

The phase of inspection intended as product-by-product control ended because it was too expensive and not very effective. The adoption of statistical methods and sample testing was introduced. In this phase, the "quality control" was born, consisting of all those actions aimed at measuring the characteristics of a product based on predefined parameters. This phase is akin to mass production, with the distinction being that the emphasis is placed on the process itself. The implementation of statistical process control yields the advantageous outcome of diminished waste and rework costs. During this phase, products are meticulously designed and manufactured, while statistical techniques are employed to effectively acquire customers.

Total Quality Management Paradigm

In this stage, the central emphasis is directed towards both the customer and the supplier. The customer plays a crucial role in defining the product. This phase is characterized by the active involvement and empowerment of employees, a strong customer orientation, continuous improvement, unwavering commitment from top management, comprehensive training, and effective teamwork. The outcome of this phase is a product of exceptional quality, achieved through low costs, swift delivery, minimal rework, and efficient production with minimal waste. It is crucial to acknowledge that within this phase, two separate approaches to quality are evident: quality control and quality assurance. Quality control is centred around the product itself, with a specific emphasis on identifying defects or ensuring adherence to the required standards. Its purpose is to verify that the final results meet the desired level of quality and that they are both complete and accurate.

Activities related to quality control include inspection, result reviews, and process testing. These activities are important for ensuring the quality of the final product. However, it is equally important to focus on quality assurance. Quality assurance is a process-oriented approach that aims to prevent defects from occurring in the first place. To achieve this, quality assurance activities should be determined and implemented before the production process begins. Examples of quality assurance activities include the use of process checklists, conducting project audits, and the development of standards, in order to achieve the highest standards of quality.

Technological-Artisanal Paradigm – Quality 4.0

This new paradigm represents a frontier in the pursuit of quality. It aims to combine the artisanal approach with efficient delivery times. To achieve this, a high degree of flexibility in production processes is required. The focus is on integrating human and machine capabilities, which necessitates process automation. Looking ahead, the future of quality lies in the emerging technologies of Industry 4.0. To excel in this new phase, it is imperative to utilize the currently available digital tools and understand how to apply them effectively.

Quality 4.0 is an evolution of the traditional approach to quality. It is important to include it at a strategic level for sustainability during digital transformation. Three approaches have been identified in literature: the first classifies Quality 4.0 as a consequence of Industry 4.0; the second considers Quality 4.0 an advancement of quality-related activities to anticipate future quality issues and achieve a higher level of excellence and stability. Ultimately, it is identified as a new trend in Quality Management. In this view, Quality 4.0 is considered a distinct and emerging trend within the field of quality management. It recognizes that the digital transformation of industries is reshaping the way quality is managed. Quality 4.0 encompasses a set of practices, methodologies, and tools that specifically address the challenges and opportunities presented by the integration of digital technologies and data into quality assurance and control processes.

The primary technologies in this field encompass artificial intelligence, computer vision, natural language processing, chatbots, personal assistants, navigation systems, robotics, intricate decision-making processes, big data analytics, and infrastructure tools like MapReduce, Hadoop, Hive, and NoSQL databases. Additionally, there are tools facilitating easier access to data sources and the management and analysis of large datasets without relying on supercomputers. Blockchain technology is utilized to enhance transparency and verification of transactions, while monitoring conditions ensures the attainment of quality objectives. Deep learning techniques are employed for tasks such as image classification, complex pattern recognition, time series prediction, text generation, sound and art creation, as well as altering images based on heuristics, such as transforming a frowning expression into a smiling one in a photograph.

Enabling technologies, such as cloud computing and 5G networks, along with convenient sensors and actuators, have paved the way for various advancements in the field of data science. These advancements include the utilization of augmented reality (AR), mixed reality, and virtual reality (VR), as well as the implementation of data streaming technologies like Kafka and Storm. Additionally, the adoption of IPv6 and the Internet of Things (IoT) has further enhanced the capabilities of data science.

Machine learning techniques, such as text analysis, recommendation systems, spam filters for email, fraud detection, and object classification, have become integral components of data science. These techniques enable the extraction of valuable insights from large and diverse datasets, allowing for accurate predictions and effective classifications.

Data science itself involves the practice of gathering heterogeneous data sets and utilizing them to make predictions, perform classifications, and identify patterns within large datasets. Furthermore, data science aims to reduce the complexity of large sets of observations by identifying significant predictors. To achieve this, valid traditional techniques such as visualization, inference, and simulation are employed to generate reliable models and solutions.

In summary, the combination of enabling technologies, machine learning techniques, and the practice of data science has revolutionized the way we analyse and utilize data. These advancements have opened up new possibilities for various industries and have the potential to drive innovation and efficiency in numerous domains.

The benefits of implementing Quality 4.0 can be succinctly summarized as follows:

- Firstly, it leads to an increase in human intelligence. This means that individuals are able to enhance their cognitive abilities, resulting in improved problem-solving skills and overall decision-making capabilities.
- Secondly, Quality 4.0 facilitates a significant boost in the speed and quality of the decision-making process. By leveraging advanced technologies and data analytics, organizations are able to make informed decisions more efficiently and effectively.
- Thirdly, the implementation of Quality 4.0 brings about improvements in transparency, traceability, and verifiability. This ensures that processes and actions can be easily monitored, audited, and validated, thereby enhancing accountability and trustworthiness.

- Furthermore, Quality 4.0 enables organizations to anticipate changes. By utilizing real-time data and predictive analytics, businesses can identify potential disruptions or opportunities in advance, enabling swift adaptation and response.
- Lastly, Quality 4.0 fosters the evolution of relationships, organizational boundaries, and the concept of trust. This, in turn, unveils opportunities for continuous improvement and the development of new business models.

In conclusion, the implementation of Quality 4.0 offers numerous advantages, including the enhancement of human intelligence, the acceleration of decision-making processes, the improvement of transparency and traceability, the anticipation of changes, and the evolution of relationships and business models.

Pioneering Quality Gurus

Ronald Fisher

Fisher, while not widely recognized as a quality guru, has indeed made significant contributions to the field of statistics. In the 1930s, he laid the groundwork for several robust statistical methods, including Design of Experiments (DOE) and Analysis of Variance (ANOVA). These methods have since become indispensable tools for numerous organizations in their pursuit of problem-solving and process improvement. It is worth noting that Fisher's work on analysis of variance gained considerable recognition following its inclusion in his seminal book *Statistical Methods for Research Workers*, published in 1925. Furthermore, Fisher's expertise in experimental design led him to publish *The Design of Experiments* in 1935, further solidifying his reputation as a leading authority in the field. Lastly, his publication of *Statistical Tables* in 1947 served as a valuable resource for researchers and statisticians alike.

W. Edwards Deming

According to this influential scholar, increased productivity leads to long-term competitiveness in the market that results from achieving superior quality in all processes. His theory is that updated quality standards lead to lower costs and improved productivity (number of defect-free products). The optimization of time and resources can be achieved through the reduction of labour, particularly in the form of rework, as well as the minimization of errors and delays. By implementing these measures, one can expect a more efficient utilization of available resources. Gaining a larger market share to stay in business is possible when a company provides better quality at a lower price.

Deming's Philosophy

According to Deming, it is necessary to have a deep understanding of the system in which one operates. Therefore, he developed the Theory of Profound Knowledge, which is based on four parts that interact with each other:

- The theory of optimization suggests optimization for the entire system. The goal of an organization or company is to optimize the entire system, which includes all parts and not just the optimization of individual subsystems. The complete system includes all components: customers, employees, suppliers, distributors, manufacturers, shareholders, communities, and the environment. The long-term goal of a company is to create an advantageous situation for all its partners. The use of materials available at lower costs could lead to much lower quality and excessive cost increases during rework in production and assembly. All parts of a system and their interrelationships need to be analysed, otherwise, each part will have its own direction. The more complex a system is, the more communication is required among the parts, and the greater the possibilities for optimization.

- The theory of variance, according to which productivity increases when process variance decreases. Deming estimated that variation was one of the main causes of poor quality. In the realm of mechanical assemblies, it is worth noting that even slight variations in the dimensional specifications of individual parts can have a profound impact. Specifically, such variations can lead to increased wear and, consequently, result in premature failures and inconsistent performance. It is crucial to acknowledge that customers, being discerning individuals, possess the power to significantly influence the commercial reputation of any organization. This influence is particularly potent when it stems from experiences with inconsistent service. Therefore, it becomes imperative to possess a deep understanding of statistical theories and statistical process control. These tools enable us to identify the causes of variance and, in turn, verify the stability of the system. By employing such knowledge, we can ensure that our organization remains steadfast in delivering consistent and reliable performance, thereby safeguarding our commercial reputation. The causes of variance can be of two types: "common" (intrinsic variances in the process that continually shift the process itself, which, according to Deming, are attributable to poor management) and "special" (causes external to the process that sporadically affect variance and can be well identified).
- The theory of knowledge, according to which knowledge is a continuous process of collecting, analysing, and interpreting data derived from experimentation. Deming believed that the generation of knowledge was possible only with theory and experience together.
- The scientific study of psychology aids in comprehending individuals and their interactions in various circumstances and conditions, as well as their interactions within the hierarchy of any management system. Psychology connects the people who make up the management system. Therefore, people management requires knowledge of psychology and knowledge of what motivates people. With proper motivation, they can work more efficiently. Job satisfaction and motivation to excel are essential components of an individual's psychological and emotional well-being within the context of their professional endeavours. These elements are crucial in promoting a sense of fulfilment and drive, thereby propelling individuals towards achieving their highest potential. Reward and recognition are extrinsic factors, part of motivation. Therefore, management must create the right mixture of intrinsic and extrinsic factors to motivate employees. Deming provided the concept of quality control.

Deming and the 14 Points of Total Quality Management

Monitoring and preparing long-term action plans/business plans require the development of knowledge from the workplace. Therefore, Deming created 14 points for it, in which his thoughts can be synthesized.

1. To begin, it is imperative to establish a sense of unwavering commitment towards enhancing both the product and service. This commitment is driven by the ultimate objective of attaining competitiveness, ensuring the longevity of the business, and facilitating job creation.
2. Adopt the new philosophy. Management must assume its responsibilities and have the leadership for change.

3. Create quality in the product from the beginning.
4. Minimize total cost.
5. Continuously improve the production and service system to improve quality and reduce waste.
6. Introduce training in the company.
7. Establish leadership with the aim of guiding and helping people to perform their tasks better.
8. Create a comfortable working environment in which to work effectively for the company.
9. Get rid of barriers that deprive people of pride in workmanship.
10. Remove any slogans, exhortations, and targets that may be present in the workforce's requests for zero defects and increased productivity. These elements can be counterproductive and may hinder the overall performance of the workforce. By eliminating them, you will create a more focused and efficient work environment.
11. Eliminate numerical goals and annual merit rating.
12. Stimulate the pride of working.
13. Introduce a dynamic program of instruction and retraining for both management and workforce.
14. Management and workforce must work together.

Joseph Juran

The quality trilogy is a fundamental concept that requires significant focus. It centres on three essential elements: advanced quality planning, quality control, and quality enhancement. Firstly, advanced quality planning is an indispensable aspect of the quality trilogy. It entails meticulous preparation and strategizing to ensure that quality standards are met and exceeded. By engaging in this proactive approach, companies can detect potential risks and implement pre-emptive measures to minimize them. This, in turn, leads to enhanced efficiency and effectiveness in the overall quality management process. It aims to help operators profitably produce goods and services to meet customer expectations and needs. It also provides a method that will be implemented for customer satisfaction. In the quality planning phase, it is important that customers and their needs are identified by the organization.

Next, quality control holds a crucial position within the quality trilogy. It encompasses the structured inspection and assessment of products, processes, and services to ensure their compliance with predefined quality criteria. Through rigorous inspections, testing, and analysis, organizations can identify any deviations or non-conformities and take corrective actions promptly. This thorough examination ensures that only top-quality products and services are provided to customers, enhancing satisfaction and loyalty.

Lastly, quality improvement is an imperative component of the quality trilogy. It emphasizes the continuous enhancement of processes, systems, and practices to achieve superior quality outcomes. By embracing a culture of continuous improvement, organizations can identify areas for optimization, implement innovative solutions, and drive positive change. This relentless pursuit of excellence ensures that organizations remain at the forefront of their respective industries, consistently delivering exceptional products and services.

In conclusion, the quality trilogy is an indispensable framework that demands our unwavering attention. By embracing advanced quality planning, quality control, and quality improvement, organizations can elevate their performance, exceed customer expectations, and secure a competitive advantage.

Philip Crosby

Philip Crosby analysed the meaning of quality from different perspectives and taught the meaning of quality based on defects, prevention, requirements, and the price of non-conformity.

Quality is conformance, not goodness, to requirement. Quality is defined as a system in which prevention leads to quality, not evaluation. The performance standard should be "no defects", instead of "close enough to no defects".

Genichi Taguchi

Genichi Taguchi is known for his contributions in robust design in product development and optimized product production. Taguchi loss function, also known as the Taguchi method (experimental design), and other methodologies have greatly contributed to reducing variations and significantly improving the quality of engineering and productivity. This method optimizes results by taking different parameters into account, thus contributing to ensuring customer satisfaction.

Taguchi's product development comprises three phases:

- **System design**; In the system design, the non-statistical phase for engineering, marketing, and customer is determined.
- **Parameter design**; It determines how the product should perform with respect to defined parameters. For example, to optimize the result of material removal rate, different parameters such as cutting speed, feed rate, depth of cut, and so on are used.
- **Tolerance design**; It determines the balance between production costs and losses.

Armand V. Feigenbaum

Armand V. Feigenbaum is known for first elaborating the concept of Total Quality Control (TQC). In the 1950s, Feigenbaum cautioned against the adverse financial implications that arise when the production of a product or service fails to yield a satisfactory outcome on the initial attempt.

Kaoru Ishikawa

Kaoru Ishikawa is known for the Ishikawa (fishbone) diagram and developed quality circles.

Shigeo Shingo

Shigeo Shingo developed the principles of Poka Yoke (intrinsic safety), Just-in-Time, and Single-Minute Exchange of Die (SMED). Today, one of the world's most prestigious business awards is named after him: the Shingo Award for Manufacturing Excellence.

Frederick Winslow Taylor

Frederick Winslow Taylor is renowned as the pioneer of scientific management. Taylor's primary objective was to enhance production efficiency, aiming not only to minimize expenses and boost profits but also to raise workers' wages by enhancing productivity.

The fundamental principles that Taylor saw as the basis for the scientific approach to management can be summarized as follows:

- Replace work methods with methods based on a scientific study of tasks.
- Scientifically select and train each worker instead of passively promoting training.
- Cooperation is necessary to ensure the implementation of scientifically developed methods.
- Divide the work between the workforce and managers so that managers can use the principles of scientific management for work planning, while workers perform their assigned tasks.

Taylor was a supporter of productivity and salaries based on the productivity achieved by workers. He focused on studying time and motion and other techniques for measuring work.

Finally, H. James Harrington and Walter A. Shewhart have also had a significant impact in the field of quality worldwide with their contributions to improving businesses and a variety of government, military, educational, and healthcare structures.

The 9M Framework

Total Quality Control, as defined by Armand Feigenbaum, is a highly effective system that aims to seamlessly integrate and coordinate the development, maintenance, and improvement of Quality. Its aim consists in ensuring that production and service offerings are able to fully satisfy the customer, while keeping costs at a minimum. Feigenbaum, in his extensive analysis, has identified a range of components that significantly influence the widespread adoption of quality practices. These components, which he has aptly labelled "9M", encompass both internal and external factors that are pivotal to spread quality culture throughout an organization. By thoroughly understanding and addressing these components, companies can establish a solid foundation for achieving excellence in quality management:

- **Markets**: An organization is significantly influenced by the markets it serves. These influences come both from the quality requirements expressed by consumers, to which the organization must align, and from the quality offered to them by competing companies with which the organization must compete.
- **Money**: Referring to profit, which has been the driving force behind the organization since its inception.
- **Management**: Top management defines the company's vision and mission. Therefore, it determines the guiding function of the organization.
- **Men**: One of the most important factors, as it is with the knowledge, skills, and abilities of the company's human resources that quality is applied.
- **Motivation**: As we have just said, it is the work of human resources that determines the degree of quality of a company. However, it is not enough for resources to have the necessary skills, knowledge, and abilities to guarantee quality in the organization, but it is necessary that they have a degree of motivation such as to maximize their performance and, consequently, maximize the level of quality. From this, the importance of management emerges once again, which has the task of adopting an appropriate leadership style that creates resonance and motivates staff.
- **Materials**: The quality of a production system largely depends on the quality of the raw materials needed to carry out the production itself.

- **Machines and mechanization**: Similar to what has just been said for raw materials, machinery is another fundamental element for the company's production system on which its quality depends. An organization that intends to produce with quality cannot do without the use of machinery that is up to its quality aspirations.
- **Modern information methods**: The information system of an organization is a fundamental element for it. From it, in fact, depends the quality of the information circulating between the various functional areas of the company, and consequently, the total quality depends on it. Specifically, an information system must be reliable, timely, transparent, and clear so that information, whether it is instructions communicated by top management to staff or communications that the company makes with external subjects, reaches the recipient promptly and is easily understandable.
- **Mounting product requirements**: Companies are constantly evolving, and with them, the needs and preferences of consumers. To ensure quality, a company must be able to constantly identify consumers' preferences regarding the product characteristics and satisfy them by maximizing quality and minimizing production costs and times.

The approach to quality just outlined, starting from the 1980s, has been adopted by most companies in the Western world. However, in the Eastern world, particularly in Japan, there was a different approach to quality, a different philosophy.

Total Quality Management – From Inspection to TQM

Total Quality Management (TQM) is a revered business methodology that meticulously scrutinizes not only the goods and services offered by a company but also the intricate processes involved in their development, as well as the dedicated workforce responsible for ensuring that the final outcomes consistently meet the stringent standards of valued customers. TQM seamlessly integrates all aspects of a company's complex operations, spanning marketing, finance, design, engineering, production, and customer service, with the primary aim of effectively fulfilling the diverse needs of valued customers while simultaneously attaining organizational objectives. It is an unwavering and perpetual endeavour to attain a competitive advantage by ceaselessly enhancing and refining all aspects of a company's multifarious activities.

Many believe that the concept of TQM emerged from a misinterpretation of Japanese terminology regarding "control" and "management". According to esteemed quality scholar and consultant William Golimski, Koji Kobayashi, former CEO of Nippon Electrical Company (NEC), introduced the term TQM during his acceptance speech for the prestigious Deming Prize in 1974. Furthermore, the American Society for Quality notes that in 1984, the US Naval Air Systems Command adopted the term to describe their management approach, influenced by Japanese practices, aimed at improving quality. This decision was made in order to replace the previously used term "total quality control", as the word "control" was deemed unfavourable by the command.

These historical accounts provide us with valuable insights into the origins and usage of the term TQM, shedding light on its evolution and widespread adoption in the field of quality management.

Nancy Warren, a staff member, introduced the term "management". This proposal coincides with the historical event where the United States Navy Personnel Research and Development Center began researching statistical process control (SPC), the contributions of Juran, Crosby, Ishikawa, and the principles of W. Edwards Deming to initiate performance improvements in 1984. The initial implementation of this approach took place at the North Island Naval Aviation Depot.

Table 1.2 Four stages of TQM.

Quality Management Stages	*Areas of Focus*	*Scope*
Inspection	Detection	Error detection Rectification Sorting, grading, reblending Decision about salvage and acceptance
Quality control	Maintaining status quo	Quality standards Use of statistical methods Process performance Product testing
Quality assurance	Prevention	Quality system (ISO 9000) Quality costing Quality planning and policies Problem solving Quality design
Total quality management	Quality as a strategy	Quality strategy Customers, employees and suppliers involvement Involve all operations Empowerment and teamwork

The initial approach to quality management was inspection-based, where a team of inspectors compared a product with a product standard. During the Second World War, a quality control system was introduced, which included product testing and documentation. The focus shifted from product quality to systems quality during the quality assurance stage, which involved quality manuals, planning, and document control. Total Quality Management represented the fourth phase of evolution, highlighting a coherent and unified vision, comprehensive staff training, exceptional customer relations, and continual enhancement. This phase sought to enhance the overall quality of the organization and eradicate barriers between departments.

Total Quality Control, Company-Wide Quality Control, and TQM

Total Quality Control (TQC) embodies an integrated approach to quality management that involves all levels of a company's structure, from management to operational staff, in continual improvement and "Kaizen" (Imai, 1986). It operates under the premise that a company's survival hinges on customer satisfaction, and as customer expectations rise, companies must strive for ongoing quality enhancement.

In contrast, the "Quality Assurance" system adopts a static view of quality, aiming to meet established requirements and document the outcomes. Total Quality Control, however, is dynamic, focusing on continuous process improvement and striving for minimal defect levels or achieving "zero defects". Consequently, rigorous control is exercised over both incoming parts and components upon arrival at the plant and throughout the production process, as well as over finished products.

Total Quality Control (TQC) was initially introduced by A. V. Feigenbaum in 1960 in a book of the same title. In the 1950s, with the end of the war, approaches developed in the United States that considered quality as the result of the entire organization and not just the production process. In this context, Total Quality Control was born, mostly designed for the Western context. It provided for the application of quality controls to all business functions, but there was no willingness to apply this theory, and for this reason, it was only accepted in the engineering field and never achieved full acceptance in the West.

In *Total Quality Control*, Feigenbaum (1960) asserts that TQC serves as an efficient system for integrating diverse quality initiatives, enabling the cost-effective execution of production and services while meeting customer satisfaction. This definition encapsulates the crux of the issue. The primary reason for TQC's lack of success in the West is predominantly attributed to Western management being misled by Feigenbaum's mention of quality departments overseeing TQC. Consequently, it was not explicitly conveyed that a crucial aspect of TQC is management's dedication to quality enhancement. While effective systems are a prerequisite, they alone are insufficient for TQC implementation. Starting from the observation that Quality would become the main factor in customer choices and, therefore, the determining argument for the survival and success of a manufacturer, Feigenbaum described Quality as the characteristics in marketing, engineering, manufacturing, and support that enable a product or service to meet consumer expectations.

Quality must, therefore, become an essential element of how a company is run, like marketing and economic management.

Total Quality Control is a way of globally managing Quality to ensure full customer (and Management) satisfaction, or, in Feigenbaum's words:

> TQC is an effective system for integrating the efforts of development, maintenance, and improvement of Quality of the various groups of an organization so that marketing, engineering, production, and assistance are carried out at the highest levels of economy, compatibly with customer satisfaction.
>
> (Feigenbaum, 1960, p. 40)

The idea of a global approach appears necessary because the quality of products and services is directly influenced by the nine fundamental "9M" factors listed earlier in this chapter. This comprehensive strategy entails establishing a Total Quality System (TQS), which is defined as the acknowledged and documented organizational framework within an integrated, efficient system. The TQS adheres to management and technical procedures, devised to direct and synchronize people, machinery, and information in the most optimal and practical manner, with the ultimate goal of attaining customer satisfaction at the lowest feasible cost to the company.

The concept of CWQC can also be described as strategic quality management, where quality is seen as a controllable tool used to achieve business objectives, particularly customer satisfaction. CWQC is characterized by three fundamental principles: (1) Using quality as a competitive tool, (2) Customer focus, and (3) Inclusion of all functional areas and employees in the quality improvement process. In Ishikawa's CWQC concept, quality is defined as a competitive parameter. According to this approach, a company that operates with a focus on quality will increase its long-term profits through the reduction of internal costs and the increase of market share.

A decisive factor in customer orientation is the shift from a production-oriented quality concept to a customer-oriented quality concept. Quality is defined by the customer and not by the company. As a result, the entire product creation process is specific to the customer. Market research, therefore, plays an increasingly important role.

Moreover, within the company, each employee in the next process is seen as a customer. The so-called "the next process is your customer" principle serves to break down departmental barriers and facilitates communication and coordination of horizontal business activities. A network of customer-supplier relationships that encompasses the entire company is created in this way. By prioritizing internal customer orientation, the aim is to fulfil the requirements of external customers. The concept of CWQC also includes quality assurance elements, such as statistical quality assurance, as well as measurement and testing technology elements.

With the help of statistical methods, data and facts are collected for continuous improvement. Thus, quality is seen not as an objective, but as a continuous process. This model implies a production process that uses less of everything. The human factor in management philosophy involves the involvement of the entire company and entails autonomy and spontaneity of the individuals involved in order to unleash the unlimited potential of human beings.

Furthermore, inter-functional management, which is characteristic of quality management, is typically Japanese and requires a fundamental modification of the bureaucratic model adopted (functional hierarchy) in traditional organizations. This form of organization, sometimes called the "chimney model", is produced by the isolation of the various functions from each other. This, in turn, results in chauvinism and other behaviours that, although optimal for a given function, are harmful to the system considered as a whole. The Japanese have developed a formal approach to address this problem, which is often called inter-functional management.

Therefore, it is characterized by the following elements:

- Product quality is improved and uniform;
- Errors are reduced;
- Product reliability is improved;
- Costs are reduced;
- Production is increased and can be streamlined;
- Unnecessary work and rework are reduced;
- Techniques are standardized and improved;
- Inspection and testing costs are reduced;
- Supply contracts are simplified;
- The sales market expands;
- Cooperation between departments is improved;
- Incorrect data and reports are reduced;
- Discussions are conducted more freely and democratically;
- Meetings are more fluid and conflict-free;
- Installation, repair, and maintenance measures on machines and equipment are carried out more efficiently;
- Human relationships are improved.

In summary, starting in the 1970s, a very innovative approach called Company Wide Quality Control (CWQC) was developed in Japan by Kaoru Ishikawa. The CWQC has a different focus than Total Quality Control (TQC); while the latter focuses only on reducing production costs, the CWQC is oriented towards total customer satisfaction achieved through the global involvement of employees. It is based on three fundamental principles: (1) use of quality as a competitive tool, (2) customer focus, and (3) involvement of all functional areas and employees. The CWQC approach considers quality as a competitive parameter, and companies that operate with a focus on quality will increase their long-term profits by reducing internal costs and increasing market share. Unlike the TQC approach, where quality is focused on the company's production cycle, the CWQC approach focuses on the customer. According to Ishikawa, to achieve total quality, it is necessary to involve all employees of the organization in order to converge the work of all business dimensions towards a single interest: quality. Therefore, the quality becomes not just a cost but a strategic investment for companies, with the customer at the centre of this management model around which companies, as active, dynamic, and flexible entities, develop new development strategies for production that meet consumer needs.

On these foundations, Total Quality Management (TQM) was born, a new way of managing the organization, tending towards long-term customer satisfaction, cantered on quality. The main points of this new managerial philosophy are developed around five main strategies: customer as the company's priority; quality as a lever for customer satisfaction; continuous improvement;

product/service quality as a result of process and system quality; and involvement of human resources. TQM represents a vision achievable only through extensive long-term planning, entailing the development and implementation of annual quality plans that progressively steer the company toward realizing its vision. Quality is intrinsic to this definition as TQM can be regarded as the apex of a hierarchy of quality concepts.

TQM does not present an inconsistent vision. In an era marked by intense competition in both domestic and international markets, an increasing number of companies recognize TQM as a strategic imperative for survival. Today's consumers have abundant options and select products based on the "highest quality-price ratio" – those offering the greatest customer satisfaction relative to price.

In comparison to TQM, CWQC is a concept more centred on employees, with quality responsibilities dispersed throughout the organization.

Compared to Feigenbaum's definition of Total Quality Control (TQC), which is focused on the production cycle of the product, CWQC is a concept that is much more focused on customer orientation, as satisfying customer needs related to the manufactured product is at the centre of the entire company and for all employees. The goal is continuous improvement and the production of quality at low cost.

TQM, which was born in 1987, represents an evolution of TQC and a Western interpretation of CWQC. It differs from the latter in the introduction of the stakeholder concept and a more engineering approach. It was defined as "a way of governing the organization focused on quality, based on the participation of all its members. It leads the organization to long-term success by satisfying customers and providing benefits to all members of the organization and the community".

The goal of TQM is to ensure that the company leadership cannot evade its responsibilities. In other words, TQM must be among the tasks of top management and the board of directors.

ISO 9001:2015 (International Organization for Standardization) – Standard and Evolution

At the international level, it is through ISO standards that we arrive at the technical definition of the Quality factor. The ISO 9000 standards date back to 1987 and were reviewed in 1994 and 2000. In 1994, ISO 9001, 9002, and 9003 were oriented towards Quality Assurance, which includes all coordinated and systematic activities to ensure that the product, process, and/or service will satisfy the quality standards. In Quality Assurance, it is crucial to objectively prove the activities carried out, the way they were executed, and the results achieved.

Subsequently, with the 2000 revision, the concept of Quality Management was introduced, which includes Quality Assurance activities and other coordinated activities such as defining quality policies and objectives, quality planning, quality control, quality assurance, and quality improvement to meet customer needs and expectations. ISO 9001:2000 contained the description of these quality management system requirements, which remained largely unchanged in the subsequent ISO 9001:2008, where continuous improvement is expected as a permanent objective.

Therefore, the 2000 revision grouped and simplified the ISO 9000 family into three standards:

- ISO 9000 Quality Management Systems – Fundamentals and Vocabulary. Enunciates the fundamental principles and concepts of Quality Management Systems.
- ISO 9001 Quality Management Systems – Requirements. Specifies the requirements for a Quality Management System. It focuses on the effectiveness of the Quality Management System in meeting customer requirements. It is the reference document that reports the minimum compliance requirements for certification.

- ISO 9004 Quality Management Systems – Guidelines for Performance Improvement. Contains guidance for a wider range of Quality Management System objectives than ISO 9001. The two standards, ISO 9001 and ISO 9004, are complementary and consistent, albeit with different objectives. ISO 9001 is focused on product and/or service quality, while ISO 9004 is focused on performance improvement. They are two independent but fully overlapping standards designed to be used together. Implementing ISO 9001 marks the initiation of the establishment of a Total Quality Management system.

The ISO 9000 standards were revised in 1994, but this revision did not bring about significant changes. In contrast, the 2000 revision, also known as Vision 2000, had a more incisive strategic relevance for the changes it introduced. In fact, the ISO model came considerably closer in content and basic concepts to TQM models, following the introduction of the process management concept. Following this, ISO 9000 underwent a revision in 2005; however, the fundamental principles of Quality Management Systems remained unchanged from the previous 2000 edition.

The vocabulary was enriched with the addition of new definitions and updated explanatory notes in line with the latest ISO 9000 family documents. In particular, it was necessary to integrate UNI EN ISO 9000 with the terms and definitions of UNI EN ISO 19011 (for example, audit, technical expert, audit field, audit plan, etc.) and UNI EN ISO 10012 on measurement management (for example, metrological confirmation, measurement management system, metrological function, etc.), as well as other standards in the ISO 10000 series, which provide guidelines on support techniques for individual quality aspects. In conclusion, ISO 9000 underwent a revision in 2015, focusing on "Quality Management Systems (QMS) – Fundamentals and Vocabulary". This standard addresses fundamental concepts pertaining to Quality Management Systems (QMS) and offers the terminology utilized in the other two standards.

Therefore, to summarize:

- ISO 9000:1994 primarily emphasized ensuring quality through pre-emptive measures rather than relying solely on the inspection of the final product. According to this framework, organizations were required to strive for quality products and enhance quality through the adoption of pre-emptive measures.
- ISO 9001:2000 marked a departure from the standards set by ISO 9000:1994. Organizations engaged in new product development necessitated a quality standard cantered on design and development. The 2000 edition shifted its focus to the concept of process management. Additionally, ISO 9000:2000 mandated the involvement of senior officials in integrating quality into the business system and prevented junior administrators from delegating quality-related tasks.

Another objective of this standard was to enhance efficiency through continuous monitoring of actions and activities. Routine process improvement and monitoring customer satisfaction became integral aspects of this standard. The requirements of this standard included:

- Approval of documents before dissemination.
- Availability of the appropriate version of documents upon request.
- Maintenance of adequate records to demonstrate compliance with all requirements.
- Establishment of a process for record maintenance.

The ISO 9000:2000 revision included five objectives:

- Meeting the needs of stakeholders.
- Creating a model applicable to all types of organizations.
- Applicability across all domains.
- Ensuring ease of comprehension.
- Aligning the quality management system with business processes.

The ISO 9000:2008 edition further aligned the eight principles of quality with the tenets of TQM. ISO 9001:2008 was essentially a revision of ISO 9001:2000, providing explanations of the current requirements of ISO 9001:2000 without introducing any new requirements. For instance, if an organization updates its quality management system, it should ensure alignment with the explanations provided in the revised version.

ISO 9000: 2015 Edition

ISO 9000:2015 is characterized by increased interaction with customers and stakeholders, greater focus on the customer and their loyalty to achieve lasting success. The main benefits of this new perspective are:

- greater value for the customer;
- greater customer satisfaction;
- customer loyalty.

Table 1.3 summarizes the seven principles of ISO 9000:2015 compared to the eight principles of ISO 9000:2008.

In the updated version, the principle of the "Process Approach" combines the previous "Process Approach" and "Systemic Approach to Management" principles, emphasizing that the Quality Management System comprises interconnected processes.

Additionally, a new aspect is introduced by the final principle, which asserts that enduring organizational success is attained when the organization effectively manages relationships with all stakeholders, encompassing both suppliers and partners.

The fourth edition of ISO 9001 was released in 2008 and replaced the third edition (ISO 9001:2000), which was modified to elucidate some themes and improve adaptability with ISO

Table 1.3 The seven quality management principles of ISO 9000:2015 versus the eight principles of ISO 9000:2008.

ISO 9000:2008	*ISO 9000:2015*
Customer orientation	Customer focus
Leadership	Leadership
Employee engagement	Active involvement of people
Process approach	Process approach
Systematic management approach	
Continuous improvement	Improvement
Data-driven decision making	Evidence-based decision making
Mutually beneficial supplier relationships	Relationship management

Source: Authors' own elaboration

14001:2004. It was also revised in 2015, "Quality Management Systems (QMS) – Requirements". This fifth edition replaces the fourth edition ISO 9001:2008. The latest edition defines the standard utilized for certification, showcasing compliance with Quality Management System (QMS) requirements to customers, regulations, and organizational needs.

Initially based on the British standard BS-5750, ISO 9000:1987 introduced three "models" for quality management systems tailored to different organizational activities:

- ISO 9001:1987 Model: Focused on ensuring quality across design, development, production, installation, and post-sales service, making it most suitable for organizations engaged in new product development.
- ISO 9002:1987 Model: Geared towards ensuring quality in production, installation, and service, making it suitable for organizations involved in production, installation, and maintenance but not in new product development.
- ISO 9003:1987 Model: Centred on ensuring quality during final inspection and testing, focusing solely on inspecting finished products without addressing the production process.

ISO 9004, initially revised in 1994 and 2000, underwent another revision in 2009 as "Quality Management Systems (QMS) – Guidelines for Performance Improvement". This edition offers guidelines for organizations to establish QMSs focused on enhancing performance, with its latest review occurring in 2018.

The ISO 9000 series has been widely adopted by most countries as a national standard, with thousands of organizations globally having their quality systems certified to the standard. For instance, in the United States, national standards are published by the American National Institute/American Society for Quality (ANSI/ASQ) under the ANSI/ASQ Q9000 series.

Initially designed for use in two-party contractual scenarios and internal audits, these standards required suppliers to develop quality systems compliant with the standards, with customers responsible for verifying the acceptability of the system. However, this approach proved costly due to multiple audits by both parties, prompting a shift to a third-party registration system.

Under the third-party registration system, a supplier's quality system is periodically evaluated and audited for adequacy by a third-party Registration Authority (Registrar). Upon compliance with the Registrar's interpretation of the standard, the supplier receives a registration certificate, reassuring customers that the supplier maintains a monitored quality system.

ISO 9000 vs ISO 9001

ISO 9000 comprises a set of global quality management standards, with ISO 9001 being one of its constituent standards. Furthermore, ISO 9000 includes a standard known as ISO 9000 within its family, which serves to define the foundation and terminology for quality management systems.

Industry-Specific Standards

The ISO 9000 system is designed to offer a straightforward framework applicable across various industries. However, specific industries such as automotive or aerospace have developed their own tailored systems, which are built upon ISO 9001 but adjusted to suit their unique requirements. Presently, there exist three additional quality systems: AS9100, ISO/TS 16949, and TL 9000.

One challenge associated with industry-specific standards is the necessity for suppliers serving clients across multiple sectors to establish quality systems that align with the distinct requirements of each industry. For instance, a packaging supplier catering to the aerospace, automotive, and telecommunications sectors must ensure their system meets not only the criteria outlined in ISO 9001 but also those of three other standards.

The Registrar Accreditation Board (RAB) has highlighted that industry standards have resulted in a demand for specialized auditors and specific training programs. Moreover, standardizing requirements beyond ISO 9001 simplifies compliance for key suppliers and facilitates implementation for major customers.

The Eight Principles of ISO 9000:2008

There are eight quality management principles on which ISO 9000 is based:

- **Customer focus**; Customer satisfaction is the primary goal of a company. By understanding the needs of customers, it is possible to identify and secure one's target market. This implies an increase in profits.
- **Leadership**; It is necessary to establish unity and coordination in the work environment. The goal of company leaders should be to motivate all parties involved in the project, minimizing the risk of misunderstandings between departments.
- **Involvement of people**; The involvement of everyone in a work group is an essential factor for its success. Involvement will lead everyone to personally invest in the project, making workers more motivated and committed, focusing on innovation and creativity and utilizing their full abilities to complete the project.
- **Quality management through a process approach**; The best results are achieved when activities and resources are managed as a single process. A quality management system through processes can reduce costs through effective use of resources, personnel, and time.
- **Systemic approach to management**; Coordinating, integrating, and aligning different management groups and related processes can lead to an efficient and effective management system.
- **Continuous improvement**; Through performance improvement, a company can increase profits and have an advantage over its competitors. If the entire production is oriented towards continuous improvement, activities will be aligned, leading to faster and more efficient development.
- **Concrete decision-making approach**; The most effective decisions are based on analysis and interpretation of data and information. By making informed decisions, an organization is more likely to make the right decision and demonstrate the effectiveness of past choices.
- **Supplier relationships**; It is important to establish a relationship with suppliers that brings mutual benefits, a relationship that creates value for both parties.

ISO 9001:2015 Requirements

The main changes introduced by the 2015 revision are:

- greater emphasis on the process approach;
- greater applicability to services;
- less importance given to documentation.

Since 2012, the International Organization for Standardization (ISO) has worked to define a single basic structure for all management system standards, including a common vocabulary of terms and definitions. From this work, ISO created Annex SL, which presents the structure known as the High Level Structure. It defines a set of terminologies, texts, definitions, titles, and their common sequence for all standards, as well as a greater focus on the concept of risk.

In particular, all management system standards have a structure divided into 11 main points and paragraphs with contents common to all standards: Scope, Normative References, Terms and Definitions, Context of the Organization, Leadership, Planning, Support, Operation, Performance, Evaluation, and Improvement. The HLS reflects the approach of continuous improvement, known as PDCA developed by Deming.

Scope

This International Standard outlines the criteria for establishing a quality management system in situations where an organization:

(a) must showcase its capability to consistently deliver products and services that meet customer expectations and comply with relevant legal and regulatory obligations, and
(b) seeks to improve customer satisfaction by effectively implementing the system, including mechanisms for system enhancement and ensuring adherence to customer needs as well as applicable statutory and regulatory prerequisites.

All provisions within this International Standard are universally applicable and can be implemented by any organization, irrespective of its nature, size, or the products and services it offers.

Normative References

The documents referenced herein, either wholly or partially, are normatively referenced in this document and are essential for its application. For dated references, only the cited edition is applicable. For undated references, the most recent edition (including any amendments) is applicable.

Terms and Definitions (Specified in ISO 9001, Not Specified in ISO 9000)

For the purposes of this document, the terms and definitions given in ISO 9000:2015 apply.

Context of the Organization

- **Understanding the organization and its context** means that: The organization is required to identify both external and internal factors that are pertinent to its mission and strategic objectives, and which may impact its ability to achieve the desired outcomes of its quality management system. Continuous monitoring and evaluation of information concerning these factors are necessary.
- **Understanding the needs and expectations of interested parties** implies that: In light of their influence or potential impact on the organization's consistency in delivering products

and services that meet customer expectations and comply with relevant legal and regulatory standards, the organization must ascertain:

(a) the relevant stakeholders and their respective requirements pertaining to the quality management system;
(b) the specific requirements outlined by these stakeholders. Continuous monitoring and assessment of information regarding these stakeholders and their demands are essential.

- **Determining the scope of the quality management system**: The organization must delineate the boundaries and applicability of its quality management system to establish its scope. In this determination process, consideration must be given to:

  (a) the external and internal factors mentioned previously;
  (b) the relevant requirements specified by interested parties as discussed earlier;
  (c) the organization's range of products and services.

All requirements outlined in this international standard must be adhered to if they are applicable to the defined scope of the quality management system. The scope of the organization's quality management system must be clearly documented and accessible. It should outline the types of products and services covered and justify any exceptions from the requirements of this international standard that the organization deems not applicable to its quality management system's scope. Declaration of compliance with this international standard is only valid if the determined requirements that are deemed not applicable do not impede the organization's ability or responsibility to ensure the compliance of its products and services and the enhancement of customer satisfaction.

Quality Management System and its Processes

The organization is mandated to establish, implement, maintain, and continually enhance a quality management system in accordance with the stipulations outlined in this international standard. This includes identifying the requisite processes and their interrelations, ensuring:

(a) Definition of the inputs and desired outputs for these processes.
(b) Sequencing and interactions among these processes.
(c) Application of criteria and methodologies (including monitoring, measurement, and relevant performance indicators) to ensure efficient operation and control of these processes.
(d) Allocation of necessary resources for these processes and ensuring their availability.
(e) Assignment of responsibilities and authorities for these processes.
(f) Mitigation of risks and capitalization on opportunities as identified per the Planning requirements.
(g) Evaluation of these processes and implementation of any necessary modifications to ensure intended outcomes are achieved.
(h) Continuous improvement of the processes and the quality management system.

As deemed necessary, the organization must:

(a) Maintain documented information to support the operation of its processes.
(b) Retain documented information to ensure confidence in the execution of the processes as planned.

Leadership

In general, Top management must exhibit leadership and dedication concerning the quality management system by:

(a) Assuming accountability for the effectiveness of the quality management system.
(b) Establishing quality policy and objectives aligned with the organization's context and strategic direction.
(c) Integrating quality management system requirements into the organization's business processes.
(d) Encouraging the adoption of the process approach and risk-based thinking.
(e) Ensuring the provision of necessary resources for the quality management system.
(f) Communicating the significance of effective quality management and adherence to quality management system requirements.
(g) Ensuring the achievement of intended outcomes by the quality management system.
(h) Involving, guiding, and supporting personnel to contribute to the effectiveness of the quality management system.
(i) Encouraging continual improvement.
(j) Supporting other managerial roles in demonstrating leadership within their respective areas of responsibility.

Customer Focus

Top management must demonstrate leadership and commitment with regard to customer focus by ensuring that:

(a) the customer and applicable legal and regulatory requirements are determined, understood, and consistently met;
(b) the risks and opportunities that can affect the conformity of products and services and the ability to enhance customer satisfaction are determined and addressed;
(c) attention to improving customer satisfaction is maintained.

Establishing the Quality Policy

Top management must establish, implement, and maintain a quality policy that:

(a) is appropriate to the purpose and context of the organization and supports its strategic direction;
(b) provides a framework for setting quality objectives;
(c) includes a commitment to satisfy applicable requirements;
(d) include a commitment to continuous improvement of the quality management system.

Communication of Quality Policy

The quality policy should:

(a) be available, up-to-date, and documented;
(b) be communicated, understood, and applied within the organization;
(c) be made available to relevant interested parties, as appropriate.

Organizational Roles, Responsibilities, and Authorities

Top management is responsible for ensuring that roles and responsibilities within the organization are clearly assigned, communicated, and comprehended. They are tasked with delegating responsibility and authority for:

(a) ensuring compliance of the quality management system with the requirements of this International Standard;
(b) ensuring processes yield the anticipated results;
(c) reporting on the performance of the quality management system and identifying improvement opportunities, particularly to top management;
(d) promoting customer focus across the organization;
(e) safeguarding the integrity of the quality management system during planned changes and implementations.

Planning

During the planning of the quality management system, the organization needs to assess the issues and requirements at hand and identify the associated risks and opportunities to:

(a) ensure that the quality management system attains the intended outcomes;
(b) amplify positive effects;
(c) prevent or mitigate negative effects;
(d) foster enhancements.

Subsequently, the organization must:

(a) develop strategies to tackle these risks and opportunities;
(b) determine how to seamlessly integrate and execute these strategies into its quality management system processes;
(c) assess the effectiveness of these strategies.

It's imperative that the actions taken to address risks and opportunities are commensurate with their potential impact on the conformity of products and services.

Quality Objectives and Designing to Achieve Them

The organization is required to establish quality objectives for the functions, levels, and processes integral to the quality management system. These objectives should:

(a) align with the organization's quality policy;
(b) be quantifiable;
(c) consider relevant requirements;
(d) directly impact the conformity of products and services and the enhancement of customer satisfaction;
(e) undergo regular monitoring;
(f) be effectively communicated;
(g) adapted as needed.

Additionally, the organization must maintain documented information pertaining to these quality objectives. When strategizing to achieve these objectives, the organization must determine:

(a) action plans;
(b) necessary resources;
(c) responsible parties;
(d) timelines for completion;
(e) criteria for evaluating outcomes.

Planning Changes

When the organization identifies the necessity for alterations to the quality management system, these changes should be executed in a methodical manner. The organization should take into account:

(a) the rationale behind the changes and their potential repercussions;
(b) the cohesion of the quality management system;
(c) the availability of resources;
(d) the allocation or reallocation of responsibilities and authorities.

Support

The organization is obligated to identify and furnish the resources essential for establishing, implementing, maintaining, and continuously enhancing the quality management system. Considerations should include:

(a) the capabilities and limitations of existing internal resources;
(b) requirements for external sourcing.

Personnel

The organization must ascertain and provide the necessary personnel for the efficient implementation of its quality management system and for overseeing and controlling its processes.

Infrastructure

The organization must determine, furnish, and uphold the infrastructure requisite for process operation and ensuring product and service conformity. Infrastructure encompasses:

(a) buildings and related facilities;
(b) equipment, encompassing both hardware and software;
(c) transportation;
(d) information and communication technologies.

Environment for the operation of processes

The organization is responsible for identifying, providing, and upholding the environment necessary for its processes' operation to ensure product and service conformity. This environment encompasses a blend of human and physical elements, including:

(a) social factors, such as fostering a non-discriminatory, tranquil, and non-confrontational atmosphere;
(b) psychological aspects, including stress reduction, burnout prevention, and emotional support;
(c) physical conditions like temperature, humidity, lighting, airflow, cleanliness, and noise levels.

These factors may vary significantly based on the nature of the products and services offered.

Monitoring and Measurement of Resources

The organization must ascertain and furnish the resources essential to ensure accurate and dependable results when monitoring or measuring product and service conformity. It must ensure that these resources:

(a) are suitable for the specific monitoring or measurement activity being conducted;
(b) are maintained to sustain their ongoing suitability for their intended purpose.

The organization must retain pertinent documented information to demonstrate the suitability of monitoring and measurement resources for their designated purpose.

Measurement Traceability

When measurement traceability is deemed necessary by the organization to validate measurement results, measuring equipment must:

(a) Be calibrated or verified at specified intervals or before use against measurement standards traceable to international or national measurement standards. If such standards are unavailable, the basis used for calibration or verification must be documented.
(b) Be clearly identified to determine their status.
(c) Be safeguarded against adjustments, damage, or deterioration that could compromise calibration status and subsequent measurement results.

The organization must assess whether the validity of prior measurement results has been compromised if the measuring equipment is determined to be unsuitable for its intended purpose and take appropriate corrective measures if necessary.

Organizational knowledge

The organization is responsible for identifying the knowledge essential for its processes' operation and to ensure product and service conformity. This knowledge must be continuously maintained and accessible as needed. To address evolving needs and trends, the organization must leverage its existing knowledge base and determine strategies for acquiring any additional necessary knowledge and updates.

Organizational knowledge may be derived from:

(a) Internal sources, including intellectual property, experiential knowledge, lessons learned from both successes and failures, undocumented knowledge and experiences, and insights gained from process, product, and service enhancements.
(b) External sources, such as industry standards, academic institutions, professional conferences, and insights gathered from external customers or suppliers.

Competence

The organization must:

(a) Assess the necessary competence levels of individuals performing tasks under its control that impact the quality management system's performance and effectiveness.
(b) Ensure that these individuals possess the required competence through appropriate education, training, or experience.
(c) Take necessary actions to attain the required competence where applicable and evaluate the effectiveness of these actions.
(d) Maintain adequate documented evidence of competence.

Awareness

The organization must ensure that individuals performing activities under its jurisdiction are informed about:

(a) The quality policy.
(b) Relevant quality objectives.
(c) Their contribution to enhancing the quality management system's effectiveness, including the benefits of improved performance.
(d) The ramifications of nonconformity with the quality management system's requirements.

Communication

The organization must identify the pertinent internal and external communications for the quality management system, determining:

(a) what information to communicate;
(b) when to communicate;
(c) whom to communicate with;
(d) how to communicate;
(e) who will be responsible for communication.

Documented Information

The organization's quality management system must encompass:

(a) Documented information mandated by this international standard.
(b) Documented information deemed necessary by the organization for the quality management system's effectiveness.

Creation and Update

During the creation and updating of documented information, the organization ensures:

(a) Adequate identification and description, including title, date, author, or reference number.
(b) Proper formatting, encompassing language, software version, graphics, and medium (e.g., paper or electronic).
(c) Review and approval to ensure suitability and adequacy.

Control of Documented Information

Documented information required by the quality management system and this international standard must be managed to ensure:

(a) Availability and utilization when and where needed.
(b) Adequate protection against loss of confidentiality, improper use, or loss of integrity.

Operational Planning and Control

The organization is tasked with planning, executing, and overseeing the processes essential to meet the requisites for delivering products and services, as well as executing the actions outlined in the planning phase for:

(a) Identifying the prerequisites for products and services.
(b) Establishing standards for processes and the acceptance of products and services.
(c) Determining the resources required to ensure compliance with product and service requirements.
(d) Implementing process control in accordance with these standards.
(e) Identifying, retaining, and managing documented information to ensure confidence in the execution of planned processes.
(f) Validating the conformity of products and services to their specifications.

The outcomes of this planning must align with the organization's operational needs. Additionally, the organization must regulate planned alterations and assess their repercussions, intervening when necessary to mitigate adverse effects. Moreover, it must exercise control over outsourced processes.

Changes to Product and Service Requirements

The organization must ensure that documented information is updated and that personnel are informed of any changes in requirements when modifications are made to the requirements for products and services.

Regarding the design and development of products and services, the organization should consider various factors to determine the phases and controls:

(a) The nature, duration, and complexity of design and development activities.
(b) The necessary process stages, including relevant design and development reviews.
(c) Verification and validation activities required for design and development.
(d) Responsibilities and authorities involved in the design and development process.
(e) Internal and external resource requirements for product and service design and development.
(f) The need to manage interfaces between parties engaged in the design and development process.
(g) Involvement of customers and users in the design and development process.
(h) Requirements for the subsequent supply of products and services.
(i) The expected level of control over the design and development process by customers and other relevant stakeholders.
(j) Documented information needed to demonstrate compliance with design and development requirements.

Designing and developing changes encompasses overseeing externally sourced processes, products, and services.

Post-delivery actions involve regulating modifications, product and service releases, and handling nonconforming outputs.

Performance entails the ongoing monitoring, measurement, analysis, and evaluation processes.

Continuous Improvement

The organization must continually improve the suitability, adequacy, and effectiveness of the quality management system. The organization must consider the results of analysis and evaluation and the outputs from management review to determine whether there are needs or opportunities that must be addressed as part of continuous improvement.

The three pillars of the ISO 9001:2015 standard are process approach, risk-based thinking, and continuous improvement.

Two of the most important objectives in the revision of the ISO 9000 series of standards are (a) to develop a simplified series of standards that will be equally applicable to small as well as medium and large organizations, and (b) the quantity and detail of the required documentation that must be more relevant to the desired outcomes of the organization's process activities.

Thus, the changes are made to:

- structure – following the 10-section structure defined in the SL appendix;
- key concepts;
- terminology.

All the requirements of the ISO 9001:2015 standard are generic and intended to be applicable to any organization, regardless of its type or size, or the products and services it provides.

ISO 9001:2015 outlines criteria for a Quality Management System when an organization:

- Seeks to showcase its capacity to consistently deliver products and services in line with customer and relevant legal and regulatory demands.
- Strives to elevate customer satisfaction by effectively applying the system, including processes for enhancing the system and ensuring compliance with customer and relevant legal and regulatory prerequisites.

Benefits of ISO 9001 Certification

Obtaining ISO 9001 certification offers numerous advantages for organizations. It enhances customer satisfaction by ensuring products consistently meet their requirements and fosters trust among stakeholders. ISO 9001 also drives operational efficiency and cost reduction through continuous improvement of processes. By complying with legal requirements and enhancing risk management practices, organizations can minimize disruptions and maintain operational integrity. Achieving ISO 9001 certification establishes credibility and unlocks business opportunities, providing a competitive edge in the market. Overall, ISO 9001 certification signifies a commitment to quality excellence and continuous improvement, leading to sustained growth and success for organizations.

ISO 14001 for Environmental Management

According to ISO 14001:2015, implementing an environmental management system can enhance a business's environmental performance. Organizations seeking to systematically manage their environmental responsibilities in alignment with sustainability goals are encouraged to adopt ISO 14001:2015. This standard assists organizations in achieving environmental management objectives, benefiting the environment, the organization itself, and stakeholders. Anticipated outcomes of implementing an environmental management system include improved environmental performance, compliance with regulations, and attainment of environmental objectives outlined in the organization's environmental policy. ISO 14001:2015 is applicable to organizations of any size, type, or industry, addressing environmental aspects across their operations, products, and services. While ISO 14001:2015 does not prescribe specific environmental performance requirements, organizations can choose to implement the standard wholly or partially to systematically enhance environmental management. However, claims of conformance to ISO 14001:2015 are only valid if the organization's environmental management system fully complies with all standard requirements consistently. It provides instructions on how to achieve continuous improvement in environmental performance and defines the prerequisites for developing an environmental management system (EMS). Regardless of size or industry, all firms can use ISO 14001, which can be applied to any part of their operations that has an influence on the environment (Raines, 2002).

Commitment to environmental protection, observance of relevant rules and regulations, pollution prevention, and continuous improvement are among ISO 14001's core values. The requirements of ISO 14001 include developing an environmental strategy, carrying out an extensive environmental impact assessment, creating goals and benchmarks for improvement, putting operational controls in place to reduce environmental impacts, and routinely tracking and evaluating performance. Putting ISO 14001 into practice can assist businesses in minimizing their environmental impact (Gattiker and Carter, 2010).

There are various steps in the ISO 14001 implementation process. Organizations must first create an environmental strategy that complements their corporate goals and legal obligations. Finding potential environmental dangers and possibilities is part of this. To understand the organization's environmental performance at the present, a thorough environmental impact assessment should be conducted. This evaluation aids in locating areas that can be improved. Organizations can set targets and benchmarks for improvement following the evaluation, which will give their environmental management a clear direction (Sambasivan and Fei, 2008).

Environmental management techniques must be successfully implemented with the support of top management and employee participation. Top management sets the tone for the entire organization and motivates staff to actively participate in environmental projects when they demonstrate a strong commitment to environmental sustainability. This involvement can take the shape of including environmentally friendly practices in routine business operations or involving staff members in decision-making procedures regarding environmental performance. The organization's environmental management system becomes more efficient and sustainable due to the collaborative efforts of both upper management and staff members (Bresciani et al., 2022).

Resistance to change, a lack of awareness or understanding among staff, and the initial costs associated with implementing new practices or technologies are some of the challenges and barriers encountered during the implementation process of incorporating environmentally friendly practices into routine business operations or involving staff members in decision-making procedures regarding environmental performance. These problems, however, can be solved with good training, communication, and top management support.

ISO/IEC 27001

ISO/IEC 27001 is a global standard that specifies the standards for establishing, implementing, maintaining, and continuously improving an information security management system (ISMS). This standard establishes a framework for companies to manage and safeguard sensitive information, assuring data confidentiality, integrity, and availability. Companies may successfully manage risks, preserve precious assets, and build a security culture within their operations by adhering to ISO/IEC 27001 principles. ISO/IEC 27001 also assists enterprises in meeting legal, regulatory, and contractual information security requirements. Furthermore, by adhering to this standard, businesses may demonstrate their dedication to protecting client data while gaining a competitive advantage in the market.

The ISO/IEC 27001 standard establishes the structure for creating an ISMS and assists businesses in managing and safeguarding their information assets. By implementing an ISMS based on ISO/IEC 27001, organizations can identify and assess potential risks to their information assets, allowing them to adopt appropriate controls and procedures to mitigate these risks. This helps to maintain the confidentiality, integrity, and availability of sensitive information, which increases consumer trust and fosters commercial partnerships. Furthermore, ISO/IEC 27001 provides a framework for continuous evolution, allowing firms to regularly examine and update their information security policies in order to react to new threats and technology (Accerboni and Sartor, 2019).

ISO/IEC 27001, an internationally recognized standard for information security management systems (ISMS), serves as a comprehensive framework for organizations to safeguard their sensitive information and effectively manage associated risks. By adopting ISO/IEC 27001, organizations can establish, implement, maintain, and continually improve their ISMS, thereby ensuring the protection of their valuable data.

The primary purpose of ISO/IEC 27001 is to provide organizations with a systematic approach to managing information security. By adhering to this standard, organizations can effectively identify, assess, and mitigate risks related to their information assets. This systematic approach enables organizations to establish robust controls and procedures, ensuring the confidentiality, integrity, and availability of their information.

It is worth mentioning that ISO/IEC 27001 applies to organizations regardless of their size, industries, or sectors. Whether a public or private sector organization, the standard can be implemented to enhance information security practices. This inclusivity allows organizations across various domains to benefit from the guidance provided by ISO/IEC 27001.

In terms of structure, ISO/IEC 27001 follows a similar format to other ISO management system standards, such as ISO 9001 for quality management. The standard consists of several clauses, each addressing specific aspects of information security management. These clauses encompass various subjects such as risk assessment, security policy, and asset management, access control, and incident management. By adhering to these clauses, organizations can establish a robust and comprehensive ISMS that aligns with their overall business objectives.

In summary, ISO/IEC 27001 offers organizations a globally recognized framework to protect their sensitive information and manage associated risks effectively. The structured nature of ISO/IEC 27001, with its various clauses, ensures that organizations can implement comprehensive controls and procedures to safeguard their valuable information assets. The framework comprises of several essential clauses that are crucial for the effective implementation of an Information Security Management System (ISMS). These clauses are meticulously designed to ensure the comprehensive protection of sensitive information within an organization. Let us delve into each clause in detail:

1. **Context of the organization**: This clause emphasizes the significance of understanding the organization and its context. By thoroughly comprehending the internal and external factors that influence the organization, one can effectively identify the interested parties and their specific requirements. This knowledge serves as a solid foundation for the successful implementation of the ISMS.
2. **Leadership**: Demonstrating unwavering commitment to information security is of paramount importance. This clause emphasizes the need for strong leadership within the organization. By defining clear roles and responsibilities, individuals can be held accountable for the security of information. This ensures that all members of the organization are aware of their obligations and actively contribute to the protection of sensitive data.
3. **Planning**: The planning clause plays a pivotal role in establishing a robust ISMS. It involves determining the scope of the system, setting information security objectives, conducting meticulous risk assessments, and developing a comprehensive risk treatment plan. These meticulous planning activities lay the groundwork for effective implementation and management of information security measures.
4. **Support**: Adequate support is essential for the successful implementation of an ISMS. This clause emphasizes the provision of necessary resources, competence, awareness, communication, and documentation to support the ISMS. By ensuring that all required elements are in place, organizations can effectively address information security challenges and mitigate potential risks.
5. **Operation**: This clause focuses on the operational aspects of managing identified risks. It involves meticulous planning, implementation, and control of security measures. By effectively managing and monitoring security measures, organizations can proactively address potential vulnerabilities and safeguard sensitive information.
6. **Performance evaluation**: Regular monitoring, measuring, analysing, and evaluating the performance of the ISMS is crucial for its continuous improvement. This clause emphasizes the importance of conducting thorough evaluations to identify areas of improvement and address any shortcomings. By analysing the performance of the ISMS, companies can make well-informed decisions and implement necessary actions to enhance information security.
7. **Improvement**: The final clause emphasizes the need for continuous improvement of the ISMS. Based on the results of evaluations and changing circumstances, organizations should strive to enhance the effectiveness of their information security measures. Through fostering a culture of ongoing enhancement, organizations can effectively respond to changing threats and safeguard sensitive information in the long term.

The comprehensive framework of the ISMS comprises of several essential clauses that collectively contribute to the effective protection of sensitive information within an organization. By adhering to these clauses and implementing the necessary measures, organizations can confidently safeguard their valuable data and maintain the trust of their stakeholders.

ISO/IEC 27001 places a strong emphasis on risk management. It is of utmost importance for organizations to meticulously assess and treat information security risks in a systematic manner. This ensures that information assets are adequately protected, safeguarding the integrity and confidentiality of sensitive data.

Organizations have the opportunity to seek ISO/IEC 27001 certification from accredited certification bodies. This certification process entails a rigorous audit, meticulously evaluating whether the organization's Information Security Management System (ISMS) aligns with the stringent requirements set forth by the standard. By undergoing this thorough assessment, organizations can demonstrate their commitment to maintaining the highest standards of information security.

The attainment of ISO/IEC 27001 certification can yield numerous benefits for organizations. Firstly, it enhances information security measures, fortifying the organization's ability to protect valuable data from unauthorized access or breaches. Additionally, achieving certification instils confidence and trust among stakeholders, assuring them that the organization has implemented robust security controls. Furthermore, compliance with legal and regulatory requirements is ensured, mitigating potential risks and liabilities. Lastly, ISO/IEC 27001 certification provides a competitive advantage in the marketplace, distinguishing the organization as a trusted and reliable partner in the realm of information security.

ISO/IEC 27001 fosters a culture of perpetual enhancement in the realm of information security management. It strongly encourages organizations to engage in regular and systematic evaluations of their Information Security Management Systems (ISMS) in order to effectively adapt to the ever-evolving landscape of threats and vulnerabilities.

ISO/IEC 27001 seamlessly integrates with other respected management system standards like ISO 9001 (quality management) and ISO 14001 (environmental management), forming a comprehensive management system that effectively addresses diverse aspects of an organization's operations and safeguards information security optimizing the overall operational efficiency.

Towards Quality 5.0 – A Glimpse into the Future

Quality 4.0 is a contemporary notion stemming from the Industry 4.0 movement, highlighting the utilization of technology to elevate quality management methodologies. Coined in the ASQ Future of Quality Report of 2015, this concept anticipates a revitalization of quality management tools and techniques, cantered around four key themes: (1) quality through inspection; (2) quality via design; (3) quality through empowerment, entailing the adoption of comprehensive quality approaches like TQM and Six Sigma to foster greater accountability and empowerment for continual enhancement; and (4) quality through discovery, entailing the use of adaptive and intelligent environments to address challenges and issues. The implementation of Quality 4.0 involves the integration of advanced technologies such as artificial intelligence, the Internet of Things, and big data analytics into quality control processes to boost effectiveness, minimize expenses, and enhance customer contentment. The adoption of Quality 4.0 has the potential to transform quality management from a reactive to a proactive approach, with real-time data and insights enabling predictive maintenance and quality assurance. Several studies have shown that the adoption of Quality 4.0 can result in notable enhancements in quality performance, including reduced defects, improved reliability, and increased customer satisfaction (Jadhav and Deshmukh, 2022; Zeng et al., 2024). Furthermore, the application of Quality 4.0 can facilitate compliance with regulatory requirements and provide organizations with a competitive advantage in the market (Dey et al., 2020). Professor Antony, a well-known quality expert, has also discussed the concept of Quality 4.0. He highlights the importance of Quality 4.0 in addressing the challenges faced by traditional quality management systems in the digital age. According to Professor Antony, Quality 4.0 can help organizations overcome the limitations of manual quality management systems by providing real-time data and insights to improve decision-making, and enabling proactive and predictive quality management. He also emphasizes the need for organizations to embrace Quality 4.0 as a means of achieving competitive advantage and sustainability (Antony et al., 2019). Overall, the concept of Quality 4.0 represents a significant shift in quality management, with the potential to revolutionize quality control practices and enhance organizational performance. Radzwill (2018) highlighted six value propositions for Quality 4.0 initiatives, placing the enhancement of human intelligence as the top priority. Esko Kilpi, famous researcher and strategist, as cited by Radzwill in 2018, proposed that the future of work should prioritize augmentation over automation within a post-industrial framework rather than pursuing automation within the traditional industrial model. These concepts have influenced the development of Quality 4.0 Tools, which encompass various artificial intelligence capabilities like natural language processing and sophisticated decision-making, a robust Big Data infrastructure facilitating seamless access to data sources, management, and analysis, blockchain technology to enhance transparency and auditability of transactions while monitoring conditions to meet quality objectives, deep learning for tasks such as image classification, intricate pattern recognition, time series forecasting, text generation, and recognition, alongside facilitating technologies such as sensors, open-source software, and 5G networks. Machine learning is also incorporated, with applications such as text analysis, recommendation systems, fraud detection, and forecasting. Additionally, data science techniques like predicting, simulation, classification, and inference are employed to generate viable models and solutions. As outlined by Radzwill (2018), professionals in the quality domain are tasked with spearheading the shift from traditional methodologies to Quality 4.0. This transition necessitates proficiency in several key areas, including systemic thinking, data-informed decision-making, fostering leadership conducive to organizational learning, instituting protocols for ongoing enhancement, and comprehending the broader societal impacts of decisions on individuals, relationships, communities, well-being, health, and society as a whole.

Conversely, Quality 5.0 is a relatively new concept that builds upon the principles of Quality 4.0, taking into account the evolving needs of organizations in the digital age. Quality 4.0 emphasizes the integration of advanced technologies into quality management practices, whereas Quality 5.0 places a greater emphasis on the human element of quality control, recognizing the importance of people and their knowledge and skills in ensuring quality. The notion of Quality 5.0 underscores the importance of organizations adopting a people-centred strategy towards quality management, prioritizing the cultivation of a quality-oriented culture and empowering staff to assume responsibility for quality assurance procedures. This perspective acknowledges that while technology plays a role, it alone cannot guarantee quality, emphasizing instead the indispensable role of human insight and discernment in maintaining quality within intricate and evolving contexts (López-Fresno et al., 2021). However, the concept of Quality 5.0 is still evolving, it represents an important development in quality management, reflecting the changing needs of organizations in the digital age.

Quality 5.0 represents a tier of quality management practices that integrates technology and data-driven decision-making. It underscores the utilization of automation, AI, and machine learning to elevate the standards of products and services. In the ensuing phase of the industrial revolution, dubbed Industry 5.0, the emphasis is on the symbiotic relationship between humans and machines, placing their coexistence at the forefront. By utilizing the distinct advantages of both people and robots, this strategy strives to promote innovation and productivity. Industry 5.0 makes it possible to create new solutions and improve existing processes by fusing human creativity, problem-solving skills, and emotional intelligence with cutting-edge technologies (Tadić, 2022).

Industry 5.0 has the potential to greatly increase productivity and creativity across a range of industries. For instance, the coexistence of people and robots can result in quicker and more effective production processes in the manufacturing industry. Humans can concentrate on tasks that call for creativity, problem-solving abilities, and emotional intelligence, while robots can execute repetitive and physically taxing duties with accuracy and speed. This interaction between humans and robots has the potential to generate fresh ideas, enhance product designs, and boost overall productivity. Furthermore, by enabling predictive analytics and data-driven decision-making, the incorporation of cutting-edge technology, such as artificial intelligence and machine learning, can further increase innovation (Souza et al., 2022).

Companies can improve the accuracy and efficiency of their operations by embracing Quality 5.0 practices. Automation and artificial intelligence in quality control procedures are examples of quality 5.0 techniques that can assist businesses. Automation can speed up and simplify the quality control procedure, requiring less manual inspection time and effort. More accurately than human inspectors, artificial intelligence can analyse vast amounts of data to find patterns and spot flaws. This may lead to higher customer satisfaction, lower error rates, and better product quality.

These cutting-edge technologies can be used in conjunction with Industry 5.0 principles, which place a premium on human creativity and problem-solving skills, to create novel solutions and raise the calibre of all products (Sami et al., 2022).

However, implementing Quality 5.0 methods in quality control processes, such as automation and artificial intelligence, may present certain issues. Companies, for example, may need to invest in new technology and infrastructure, which may necessitate large financial resources. Employees who are concerned that automation would take their jobs may also be resistant. To address these concerns, businesses can offer training and re-skilling programs to empower individuals with the skills they need to work alongside technology and reap its benefits. It is critical for organizations to highlight the positive impact of Quality 5.0 standards on job roles and to encourage collaboration over replacement. Companies can ensure an easier transition to Quality 5.0 practices by proactively addressing these difficulties.

Defining Total Quality Management

Total Quality Management (TQM) aims to optimize the effectiveness and efficiency of all stages in product or service creation, considering overall system performance. It comprises quality control, assurance, and improvement. To achieve a more consistent level of quality, TQM utilizes quality assurance, as well as process and product control.

TQM is predicated upon the comprehensive engagement of all employees within the organization. In order to diligently pursue the objective of quality, an organization must meticulously scrutinize it at three distinct levels: the organizational level, the process level, and the execution level, also known as the work level or activity design level. At the organizational level, the primary focus is directed towards the fulfilment of the external customer's desired requirements pertaining to the product or service. Consequently, it is imperative for an organization to consistently solicit customer input in order to ensure utmost satisfaction.

At the process level, organizational units are classified as functions or departments. These include marketing, design, product development, operations, finance, procurement, and billing. It is worth noting that most processes are inter-functional, meaning that those who are responsible for specific organizational units have the opportunity to optimize activities under their control.

At the execution level, it is imperative to establish production standards that are firmly rooted in the quality and service requirements as defined by the esteemed customer. These requirements are meticulously monitored and upheld at both the organizational and process levels. The standards in question encompass a wide range of criteria, including but not limited to innovation, timeliness, completeness, accuracy, and cost. The overarching objective of TQM is to seamlessly integrate all organizational functions, thereby directing their collective efforts towards the fulfilment of customer needs and the attainment of organizational goals.

It is defined as:

- Total = Made up of the whole.
- Quality = Degree of excellence of the product/service.
- Management = The art of dealing with, controlling, and directing.

According to the TQM approach, an organization is seen as a set of processes. Organizations should constantly enhance processes by incorporating employee knowledge and experiences. The simple goal of TQM is "*To do things right, the first time, every time*". TQM is infinitely variable and adaptable. This management approach covers different areas, although it was originally applied to production operations. For many years, it was used only in this field. Later, TQM was identified as a generic management tool that also applies to services (health and safety), industries, and the public sector.

Core Principles of TQM

The core principles of TQM comprise:

(a) Management commitment: This principle highlights the crucial role of management in leading, directing, and actively participating in quality improvement efforts. It involves planning, executing, reviewing, and continuously revising processes to ensure quality standards are met.
(b) Employee responsibility: TQM emphasizes that every employee is responsible for maintaining and improving quality. This involves providing adequate training, fostering teamwork, setting measurable goals, and recognizing employees for their contributions to quality improvement.
(c) Continuous Improvement: TQM promotes the idea of continuous improvement as a fundamental aspect of quality management. This includes systematically measuring performance, forming teams dedicated to excellence, managing processes across different functions, and striving to achieve, maintain, and enhance quality standards over time.
(d) Customer Focus: TQM places a strong emphasis on understanding and meeting customer needs and expectations. This involves forming partnerships with suppliers, fostering service-oriented relationships with internal customers, setting customer-oriented standards, and ensuring that quality is never compromised.

Overall, these core principles of TQM emphasize the importance of leadership commitment, employee involvement, continuous improvement, and a customer-centric approach to achieving and maintaining high-quality standards in an organization.

Benefits of TQM and Essential Requirements for Successful TQM Implementation

Implementing TQM brings forth a multitude of advantages for an organization, shaping its operational landscape and fostering a culture of excellence. Chief among these advantages are:

- Enhanced Organizational Knowledge and Quality Culture: TQM instils a deep-rooted understanding and appreciation for quality maintenance across all levels of the organization. It serves as a catalyst for cultivating a culture where quality is not just a goal but a way of life. Through comprehensive training and unwavering support, management leads the charge in removing barriers to continuous quality improvement and fostering an environment conducive to attitudinal shifts.
- Emphasis on Teamwork: TQM places particular emphasis on the power of teamwork. Recognizing that quality improvement is a collective effort, the organization prioritizes

collaboration and synergy among its members. By fostering an environment where every individual's contribution is valued, TQM strengthens the fabric of the organization and drives collective success.

- Commitment to Continuous Improvement: Central to TQM is the unwavering commitment to continuous improvement. Understanding that perfection is an ongoing journey rather than a destination, the organization embraces a mindset of perpetual enhancement. This commitment is not only driven by internal motivations but also adapts to changing customer and supplier needs and expectations.

Essential requirements for successful TQM implementation include:

(a) **Commitment**: TQM requires wholehearted commitment from every member of the organization, with management taking a proactive stance in facilitating and promoting quality improvement initiatives. This involves removing barriers, providing necessary resources, and fostering an environment conducive to change through expanded training and support mechanisms.
(b) **Cultivating a quality culture**: Before tangible changes can be realized, the organization must invest in comprehensive training and awareness programs to instil a culture where quality is ingrained in every aspect of operations and attitudes.
(c) **Embracing continuous improvement**: Continuous improvement should not be viewed as a one-off effort, but rather as a perpetual journey guided by the demands of customers and suppliers, underscored by a resolute dedication to surpassing expectations.
(d) **Customer focus**: TQM places paramount importance on understanding and meeting the needs of customers, whether they are internal or external stakeholders. By prioritizing customer satisfaction and striving for perfection without defects, the organization aligns its goals with those of its end-users.
(e) **Vigilant control**: Regular monitoring and verification are essential components of TQM, ensuring that processes adhere to the planned implementation course and deviations are promptly addressed. This exercise of control safeguards against lapses and reinforces the organization's commitment to quality excellence.

TQM embodies a harmonious blend of diverse activities aimed at achieving organizational excellence. To effectively implement TQM, organizations must prioritize eight key elements, categorized into four functional groups, as described below.

The Foundation: Ethics, Integrity, and Trust

At its core, TQM is built upon a foundation of ethical conduct, integrity, and trust. Ethics serve as the guiding principle, instilling discipline and professionalism among all employees. Integrity, encompassing honesty, morality, and respect for facts, is paramount in fostering trust both within the organization and with external stakeholders. Trust, as a byproduct of integrity and ethical behaviour, forms the cornerstone of TQM, facilitating full participation, accountability, and commitment from all members.

The Pillars: Training, Teamwork, and Leadership

Training plays a pivotal role in enhancing employee productivity and accuracy in implementing TQM principles. Through regular training programs, employees acquire the necessary knowledge and skills to drive TQM initiatives forward. Teamwork is indispensable for leveraging resources effectively and solving problems efficiently. By fostering collaboration and synergy among team members, organizations can achieve faster and more enduring improvements in processes and operations. Leadership emerges as the linchpin of TQM, providing direction, inspiration, and strategic guidance to propel the organization toward its quality objectives. Effective leaders define clear goals, articulate compelling visions, and engage employees at all levels to realize the shared vision of excellence.

The Engine: Communication

Communication acts as the engine driving TQM, serving as the dynamic link that unites all elements of the organizational framework. From the foundation to the pinnacle of TQM, effective communication fosters connectivity and cohesion among stakeholders. It facilitates the exchange of ideas, feedback, and information essential for aligning efforts and driving continuous improvement. Strong relationships between senders and receivers ensure the smooth flow of communication, both internally and externally, among all parties involved in the TQM process.

The Roof: Recognition

Recognition serves as the crowning element of TQM, motivating and reinforcing desired behaviours and outcomes. Acknowledging and rewarding individuals and teams for their contributions fosters a culture of excellence and incentivizes continued efforts toward quality improvement. Timely recognition boosts morale, enhances self-esteem, and inspires greater commitment and dedication among employees. By celebrating achievements and successes, organizations nurture a positive and supportive environment conducive to sustained excellence.

In summary, by embracing these eight elements of TQM, organizations can cultivate a culture of quality, encourage cooperation and creativity to attain enduring success in today's competitive environment.

Key Takeaways

- **Quality management principles**: The chapter delineates the fundamental principles of quality management, which form the bedrock of Total Quality Management (TQM), placing particular emphasis on customer-centricity, perpetual enhancement, and active participation of employees.
- **Quality paradigms**: The concept of quality paradigms is introduced, explaining different approaches to quality management over time, including inspection, Total Quality Control, Company-Wide Quality Control, and TQM.
- **The 9M framework**: The 9M Framework is introduced as a tool for assessing and managing various factors affecting quality, including materials, manpower, methods, and more.
- **ISO 9001:2015**: The ISO 9001 standard and its evolution are discussed, emphasizing its role in ensuring quality management within organizations.
- **Environmental management (ISO 14001)**: An overview of ISO 14001 for environmental management is provided, showcasing the importance of integrating environmental concerns into quality management.

- **Information security (ISO/IEC 27001)**: The chapter briefly touches on ISO/IEC 27001, a standard for information security management.
- **Quality 5.0**: A glimpse into the future of quality management, referred to as Quality 5.0, is presented, highlighting the evolving nature of quality practices.

Review Questions

1. What is the concept of quality, and why is it important in the context of business and manufacturing?
2. Describe the evolution of quality management from its early stages to TQM practices. What were the key milestones?
3. Discuss the different definitions of quality presented in the chapter and their significance in enhancing our comprehension of the concept.
4. List and briefly explain the core quality management principles outlined in the chapter. Why are these principles important in TQM?
5. Discuss the different quality paradigms mentioned in the chapter, such as inspection, Total Quality Control, Company-Wide Quality Control, and TQM. How do they differ from each other?
6. Who are some of the pioneering quality gurus mentioned in the chapter, and what were their contributions to quality management?
7. What is the 9M Framework, and how can it be used to assess and manage factors affecting quality?
8. Provide an overview of ISO 9001:2015. What is its significance in quality management, and how does it relate to TQM?
9. Briefly explain the concept of ISO 14001 and its relevance to environmental management in the context of TQM.
10. What is ISO/IEC 27001, and why is it important for information security management?
11. What is Quality 5.0, and how does it represent the future of quality management? What are the key characteristics of Quality 5.0?

2 AI Technologies in Modern Quality Management

Learning Objectives

- Understand the significance of AI technologies in quality control.
- Gain introductory knowledge of Artificial Intelligence, Machine Learning, and Digital Twins.
- Explore the various applications of AI in quality management.
- Learn how AI-driven quality control systems operate.
- Understand how AI can be used for predictive maintenance and fault detection.
- Discover how AI can enhance process efficiency in quality management.
- Learn about the integration of AI with Total Quality Management (TQM) and Lean Thinking 5.0.

The Significance of AI Technologies in Quality Control

Quality management is an exceedingly crucial aspect across a multitude of industries, as it serves the purpose of guaranteeing that products and services not only meet, but surpass, the established standards, thereby exceeding the expectations of customers. With the continuous advancement of technology, the integration of artificial intelligence (AI) has emerged as a revolutionary force, reshaping traditional business processes (Dieniezhnikov and Pshenychna, 2021). Thus, we shall delve into the profound role that AI plays in quality management, with a particular focus on its potential to enhance efficiency, accuracy, and decision-making capabilities within the quality process. By leveraging AI, companies can modernise their operations, reach higher levels of precision, and make more informed choices, ultimately leading to improved overall quality (Gorbacheva and Smirnov, 2017).

AI's automation of quality control processes stands as a paramount advancement in quality management. Through AI-powered systems, organizations streamline inspections and testing, leveraging algorithms to swiftly analyse extensive data, enabling real-time decision-making and diminishing manual efforts (Paliukas and Savanevičienė, 2018).

AI's integration into quality assurance extends to predictive analytics, utilizing historical data and machine learning algorithms to anticipate defects and deviations. This proactive strategy allows organizations to pre-emptively implement preventive measures, thwarting potential quality issues before they arise.

DOI: 10.4324/9781032726748-2

Consequently, real-time monitoring and continuous improvement become not only attainable but also highly achievable through the invaluable insights provided by AI-driven systems.

Furthermore, it is important to acknowledge the pivotal role played by AI-based sensors and monitoring systems in quality management. These advanced technologies provide organizations with the means to monitor and track quality-related parameters with utmost precision and accuracy. By employing AI-based sensors and monitoring systems, organizations can ensure that quality standards are consistently met, thereby enhancing overall product or service quality.

In conclusion, the integration of AI into quality management processes offers numerous advantages. From the automation of quality control processes to the implementation of predictive analytics and the utilization of AI-based sensors and monitoring systems, organizations can significantly enhance their quality assurance practices. By embracing these AI-driven solutions, organizations can reach advanced levels of efficiency, accuracy, and ultimately, customer satisfaction (Vela-Valido, 2021). These systems possess the capability to gather data in real-time, meticulously observe quality parameters, and promptly recognize any irregularities. Through the utilization of adaptive algorithms, artificial intelligence has the capacity to discern intricate patterns, prevailing trends, and deviations that may prove challenging for human operators to discern. This ensures that any concerns regarding quality are promptly addressed, thereby facilitating enhanced product consistency and customer satisfaction.

AI has the remarkable ability to eliminate the possibility of human error in the quality process. Through the automation of tasks and the utilization of advanced algorithms, AI greatly enhances the accuracy and reliability of quality control activities. One of the key advantages of AI systems is their exceptional precision in detecting and analysing defects, resulting in consistent and reliable quality outcomes (Zhang et al., 2021).

Moreover, integrating AI-driven automation and data processing significantly boosts efficiency and productivity in quality management. Automating manual tasks accelerates data processing and analysis, thereby shortening the time needed for inspections. This decrease in manual labour not only saves time but also cuts costs, enabling more effective resource allocation.

AI bestows upon organizations the invaluable gift of enhanced decision-making capabilities in the domain of quality control. Through the provision of data-driven insights, AI facilitates the implementation of proactive quality control measures. By adeptly identifying potential risks and areas for process optimization, AI lends its unwavering support to the decision-making process at various hierarchical levels within the organization. Artificial intelligence has also the capability to offer valuable recommendations in the realm of risk mitigation (Wei and Pardo, 2022). This, in turn, empowers organizations to make well-informed decisions that have the potential to enhance the overall quality outcomes they strive to achieve. The utilization of AI in this context is undoubtedly advantageous, as it provides a reliable and objective source of guidance. By leveraging the insights provided by AI, organizations can confidently navigate the complexities of risk management and take appropriate measures to mitigate potential hazards.

Introduction to Artificial Intelligence, Machine Learning and Digital Twins

The simulation of human intelligence in robots that are designed to think and learn like humans is known as artificial intelligence, or AI. It includes a number of subfields, including expert systems, computer vision, and natural language processing. Machine learning (ML), a subset of AI, empowers machines to learn from data without explicit programming. Through ML algorithms, computers analyse vast datasets, identify trends, and make conclusions or predictions.

The modern world has grown to rely more and more on AI and ML because of their capacity to automate processes, increase productivity, and make precise forecasts. Applications for these technologies can be found in a number of industries, including customer service, banking, healthcare, and transportation. AI and ML, for instance, can be utilized in the medical field to diagnose illnesses by analysing medical images. AI systems in finance are capable of predicting market movements and making investment choices. This concise synopsis should help the reader understand the vast array of opportunities and advantages that AI and ML present. Applications of AI and ML in the real world are already many. Natural language processing, for example, is used by voice assistants such as Siri and Alexa to comprehend and react to human orders. Self-driving cars employ computer vision algorithms to recognize and understand objects they encounter on the road. Recommendation systems, such as those offered by Netflix and Amazon, employ machine learning algorithms to provide customers with individualized content and merchandise recommendations. The reader would be better able to see how AI and ML are already affecting their daily life if these instances were included. Despite their enormous potential, AI and ML have some drawbacks and difficulties. One difficulty is that machine learning algorithms require vast quantities of high-quality data to be trained. Occasionally, getting such information could be expensive or challenging. Potential prejudice in AI systems is another problem. Discriminatory results may result from biased training data used to create an AI system. The application of AI and ML raises additional ethical questions, such as privacy and employment displacement. An examination of these restrictions and difficulties would offer a more complex viewpoint on the subject. ML is one of the many subfields that make up the large field of artificial intelligence. AI encompasses more than just machine learning; it also includes computer vision, natural language processing, and expert systems. ML plays a pivotal role in the field of AI as it empowers robots with the ability to analyse data and derive insightful conclusions or predictions based on patterns. By elucidating this connection, the reader will acquire a more comprehensive comprehension of the hierarchical structure and vast scope encompassed within these domains.

Digital Twins are virtual representations of physical assets, processes, or systems that enable real-time monitoring, analysis, and optimization. By creating a digital replica, businesses can simulate various scenarios, predict outcomes, and optimize performance, leading to improved efficiency, reduced costs, and enhanced innovation.

Digital twins have been a topic of interest in scientific literature across various fields such as engineering, manufacturing, healthcare, and more. The aim of digital twins is to create virtual models of physical systems or processes that can be used for simulation, analysis, and optimization. Here is an overview of the structure and case studies of digital twins in scientific literature:

Thus, digital twins are a powerful tool for modelling and optimizing complex systems across various fields. With ongoing technological advancements, it is probable that digital twins will become more sophisticated and widely used in scientific research and industry.

The structure of a digital twin typically consists of three components: the physical system, the virtual model, and the data exchange interface. The physical system is the real-world object or process that is being modelled, while the virtual model is the digital picture of the material system. The data exchange interface is the mechanism that enables the exchange of data between the physical system and the virtual model.

Table 2.1 Overview on the use of digital twins.

Field of Application	*Contribution*
Engineering	In the field of engineering, digital twins are used to model complex systems such as aircraft engines, power plants, and manufacturing processes. For example, General Electric (GE) has developed a digital twin of its aircraft engines that can predict maintenance needs and optimize performance.
Healthcare Industry	Digital twins are also being used in healthcare to model patient physiology and predict outcomes. For example, The concept of the digital cardiac twin has the potential to significantly advance personalized medicine in the near future. By utilizing clinical data to create a digital replica of the heart that encompasses all relevant clinical observations, this concept paves the way for personalized medical approaches.
Urban planning	Digital twins are also being used in urban planning to model cities and optimize infrastructure. For example, the city of Helsinki has developed a digital twin of the city that can be used to simulate the effects of different urban development scenarios on air quality, traffic flow, and energy consumption.
Manufacturing sector	In manufacturing, digital twins can be used to model production processes and optimize manufacturing lines. For example, Siemens has developed a digital twin of a production line that can simulate different production scenarios and predict the impact of changes on efficiency and productivity.
Energy	Digital twins are also being used in the energy sector to model power grids and optimize energy usage. For example, researchers at the University of California, Berkeley have developed a digital twin of the California power grid that can be used to predict the effects of renewable energy integration and energy storage implementation.
Transportation	Digital twins have been employed to simulate traffic flow, optimize logistics, and enhance vehicle safety.
Aerospace industries	Digital twins are adopted to enhance aircraft performance, predict maintenance needs, and optimize fuel consumption.
Construction sector	Digital representations of physical concrete elements enable enhanced monitoring, analysis, and optimization of various aspects related to construction and maintenance.

Source: authors' own elaboration

Artificial Intelligence: Explanation of AI, Types and its Applications

The simulation of human intelligence in robots that are designed to think and learn like humans is known as artificial intelligence, or AI. It covers a broad spectrum of uses, such as expert systems, computer vision, and natural language processing. AI has grown in significance across a range of industries due to its capacity to transform workflows and boost productivity.

AI is the emulation of human intelligence in machines built to think and learn like people. It can be applied to a wide range of tasks, including natural language processing, computer vision, and expert systems. Because AI may change workflows and increase production, its importance has expanded across a variety of industries. The goal of this book is to provide a general overview of AI and its applications, highlighting the technology's significance in a range of sectors, such as finance, healthcare, and transportation. In addition, the book explores the ethical issues facing AI, including

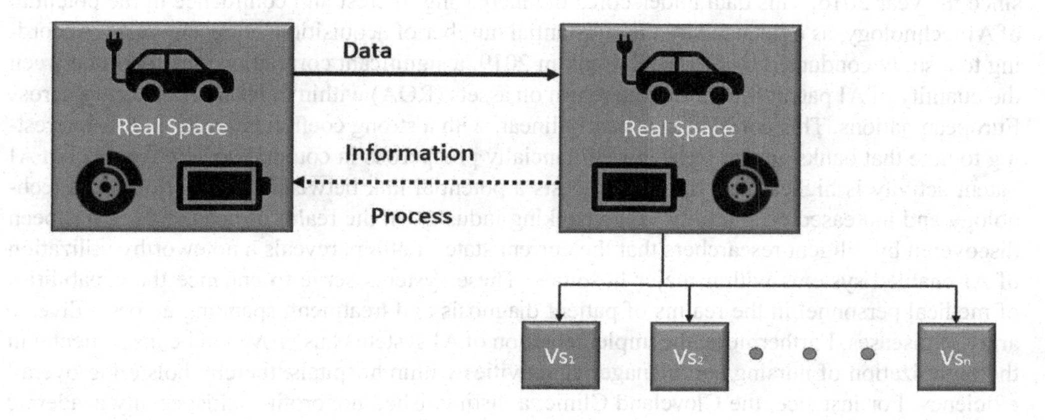

Figure 2.1 Data exchange interface in Digital Twins.

worries about privacy and employment displacement. Furthermore, it delves into the potential applications of AI in the future, including how it can propel developments in autonomous vehicles.

AI systems come in a variety of forms. Narrow AI is tailored to a single task; general AI is intelligent in many domains, comparable to human intelligence; and superintelligent AI is AI that is smarter than humans. It talks about the strengths and weaknesses of each kind, illuminating how they might affect society and the development of technology. It also looks at the current discussions and research around the creation of these AI systems and their potential effects on humans.

Artificial general intelligence (AGI), strong AI, and weak AI are all types of AI. AI systems that are constrained to their intended duties and have limited skills outside them are referred to as weak AI. Conversely, strong AI describes AI systems that are intelligent enough to be considered human-like and that have the capacity to comprehend and learn in a variety of contexts. The ultimate goal of AI research is AGI, which seeks to develop a system that is smarter than a human in every way. Every kind of AI has a unique set of strengths and weaknesses.

This book seeks to give a broad introduction of AI and its uses, emphasizing the technology's importance across various sectors, encompassing banking, healthcare, and transportation. In the banking industry, AI has the potential to revolutionize customer service by providing personalized recommendations, detecting fraud, and automating routine tasks. Banks are currently utilizing AI to cultivate groundbreaking business models. This research unequivocally demonstrates that the meticulous implementation of AI technology facilitates a profound reconfiguration of the conventional banking landscape. Consequently, financial institutions are empowered to fortify their managerial prowess, resulting in significant cost reductions and increased operational efficiency, and enhanced competitiveness within the industry (Gómez and Heredero, 2020). In healthcare, AI can assist in diagnosing diseases, analysing medical images, and improving patient outcomes through predictive analytics. In the transportation sector, AI can optimize route planning, enhance traffic management systems, and enable the development of autonomous vehicles. These examples illustrate the transformative power of AI and its ability to revolutionize various industries. The growth observed in venture capital investments during the preceding two to three years has been remarkably robust. It is worth noting that AI firms have experienced a notable surge in becoming targets for acquisition. Over the course of the past two decades, a total of 434 companies operating within the AI sector have been acquired, with a significant portion of 220 of these acquisitions occurring since the year 2016. This data underscores the increasing interest and confidence in the potential of AI technology, as evidenced by the substantial number of acquisitions in recent years. According to a study conducted by Deutsche Bank in 2019, a significant correlation was found between the quantity of AI patent filings and the return on assets (ROA) within the banking industry across European nations. This correlation is nearly linear, with a strong coefficient of 80%. It is interesting to note that banks appear to be more financially prosperous in countries where the level of AI patent activity is higher. This finding suggests a potential link between the adoption of AI technology and increased profitability in the banking industry. In the realm of healthcare, it has been discovered by diligent researchers that the current state of affairs reveals a noteworthy utilization of AI-enabled systems within major hospitals. These systems serve to enhance the capabilities of medical personnel in the realms of patient diagnosis and treatment, spanning across a diverse array of diseases. Furthermore, the implementation of AI systems has proven to be instrumental in the optimization of nursing and managerial activities within hospitals, thereby bolstering overall efficiency. For instance, the Cleveland Clinic, a distinguished nonprofit multispecialty academic medical centre located in the vibrant city of Cleveland, Ohio, made the astute decision to incorporate Microsoft's cutting-edge AI digital assistant, Cortana, into their esteemed e-Hospital system in the year 2016. This remarkable integration has allowed the Cleveland Clinic to harness the power of predictive and advanced analytics, enabling them to effectively identify potential at-risk patients who are under the meticulous care of their intensive care units (ICUs). With unwavering dedication, Cortana diligently monitors a total of 100 beds spread across six ICUs, diligently carrying out its duties from the hours of 7 p.m. to 7 a.m. Truly, this innovative utilization of technology has proven to be an invaluable asset to the Cleveland Clinic, further enhancing their ability to provide exceptional care to their patients (Lee and Yoon, 2021). These illustrative instances effectively showcase the concrete advantages and practical implications of artificial intelligence within distinct industries. It is worth noting that AI has proven to be a valuable asset in various sectors, yielding substantial benefits and generating significant impact. For instance, in the healthcare industry, AI-powered diagnostic tools have demonstrated remarkable accuracy in detecting diseases at an early stage, thereby enabling timely interventions and improving patient outcomes. Similarly, in the manufacturing sector, AI-driven automation systems have streamlined production processes, resulting in enhanced efficiency and reduced costs. These examples serve as compelling evidence of the tangible benefits and real-world impact that AI brings to specific industries.

Challenges and Ethical Considerations in AI Development

It is crucial to recognize that AI also presents challenges and constraints. Some of the key concerns include ethical implications, potential job displacement, and biases in AI algorithms. These challenges need to be carefully addressed to ensure the responsible and equitable use of AI in society. Difficulties and moral issues with AI development include managing the possible effects on employment and society, protecting data security and privacy, and resolving biases and fairness in AI algorithms. Regulations and policies must also be put in place to control the ethical application of AI technology and to allay worries about its potential for abuse or harm. Researchers and developers must actively engage in discussions about social and ethical issues if they hope to ensure that AI is created and applied in a way that benefits all people.

The process of teaching workers AI-related skills so they can collaborate with AI systems to improve productivity and decision-making is known as "AI upskilling".

AI upskilling is essential for organizations to stay competitive in the rapidly evolving digital landscape and to address the increasing demand for AI-enabled services. Consequently, the future growth of companies will depend on the capacity of their employees to upskill and adapt, resulting in a shift towards new roles that rely more on 'feeling skills' rather than job displacement. Nevertheless, there have been recent arguments from certain authors suggesting that even 'thinking skills' may not safeguard service workers from being replaced by artificial intelligence.

Supporters of AI have envisioned a future where intelligent machines take over mundane tasks traditionally performed by humans, allowing them to focus on more creative endeavours. Despite concerns about potential job displacement, organizational experts advocate for a collaborative integration of human and machine capabilities. By incorporating theories such as dynamic skills, neo-human capital, and AI job replacement, the implementation and acceptance of AI necessitate employees to enhance their skill sets through upskilling.

With the rise of AI and ML, the skill requirements in various industries, including IT, are undergoing significant transformations (World Economic Forum, 2016). This rapid rate of change has resulted in skills instability (World Economic Forum, 2016). The aforementioned report emphasizes the need for organizations to invest efforts in reskilling or upskilling their employees. It is essential for individuals and organizations to adapt to the rapid advancements in AI in order to remain competitive and take advantage of the opportunities it presents.

A literature review on AI upskilling reveals several key findings. Firstly, there is a need for individuals, particularly entrepreneurs, to accept and adopt AI technologies for digital entrepreneurship (Upadhyay et al., 2021). Factors such as performance expectancy, openness, social influence, and hedonic motivations influence the intention of entrepreneurs to accept AI (Upadhyay et al., 2021). Additionally, the introduction and adoption of AI in organizations call for employees to upskill themselves to effectively work with AI technologies (Jaiswal et al., 2021).

The impact of AI on workers' skills and workplaces is a crucial area of study. It has been found that AI will revolutionize professional skill sets and work environments, underscoring the transformation and the importance of transversal skills and the need for upskilling and reskilling strategies (Morandini et al., 2023). The integration of AI in healthcare also requires significant training and upskilling of healthcare providers, along with ethical considerations and privacy concerns (Rao et al., 2023).

The limitations of AI can also drive innovative work behaviours. Workers may be stimulated to engage in innovative work behaviours due to the mistakes made by robots, the fear of AI, the need for upskilling and reskilling, and the interface between workers and AI technologies (Zirar et al., 2023). These limitations can serve as catalysts for workers to think creatively and find new ways to overcome challenges.

To address the need for AI upskilling, various strategies and approaches have been proposed. The United States Department of Defense (DoD) and its ICT UARC are using strategies to quickly upskill their workforce of over 2 million adult learners (Brawner et al., 2023). Organizations can combine multiple strategies simultaneously, such as incorporating online learning in the curriculum, upskilling and reskilling in new technologies, and providing support and guidance to workers (Morandini et al., 2023).

In conclusion, AI upskilling is crucial for individuals and organizations to adapt to the advancements in AI and effectively utilize AI technologies. It requires individuals to accept and adopt AI, acquire new skills and knowledge, and overcome the limitations and challenges posed by AI. Strategies such as online learning, upskilling and reskilling programs, and organizational support can facilitate the process of AI upskilling. It is important for stakeholders, including entrepreneurs, employees, and policymakers, to recognize the significance of AI upskilling and take proactive measures to ensure a smooth transition into the AI-driven future.

Machine Learning: Definition of ML and its relationship with AI

ML is an incredibly fascinating field of study. It involves the creation of models and algorithms that empower computers to learn and make decisions without the need for explicit programming. Through the meticulous analysis and interpretation of large-scale data, ML algorithms are able to identify trends, extract valuable insights, and continuously improve their functionality over time. This remarkable technology has been adopted across diverse sectors including healthcare, finance, marketing, and transportation, delivering significant benefits through its applications.

It is important to note that ML is a subset of AI. This particular branch of AI utilizes statistical models and algorithms to teach computers how to learn from experience and progressively enhance their performance without the need for explicit programming. By harnessing the capabilities of ML, computers acquire knowledge and improve their decision-making abilities, obtaining more efficient and effective outcomes.

The impact of ML is of utmost importance and should not be underestimated. Its remarkable capability to independently acquire knowledge and adjust its behaviour through data analysis is truly extraordinary. By wholeheartedly adopting this technology, the manner in which decisions are made and problems are resolved can be completely transformed. It is highly recommended that individuals and organizations alike seize this opportunity to harness the power of machine learning and revolutionize their respective fields.

This technology uses big dataset analysis to find trends, forecast outcomes, and automate decision-making procedures.

The essential principles of ML include data preprocessing, feature selection, model training, and evaluation. ML algorithms are designed to iteratively learn from data, adjust their parameters, and optimize their execution gradually. ML involves the development of algorithms and models that enable computers to learn from data without being explicitly programmed. It employs statistical methods and mathematical models to analyse extensive datasets, detect patterns, and formulate predictions or decisions. Core ML principles encompass supervised learning, where models are trained with labelled data to predict or classify new data, unsupervised learning, which uncovers patterns in unlabelled data, and reinforcement learning.

Since no predetermined labels or results are needed, the algorithm is free to find hidden links or groupings within the data. Contrarily, reinforcement learning entails an agent learning by making mistakes while interacting with its surroundings and getting feedback in the form of incentives or punishments. Machine learning is a strong tool for resolving difficult issues and reaching well-informed conclusions because it is based on these three types of learning, which allow computers to learn from experience and become better at it.

In the present context, it is imperative to acknowledge the close relationship between ML and AI. These two domains are intricately intertwined, with ML serving as a fundamental component of AI. ML, as a subfield of AI, focuses on the development of algorithms and models that enable computer systems to learn and make predictions or decisions without explicit programming. This process involves the analysis of enormous quantities of data, allowing the system to recognise patterns and extract meaningful acumens. Consequently, ML plays a crucial role in the advancement of AI, as it empowers machines to exhibit intelligent behaviour and adapt to changing circumstances. Therefore, it is crucial to recognize the interdependence of ML and AI, as their synergy drives innovation and propels us towards a future where intelligent machines are capable of enhancing various aspects of our lives. While AI refers to a wider spectrum of technologies that seek to emulate human intelligence, ML focuses on the creation of learning algorithms and models. The capacity of AI systems to glean insights from data and render astute decisions is made possible by ML, which propels the development of AI capabilities. When combined, ML and AI have the power to completely transform whole industries and influence technology going forward.

Supervised Learning, Unsupervised Learning, and Reinforcement Learning

Supervised learning, unsupervised learning, and reinforcement learning are some of the several kinds of machine learning algorithms and their uses.

Supervised learning is the study of algorithms that are trained on well labelled data and learn from examples that are supplied by a human expert. Tasks like spam detection and image categorization are common uses for this specific kind of learning. On the other hand, Semi-Supervised Learning (SSL) is a genuinely novel and revolutionary approach. When compared to the techniques used in supervised learning (SL), this method achieves very competitive results with only a little amount of labelled data.

This claim was supported by a study that conducted a wide range of experiments that conclusively show the value of semi-supervised learning models in explaining human behaviour when exposed to both labelled and unlabelled data (Al-Azzam and Shatnawi, 2021). This approach has great potential to advance our understanding of the subject matter, hence it is imperative that researchers and practitioners alike adopt it.

Conversely, unsupervised learning works with unlabelled data and concentrates on identifying structures or patterns within the data. It is frequently used for tasks involving clustering or dimensionality reduction. Experiments done in a study by Leordeanu et al. (2009) show that unsupervised learning avoids the laborious manual labelling of ground truth correspondences and compares favourably to the supervised case in terms of efficiency and quality.

Last but not least, reinforcement learning entails an agent interacting with a setting and picking up knowledge from rewards-based feedback. Thus, reinforcement learning algorithms interact with their environment to optimize rewards through a trial-and-error learning process. These algorithms can be utilized for tasks like disease diagnosis, stock price prediction, and complex environment navigation across different sectors, encompassing healthcare, finance, and autonomous vehicles. The primary contribution of Cunha et al.'s work from 2021 is showing that reinforcement learning has the potential to replace conventional approaches whenever a speedy solution is required because it can solve any problem in a matter of seconds and produce high-quality results that are close to optimal solutions.

Furthermore, it is worth noting that Reinforcement Learning (RL) holds great promise as a model-free supervisory control method for optimizing the operation of HVAC systems, which stands for Heating, Ventilation, and Air Conditioning. This particular approach is particularly well-suited for addressing small-scale operation optimization problems (Yuan et al., 2021).

Common ML Algorithms and Techniques

They are extensively employed in many different domains, including recommendation systems, natural language processing, and computer vision. These algorithms use a lot of data to find patterns, forecast outcomes, or categorize objects. In order to accomplish challenging tasks and get greater precision, they are continuously being worked upon and improving. They are made up of algorithms that don't need to be explicitly written in order for them to automatically learn from experience and get better at it. These algorithms can be trained on a variety of datasets to address a broad range of issues. They can also be used to derive valuable insights from data. They also frequently use statistical methods and sophisticated mathematical models to maximize their efficiency and provide precise forecasts or classifications. Algorithms like image recognition, natural language processing, and recommendation systems can be handled by these algorithms since they are made to resemble the way the human brain processes and analyses data. They may also change and adapt as new data becomes available, guaranteeing their continuous efficacy in changing contexts.

Neural networks, support vector machines, random forests, and decision trees are a few examples of these techniques. These algorithms examine patterns in data and provide predictions or judgments based on those patterns using a variety of mathematical models and methods. They can be taught using unsupervised learning approaches, in which the algorithm finds patterns on its own, or labelled data, in which the desired outcome is known. These algorithms have been extensively utilized in a variety of industries, including marketing, finance, and healthcare, to enhance decision-making procedures and extract insightful information from big datasets.

Decision Tree

A decision tree is a kind of mathematical model that represents decisions and their potential outcomes using a structure resembling a tree. The branches of the tree indicate the potential results of each node, which in turn indicates a decision. In classification problems, where the objective is to assign an input to one of several predetermined classes, decision trees are frequently utilized. Given that they can handle both numerical and categorical data, they are very helpful when working with complicated datasets. Decision trees are a common choice for reasoning explanations since they are very simple to interpret. Here is an illustration of a decision tree that illustrates the method used by a telecom operator to decide which customers are likely to leave.

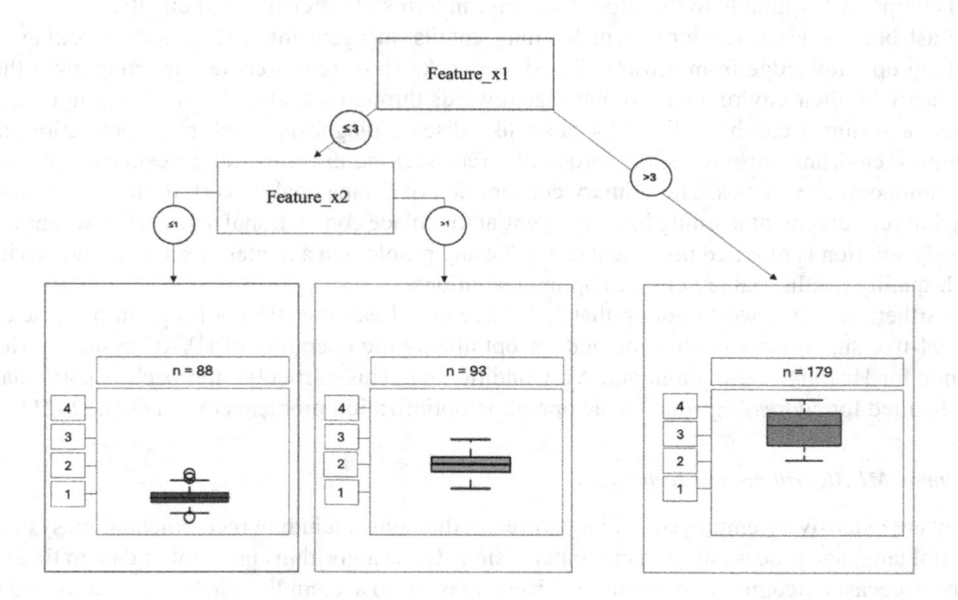

Figure 2.2 Artificial data in a decision tree. Node 5 is the destination for instances of feature x1 with a value greater than 3. Depending on whether the values of feature x2 exceed 1, the remaining cases are assigned to either node 3 or node 4

Source: authors' own elaboration

Trees can be grown by a variety of algorithms. They diverge in terms of the potential tree structure (number of splits per node, for example), the standards for identifying splits, the point at which splitting should end, and the methods for estimating the basic models inside leaf nodes. The branches of the tree indicate the potential outcomes based on the nodes, which each represent a distinct trait or feature, such as monthly fees or customer tenure. Following the branches makes it simple to see how several elements come together to determine a customer's likelihood of churning or not. Because of its interpretability, organizations are able to identify patterns and relationships and use this information to make insightful decisions. The most widely used technique for tree induction is most likely the classification and regression trees (CART) algorithm (Friedman et al., 2009).

Neural Networks

Artificial intelligence that mimics the way the human brain functions is known as neural networks. They are made up of networked nodes, known as neurons, that process and transfer data via weighted connections. Neural networks are highly accurate in recognizing patterns, making predictions, and carrying out intricate tasks by learning from enormous volumes of data.

The history of neural networks is extensive, having begun in the 1940s. Warren McCulloch and Walter Pitts initially presented the idea by putting forth a computational model of brain networks. Frank Rosenblatt's perceptron learning algorithm and the development of perceptron did not, however, yield any real advances until the 1950s and 1960s. Since then, neural networks have undergone various advancements, such as the introduction of backpropagation in the 1980s, which greatly improved their ability to learn and solve complex problems. Neural networks are now widely employed in domains including natural language processing, autonomous cars, and image and audio recognition. Their capacity to analyse vast volumes of data and generate precise forecasts has transformed sectors such as healthcare, finance, and technology. Neural networks are pushing the frontiers of artificial intelligence with continued research and development, and they could have a significant future impact on how we live our daily lives.

Artificial neurons arranged in layers as interconnected nodes make up a neural network. An input layer, one or more hidden layers, and an output layer are some examples of these layers. Every network node receives input signals, applies activation functions to process them, and then forwards the output to the subsequent layer. Through a procedure known as training, the weights and biases applied to each link between nodes are changed, enabling the network to gain performance improvements over time by learning from data.

In neural networks, activation functions are essential because they give the model non-linearity. They use the weighted total of inputs and biases to determine a node's output. Regular activation functions consist of ReLU or Rectified Linear Unit, tanh, and sigmoid. With the aid of these functions, the network is able to learn and generate precise predictions by comprehending intricate correlations between inputs and outputs. The performance and rate of convergence of the network during training can be significantly impacted by the activation function selection, which is based on the type of problem being solved.

Biases and weight initialization are important aspects of neural network training. The network's capacity to learn can be hampered by problems like vanishing or bursting gradients, which can be avoided with proper weight initialization. Conversely, biases offer adaptability in adjusting the activation function, enabling the network to more closely match the data. As a result, selecting the right starting weights and biases is crucial to attaining peak performance and accelerating training convergence.

One kind of artificial neural network is called a feedforward neural network, in which data only moves from the input layer to the output layer. They are made up of several layers of networked nodes, or neurons, with each layer's neurons connected to all of the layers below it. A feedforward neural network's architecture normally consists of an input layer, one or more hidden layers, and an output layer. After the initial data is received by the input layer, activation functions are used to process it through the hidden layers.

By adding feedback connections, recurrent neural networks (RNNs) are a kind of neural network that can handle sequential data. RNNs may store and process information from prior inputs because, in contrast to feedforward neural networks, they feature connections that permit information to flow in loops. Because of this, RNNs work especially well for applications like time series analysis, speech recognition, and natural language processing.

One kind of neural network that is particularly good at processing grid-like input, including photos and movies, is the convolutional neural network (CNN). They employ convolutional, pooling, and fully connected layers to automatically learn and extract features from input data. In the field of image and video processing, CNNs have revolutionized object recognition, image classification, video analysis, and even medical imaging. They are an essential tool in the training of neural networks because of their capacity to capture spatial hierarchies and identify intricate patterns. It computes the gradient of the loss function in relation to the network's weights and biases, enabling effective gradient descent optimization. The network learns to reduce error and enhance performance on the given task by iteratively modifying the weights and biases depending on this gradient. This iterative process of forward propagation and its impact on the training process are important topics in the field of neural networks. The methods used by these algorithms to adjust the weights and biases during training vary; stochastic gradient descent is a well-liked option for big datasets because of its effectiveness in handling mini-batches of data. In order to further increase convergence speed and performance, researchers have also looked at more complex optimization algorithms like Adam and RMSprop, which adaptively modify the learning rate dependent on the magnitude of the gradients.

Natural language processing tasks including sentiment analysis, machine translation, and question answering have made extensive use of neural networks. The accuracy and fluency of language production and interpretation systems have improved because to these models, which have produced encouraging results.

Deep Learning

The creation and usage of artificial neural networks is the main focus of the machine learning branch known as "deep learning". It entails teaching these neural networks to acquire new skills and make judgment calls without explicit programming. Due to its capacity to deliver state-of-the-art performance in a variety of tasks, including speech recognition, image recognition, and natural language processing, deep learning has attracted a lot of interest and grown in popularity in recent years. By providing cutting-edge skills like fraud detection and medical diagnosis, it has completely changed a number of industries, including healthcare, finance, and autonomous cars. For instance, a study conducted by Xue et al. (2017) presents compelling evidence regarding the potential of deep learning in the field of intelligent medical image diagnosis practice. The focus of this study was on the hip joint, specifically exploring the diagnostic value of deep learning in hip osteoarthritis. To achieve this, a deep convolutional neural network (CNN) was meticulously trained and tested on a dataset consisting of 420 hip X-ray images. The results were remarkable, with the CNN model demonstrating a remarkable balance between high sensitivity (95.0%) and high specificity (90.7%), resulting in an overall accuracy of 92.8%.

These findings were compared to the assessments made by the chief physicians, further highlighting the effectiveness of the CNN model. Despite these applications.

Applications of AI in Quality Management

Using machine learning techniques to examine huge datasets and spot trends or abnormalities that might point to quality problems is one of them. AI can also be used to automate quality control tasks like checking items for flaws or keeping an eye out for violations from standard operating procedures on production lines. It may also assist in improving quality management systems by anticipating maintenance requirements and allocating resources optimally to guarantee constant product quality.

It is of utmost importance to provide concrete examples of application in order to illustrate the subject matter at hand. In their article, Bartels et al. (2022) discuss their study on the development and implementation of an Artificial Intelligence/Machine Learning-Clinical Decision Support (AI/ML-CDS) tool called Sleep Well Baby (SWB). The case study focuses on the details of this tool and its application in the field. This tool was specifically designed and put into practice at the University Medical Center Utrecht (UMC Utrecht), an esteemed institution renowned for its commitment to academic excellence and medical education. It is worth noting that UMC Utrecht stands as one of the largest academic teaching hospitals in the Netherlands.

The SWB project, which initially emerged as a grassroots initiative, garnered significant recognition and acclaim, ultimately being awarded the prestigious best innovation prize at the esteemed Dutch Hacking Health event in 2019. The primary objective of this project was to explore the potential of utilizing clinical-care data in conjunction with AI/ML-CDS technology to facilitate personalized care. By harnessing the potential of advanced algorithms and machine learning techniques, the SWB tool aimed to enhance the quality and efficiency of healthcare delivery, ultimately leading to improved patient outcomes.

The development of the SWB tool involved a multidisciplinary approach, including clinical experts, data scientists, end-users, product/service designers, software engineers, ethicists, legal experts, financial/business development experts, and change management experts. This approach ensured that the tool was developed with input from various stakeholders and considered different perspectives.

During the development phase, the article highlights the importance of quality management in AI/ML-CDS tools. It suggests that the organizational structure of medical laboratories and ISO15189, which provides requirements for quality and competence in medical laboratories, can inspire healthcare institutes in building an effective and sustainable Quality Management System (QMS) for AI/ML usage in clinical care. ISO13485, which provides requirements for quality management systems for medical devices, can also be used as a reference for QMS in AI/ML-CDS tools.

The article emphasizes the need for a QMS that includes processes for selection, clearance, and performance verification of AI/ML-CDS tools, as well as appropriate Standard Operating Procedures (SOPs) and service agreements with manufacturers for monitoring and change management. It also highlights the importance of user expertise in auditing and validating AI/ML-CDS tools to ensure their safe and effective use in clinical practice.

The benefits of implementing AI in quality control systems include improved accuracy and efficiency in detecting defects or anomalies, increased productivity, and the capacity to analyse huge quantities of data rapidly. AI can also help recognise patterns and trends that may not be straightforwardly detectable by individuals, leading to better decision-making and problem-solving.

However, there are also drawbacks to implementing AI in quality control systems. One of the challenges is the need for high-quality and diverse data to train AI models effectively. Data quality issues, such as bias and incomplete or inaccurate data, can affect the performance and reliability of AI systems. Another challenge is the interpretability and explainability of AI algorithms.

In addition to the healthcare sector, it is worth noting that AI has found applications in quality management across various industries. A recent study conducted by Lee et al. in 2021 provides an insightful example of this. The study focuses on the collaboration between the Big Data Research Center of Kyung Hee University and Benple Inc., who worked together to develop and implement an AI system for Frontec, a small and medium-sized enterprise (SME) specializing in the manufacturing of automobile parts.

The primary objective of this project was to automate the quality management process at Frontec using AI technology. To achieve this, the team had to carefully consider a range of constraints. For instance, they needed to account for the response time requirements, ensuring that the system could provide timely feedback. Additionally, the team had to work within the limitations of the available computing resources, ensuring that the AI system could operate efficiently.

By leveraging AI, Frontec was able to streamline their quality management process, resulting in improved efficiency and accuracy. The successful execution stands as evidence of the potential advantages that AI can bring to quality management in various sectors beyond healthcare.

The purpose of the defect finding system is to accurately classify weld nuts within a remarkably short time frame of 0.2 seconds, while maintaining an impressive accuracy rate of over 95%. To achieve this, the system utilizes the Circular Hough Transform for preprocessing, which effectively prepares the images for further analysis. Additionally, researchers implemented an adjusted VGG (Visual Geometry Group) model, which serves as the backbone of our convolutional neural network (CNN) system.

The results of our CNN system are truly remarkable, boasting an accuracy rate of over 99% and a swift response time of approximately 0.14 seconds. In order to seamlessly integrate our CNN model into the factory environment, we took the necessary steps to reimplement the preprocessing modules using LabVIEW. This allowed for smooth communication between the classification model server and an existing vision inspector.

Throughout the development and embedding process of our deep learning framework, we encountered various real-world challenges. However, we successfully overcame these obstacles without the need for any hardware modifications. We believe that our experience in this endeavour can provide valuable insights to others in the field. Therefore, we are eager to share the procedure we followed, as well as the specific issues we encountered, in order to contribute to the collective knowledge of developing and embedding deep learning frameworks in existing manufacturing environments.

AI-Driven Quality Control Systems

AI-powered quality control solutions are automating and optimizing the inspection process, transforming the manufacturing sector. Using sophisticated algorithms and machine learning methods, these systems ensure that only top-notch products are introduced to the market by detecting flaws and irregularities in products. AI-driven quality control systems can detect patterns and trends that human inspectors might overlook because of their capacity to evaluate enormous volumes of data in real-time. This improves the accuracy and efficiency of identifying and addressing quality concerns. In stark contrast to previous studies conducted on algorithm aversion, Keding and Meissner (2021) find a substantial and constructive effect on choice behaviour and the perception of decision quality when utilizing AI-based advisory systems. Furthermore, they provide evidence to elucidate the reasons behind this phenomenon, which can be attributed to a heightened level of trust in the advisor and the implementation of a more structured decision-making process.

Their findings demonstrate that the incorporation of AI technology in decision-making processes yields favourable outcomes. By relying on AI-based advisory systems, individuals exhibit a greater propensity to make choices and perceive the quality of their decisions as being enhanced. This can be attributed to the establishment of a higher degree of trust in the advisor, as well as the implementation of a more structured decision-making process.

The increased trust in the advisor stems from the reliable and consistent guidance provided by the AI-based system. This technology is designed to analyse vast amounts of data, enabling it to offer well-informed recommendations. Consequently, individuals feel more confident in the advice provided, leading to a greater willingness to follow the system's suggestions.

Moreover, the implementation of a more structured decision-making process contributes to the positive outcomes observed. The AI-based advisory systems offer a systematic approach, ensuring that all relevant factors are considered and evaluated. This structured process instils a sense of confidence in individuals, as they perceive their decisions to be more thorough and well-informed.

In conclusion, their research highlights the beneficial effects of employing AI-based advisory systems on choice behaviour and decision quality perception. The increased trust in the advisor and the implementation of a more structured decision-making process contribute to these positive outcomes. These findings shed light on the potential advantages of integrating AI technology into decision-making processes, offering individuals a reliable and structured approach to enhance their decision-making capabilities.

AI-Based Predictive Maintenance and Fault Detection

A state-of-the-art technique called AI-Based Predictive Maintenance and Fault Detection uses artificial intelligence algorithms to examine real-time data from machinery and equipment. Through the monitoring of multiple characteristics, including temperature, vibration, and performance metrics, the system is able to precisely anticipate potential faults or failures before to their occurrence. By taking a proactive stance, companies can limit downtime, optimize maintenance plans, and lower total expenses related to unplanned repairs. Furthermore, as time goes on, the AI algorithms learn and advance continuously, increasing prediction accuracy and facilitating more effective maintenance.

In this particular matter, it is worth noting that a study was undertaken by Kiangala and Wang (2020).

The purpose of this study was to develop an experimental predictive maintenance framework for conveyor motors. The framework was designed with the intention of efficiently detecting impairments and minimizing the risk of incorrect fault diagnosis within the plant. To achieve this, a machine learning model was employed, which effectively classified observed abnormalities as either production-threatening or non-production-threatening. It is important to highlight that the accuracy achieved by this model surpassed that of traditional classification approaches.

The introduction of advanced digital technologies and the ever-increasing complexity of the global business environment necessitate the development of new quality management systems in the Industry 4.0 era (Sang et al., 2021). Based on a thorough analysis of the quality management literature and five actual instances of predictive quality management utilizing new technology, this study offers fresh concepts for predictive quality management. Based on big data analytics, AI, smart sensors, and platform development, this study's results suggest that predictive maintenance enabled by advanced technology can be implemented across multiple industries.

Predictive quality management systems of this kind have the potential to develop into living ecosystems capable of real-time, efficient decision-making, big data analytics, and cause-and-effect analysis. In the era of Industry 4.0, this study suggests several practical implications for the actual design and implementation of efficient predictive quality management systems. Nonetheless, the organizational culture that fosters stakeholder collaboration, information exchange, and goal co-creation should give rise to a living predictive quality management ecosystem.

While recent predictive maintenance studies have indicated encouraging outcomes from high-performance AI algorithms, the majority of previous research has focused on AI-only solutions and has not taken human-AI interaction into account. In this work, we specifically highlight the advantages of encounters in which AI solutions support a human inspector. A case study on predictive maintenance for wind farms is carried out, utilizing endoscopic pictures to identify bearing faults. In this experiment, 2301 photos of over 138 wind turbines were gathered, 54 technical inspectors participated, and each inspector was provided a picture and asked to identify bearing defects both with and without AI aid. The results showed that using AI to increase the technical inspector's specificity had a statistically significant impact and efficiency over time. The degree of improvement varied according to the level of experience; the generalist group outperformed the specialist group in terms of specificity and time efficiency, showing increases of 24.6% and 25.3%, respectively, compared to 4.7% and 6.4%, respectively. Regarding the intention to reuse and the value of AI aid, both groups gave favourable answers, and there was no statistically significant difference in the cognitive load (Shin et al., 2018).

Enhancing Process Efficiency with AI in Quality Management

By automating tedious operations and expediting decision-making, AI may greatly increase the efficiency of quality management procedures. Large data sets may be swiftly analysed by AI algorithms, which can also spot trends and anomalies. This makes quality assessments quicker and more precise. AI-powered systems can also foresee any quality problems in advance, giving enterprises the ability to avoid production disruptions and take preventive action.

Quality management has undergone a revolution thanks to AI technology, which streamlines procedures and boosts productivity. AI systems can swiftly uncover patterns and abnormalities by automating operations like data analysis and defect detection. This enables quicker decision-making and more precise quality control. Furthermore, as production environments change, AI-powered algorithms may learn from them and adjust accordingly, thereby increasing process efficiency. For manufacturers, these developments in AI technology have also meant huge cost savings. Through the elimination of manual labour requirements and the reduction of errors, businesses can enhance their profitability and optimize resource allocation. AI-powered quality management systems can also offer predictive analytics and real-time insights, which facilitate proactive problem-solving and stop possible disruptions before they happen.

In the context of enhancing process efficiency through quality management, a study conducted by Senoner et al. (2020) introduced a data-driven decision model tailored to the domain of manufacturing. Traditional methods in quality management face a formidable obstacle when confronted with the complex, high-dimensional, and nonlinear nature of manufacturing data. To overcome this challenge, they have leveraged explainable artificial intelligence techniques and incorporated them into the quality management framework. More specifically, they advocate the utilization of nonlinear modelling, employing Shapley additive explanations to unravel the intricate connections between a set of production parameters and the quality of a manufacturing process. Through this approach, they contribute a quantifiable metric of process significance, enabling manufacturers to effectively prioritize their efforts for quality enhancement. Rooted in established quality management principles, their decision model identifies actionable steps that target the underlying causes of quality variability.

To validate the practicality and efficacy of their decision model, the authors conducted a real-world application at a prominent manufacturer specializing in high-power semiconductors. Focusing on the improvement of production yield, they applied their model to select optimization measures for a specific transistor chip product. Subsequently, a field experiment was conducted to assess the impact of these selected improvements. Remarkably, results revealed a noteworthy 21.7% reduction in yield loss when compared to the average yield in their sample. Furthermore, outcomes from a post-experimental deployment of the decision model, which consistently led to substantial improvements in yield, are presented. These findings underscore the operational significance of explainable artificial intelligence by highlighting that pivotal factors influencing process quality may remain concealed when traditional methods are employed. In summary, this study underscores the pivotal role of artificial intelligence in quality management, particularly in deciphering intricate manufacturing processes and facilitating targeted quality enhancement efforts (Senoner et al., 2020).

Integrating AI with TQM and Lean Thinking 5.0

The operational effectiveness and quality management can both be significantly enhanced by this combination and, enterprises may streamline data analysis, detect trends, and make data-driven choices instantly (Frick and Grudowski, 2023). The ability of AI-powered systems to pinpoint the underlying causes of quality problems and provide the appropriate fixes based on past performance and industry best practices is another way that this integration can improve problem-solving abilities. Organizations can save time and effort by automating data analysis processes, such as employing machine learning algorithms to examine massive datasets. This can assist with finding patterns and trends in data that human analysts might miss. In addition, unstructured data, like customer reviews, can be analysed using natural language processing.

Businesses may make data-driven choices in real time using AI-powered technologies, increasing operational effectiveness and lowering the possibility of human error. Studies have indicated that companies who include AI into their quality management systems see notable increases in operational effectiveness. For instance, a research study conducted by Wamba-Taguimdje et al. (2020) was based on a thorough review of 500 case studies from reputable sources such as IBM, AWS, Cloudera, Nvidia, Conversica, and Universal Robots websites. The objective was to examine the impact of AI on organizational performance, specifically focusing on the business value of AI-enabled transformation projects. Findings revealed that AI has the potential to enhance processes, advance automation, optimize information, and transformation outcomes, while also identifying, forecasting, and engaging with humans. Their study has highlighted the benefits of AI in organizations, specifically in improving performance at both the organizational and process levels, including financial, marketing, and administrative aspects. By leveraging these attributes of AI, companies can boost the business value of their renovated projects. Additionally, results indicate that companies can reach performance enhancements by reconfiguring their processes using AI features and technologies. Practitioners and researchers must recognize the immense value of incorporating AI into their operations. Thus, it should be regarded as an indispensable tool, capable of serving as a support system or even as a pioneering force behind the development of a novel business model. By embracing AI, professionals can tap into its vast potential and reap the benefits it offers (Wamba-Taguimdje et al., 2020).

It is crucial to recognize that integrating AI into quality management may not always be easy. The accuracy and dependability of AI-powered systems, for example, may raise questions because of their heavy reliance on algorithms and historical data. In addition, the actual integration process itself could take a lot of time and money, not to mention staff training and maintaining the security and privacy of data. Organizations can demonstrate a thorough grasp of the possible hazards and advantages of integrating AI by addressing these issues. This combination has the potential to greatly improve both quality management and operational effectiveness. Businesses may rapidly identify trends, expedite data analysis, and make data-driven decisions by leveraging AI technologies.

This integration can also enhance problem-solving skills by enabling AI-powered systems to recognise the root causes of quality issues and recommend suitable solutions based on historical performance and industry best practices (Cao et al., 2021). By automating data analysis procedures, such as using machine learning algorithms to analyse large datasets, organizations can save time and effort. This can help identify trends and patterns in data that human analysts might overlook. Additionally, natural language processing can be used to examine unstructured data, such as customer reviews (Prentice and Nguyen, 2020). This combination has the potential to greatly improve both quality management and operational effectiveness.

Key Takeaways

- **Significance of AI technologies in quality control**: AI technologies play a crucial role in modern quality management by enabling efficient and data-driven quality control processes.
- **Introduction to artificial intelligence, machine learning, and digital twins**: Understanding the basics of AI, machine learning, and digital twins is essential for implementing AI in quality management.
- **Applications of AI in quality management**: AI finds applications in quality management across various industries, from manufacturing to healthcare, by automating tasks and improving decision-making.
- **AI-driven quality control systems**: AI-driven quality control systems use machine learning algorithms to detect defects and maintain consistent product quality.
- **AI-based predictive maintenance and fault detection**: AI helps predict equipment failures and defects, reducing downtime and maintenance costs.
- **Enhancing process efficiency with AI in quality management**: AI optimizes processes, reduces waste, and enhances overall efficiency in quality management.
- **Integrating AI with TQM and Lean Thinking 5.0**: AI can be integrated with Total Quality Management (TQM) and Lean Thinking 5.0 principles to create more agile and efficient quality control systems.

Review Questions

1. Why is AI technology significant in quality control, and how does it impact modern quality management?
2. What are the key components of AI, including machine learning and digital twins, and how are they relevant to quality management?

3. Give examples of industries where AI is applied in quality management and explain the benefits.
4. How do AI-driven quality control systems work, and what are the advantages of using them?
5. Discuss the role of AI in predictive maintenance and fault detection. How does it contribute to cost savings and productivity improvement?
6. How can AI enhance process efficiency in quality management? Provide specific examples.
7. Explain how AI can be integrated with TQM and Lean Thinking 5.0 for more effective quality control systems.

3 Lean Thinking 5.0 and Its Connection to TQM

Learning Objectives

- Understand the concept of Lean Thinking and its historical evolution.
- Define Lean Thinking 5.0 and its key characteristics in the context of modern business practices.
- Examine the origins of Lean Production, focusing on the role of Toyota Motor Company and the development of the Toyota Production System.
- Explore the seven types of waste (Muda) in a production process and learn strategies for their elimination in Lean Thinking.
- Analyse the components and principles of Toyota's House of Lean as a framework for implementing Lean Thinking.
- Differentiate between Lean Production and Total Quality Management (TQM), highlighting their unique features and identifying areas of overlap.
- Familiarize oneself with the practical tools and techniques used in Lean Production for process optimization.
- Understand the significance of UNI 11063:2017 and UNI 10147:2003 standards in the context of Lean Thinking and its applications.
- Comprehend and apply the Five Lean Principles as fundamental guidelines for achieving efficiency in Lean Thinking.
- Differentiate between the Push and Pull approaches in Lean Thinking and understand their respective advantages and applications.
- Explore the integration of digital technologies in Lean Thinking, focusing on the concept of Digital Lean 4.0.
- Understand the principles and methodologies of the Six Sigma Model as it relates to quality improvement.
- Explore the synergies between Lean Thinking and Six Sigma, understanding how their integration enhances overall organizational efficiency.

Unravelling Lean Thinking

The success of Japanese companies in global markets, particularly in the automobile industry with the major American automotive industry, prompted U.S. companies to commission a five-year research project at the Massachusetts Institute of Technology to understand the reasons for Japanese success. The research showed that the success of the Japanese production system

DOI: 10.4324/9781032726748-3

was essentially based on a style of company management called Lean Production, which is a production system that is based on minimizing inventory, a strictly calibrated flow of components and raw materials based on production needs, continuous adoption of small improvements to the product, and responsible workers who contribute to the success of the company. The results of these choices were summarized in two concepts: "few defects, low costs" and one word: "quality".

Lean production can be described as a comprehensive and widespread adoption of the Toyota Production System (TPS) in Western countries. This system, which goes beyond the boundaries of mass production techniques developed by Henry Ford and Alfred Sloan, has been and continues to be employed by the majority of Western companies. The TPS, with its emphasis on efficiency and waste reduction, has confirmed to be a valuable method for companies seeking to optimize their production processes. By implementing lean production principles, organizations can achieve higher levels of productivity and competitiveness in the global market. Toyota's constant growth, from a small player to becoming the world's largest and most valuable automotive company, focused attention on how it was able to do so, making "Lean Manufacturing" a hot topic in Management Sciences at the beginning of the twenty-first century.

The term Lean manufacturing, also known as Lean production or Lean Thinking, was first introduced in 1990 by scholars Womack and Jones in their book *The Machine That Changed the World*. In this seminal work, they meticulously outline the production system that propelled the Japanese company Toyota to achieve unparalleled success, surpassing all competitors on a global scale. Since its inception, the Lean model has garnered widespread adoption by numerous esteemed organizations worldwide, spanning across various industries and services. Notably, its applicability extends beyond strictly productive processes, encompassing logistic, administrative, product design, and development processes as well. Over the years, the Lean production model has undergone refinement, leading to the emergence of other designations such as Lean organization, Lean service, Lean office, Lean enterprise, and even Lean Thinking. It is worth noting that Lean manufacturing has become a sought-after initiative for many prominent American corporations, as they strive to maintain their competitiveness in an increasingly global market. The following definitions have been developed by various authors over the past decade, based on their extensive studies and research on the concept of Lean Manufacturing, which places emphasis on the systematic elimination of waste from an organization's activities. This is achieved through the implementation of a series of synergistic work practices, which aim to produce products and services at a price that aligns with the demand (Womack et al., 1990; Fullerton et al., 2003; Simpson and Power, 2005; Shah and Ward, 2007).

Lean Thinking represents a multidimensional concept that groups distinct bands of lean organizational practices, including just-in-time, total quality management, total preventive maintenance, human resource management, pull logic, flow of operations, low set-up, controlled processes, productive maintenance, and employee involvement (McKone et al., 1999; Swink et al., 2005; Linderman et al., 2006; Shah and Ward, 2007).

Lean manufacturing is a methodology that aims to enhance employee responsibility and engagement in waste reduction initiatives within an organization (Shah and Ward, 2003; Tu et al., 2006). By implementing a Lean orientation, companies can also effectively adopt environmental management practices that focus on waste reduction and the subsequent decrease in pollutants (Yang et al., 2010). Consequently, organizations that have experience with lean production are better equipped to embrace and implement environmental management practices.

Henceforth, Lean manufacturing can be comprehended as a collection of practices that are diligently concentrated on the reduction of waste and the elimination of activities that do not contribute any value to a company's production endeavours (Womack et al., 1990; McLachlin,

1997; Shah and Ward, 2003, 2007; Li et al., 2005; Browning and Heath, 2009). The practices of Lean manufacturing increase productivity by reducing installation times and inventory work, improving information flow, and market performances (Tu et al., 2006).

Lean manufacturing, when combined with the implementation of Six Sigma, was recognised to be an operative method for resolving complex problems within business processes. This methodology not only ensures customer satisfaction, but also enhances the overall responsiveness of the organization while reducing delivery times to customers (Shah and Ward, 2003; Ward and Zhou, 2006).

Furthermore, it is important to note that lean manufacturing has a noteworthy influence on the financial performance of an organization. By improving various organizational processes, it enables greater cost efficiency, resulting in substantial savings (Fullerton et al., 2003; Christopher and Towill, 2001; Fullerton and Wempe, 2009). Additionally, the implementation of lean manufacturing leads to increased productivity, as evidenced by studies conducted by Blackburn and Rosen (1993), Golhar and Stamm (1991), and Kinney and Wempe (2002). These improvements in productivity ultimately contribute to the overall success of the organization.

In conclusion, it is highly recommended that organizations adopt lean manufacturing practices in order to achieve optimal results, improve their problem-solving capabilities, customer satisfaction, and ultimately achieve financial success.

Seth and Gupta (2005) have presented that the goal of Lean production is to reduce waste, reduce human effort, inventory, time to market, and production space to become highly responsive to customer demands.

Some scholars define LM from a technological standpoint as a zero-waste production process, which means its main function is to eliminate all waste from the value stream (MacBryde et al., 2006).

Taj (2008) indicated that Lean thinking, which began in the manufacturing sector, represents the meaning of "production without waste" (i.e. producing with the minimum quantity of equipment, materials, components, working hours, essential for production). Sun et al. (2010) emphasized that Lean production is an effective tool for companies for their continuous improvement.

Lean, traditionally associated with production systems, is increasingly advocated to be applied holistically across the entire business framework for success (Grasso, 2005; Kennedy and Widener, 2008; McVay et al., 2013; Solomon and Fullerton 2007; Womack and Jones, 1996). The crux of Lean ideology lies in integrating all business processes and functions into a cohesive system aimed at delivering enhanced value to customers through continuous improvement and waste reduction (Grasso, 2005; Shingo Institute 2010). Given the interconnected nature of business processes, the full potential of Lean manufacturing can only be realized when it operates within an integrated context (Maskell and Kennedy, 2007).

LM practices, encompassing waste reduction, production cells, statistical process control, and just-in-time techniques (Shah and Ward, 2003), are influenced by the prevailing cultural values of both managers and workers (Khurana and Singh, 2012; Rother and Shook, 1999; Spear and Bowen, 1999). Described as a system of interdependent components aimed at enhancing operational efficiency (Demeter and Matyusz, 2011; Hofer et al., 2012), LM represents an integrated socio-technical system comprising closely linked practices targeting waste elimination (de Treville and Antonakis, 2006). Shah and Ward (2007) propose that LM, a subset of Lean Production, incorporates supplier-related and plant-related practices, each contributing to improved operational performance.

Therefore, Lean thinking is first and foremost a vision and a new way of understanding the industrial reality. Then it is a technical, organizational, and managerial model capable of achieving high performance on multiple fronts. It is a fundamental lever for changing the rules of competition and for acquiring significant competitive advantages. It is a system that can use resources in the most convenient way and achieve economies of scale through close ties between many upstream and downstream companies.

Lean Manufacturing or Lean Production (LM/LP), literally "Lean Production", is a particular production system that many large companies in the United States have tried to adopt in order to remain competitive in an increasingly global market. However, it is mostly unknown in Italy, with few companies having decided to apply it to date (Pirelli being the exception). The focus of this approach, studied and spread worldwide following the work carried out by Womack et al. (1990) and Womack and Jones (1996) and the son of the production system devised by Taiichi Ohno, is on cost reduction by eliminating activities that do not add value. Many of the tools and techniques of LM, originating from the Toyota Production System, are now widely used for discrete productions (Abdulmalek and Rajgopal, 2007). However, it should be noted that the commitment to eliminate or reduce waste has no continuity solution (perfection, as we know, is not part of this world) and requires constant improvement of business performance. By continually focusing on the reduction of muda, a company can better respond to the needs of its customers and will also be able to operate at more efficient performance levels (Ortiz, 2010).

The aim of Lean philosophy is therefore to endorse a culture of continuous improvement within a company. Definitely, it considers spending resources for any purpose other than creating

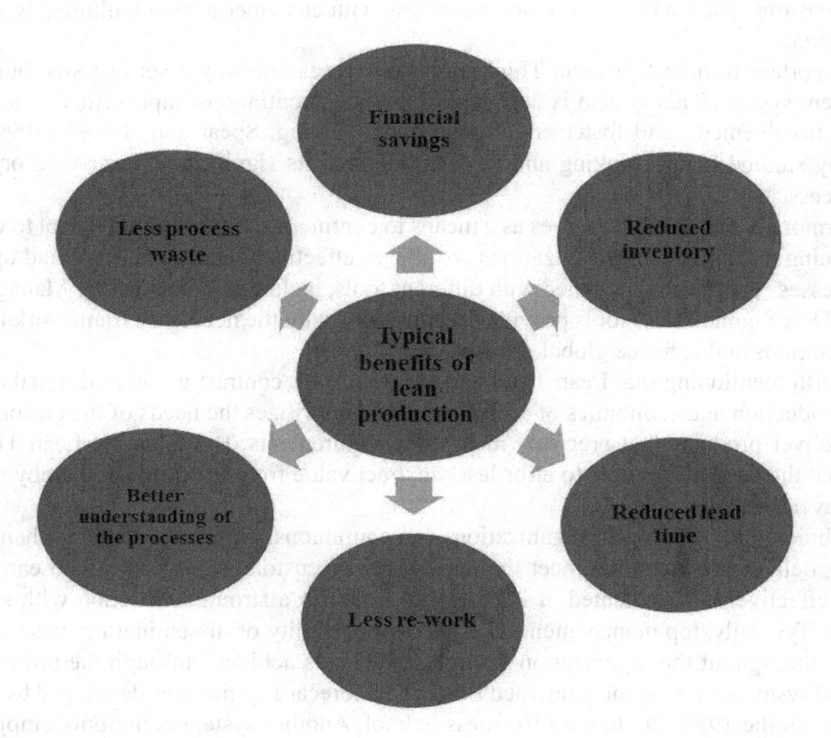

Figure 3.1 Lean Production benefits.

Source: authors' own elaboration

value for the end customer as waste. Moreover, several studies suggest that a Lean method can contribute to reducing the ecological impact of industries (Pampanelli et al., 2013).

The observed benefits are well documented (Melton, 2005):

- reduced lead-time for customers;
- reduced production waste;
- reduced inventories for manufacturers;
- improved knowledge management;
- more robust processes (fewer errors and therefore fewer reworkings).

The notion of Lean Management can be characterized as a comprehensive framework of measures and methodologies that, when implemented in unison, possess the capacity to establish a state of leanness and distinctive competitiveness, not solely within the production sector, but across the entire organization (Warnecke and Huser, 1995). The primary objective of this approach is to pinpoint the principal origins of waste, thereby facilitating the eradication of resultant inefficiencies.

Defining Lean Thinking 5.0

The National Institute of Standards and Technology (USA) has provided a comprehensive definition of Lean Thinking. According to their esteemed institution, Lean Thinking is a systematic method that aims to recognise and eradicate waste through continuous improvement. This approach ensures that the flow of the product aligns with customer demand, ultimately striving for perfection.

It is important to note that Lean Thinking encompasses not only a set of tools, but also a management system. This system is designed to facilitate continuous improvement, encourage employee involvement, and foster effective problem-solving. Spear and Bowen (1999) have extensively studied Lean Thinking and have highlighted its significance in driving organizational success.

Furthermore, Lean Thinking serves as a means to continuously improve and adapt to change. By embracing this approach, organizations are able to effectively eliminate waste and optimize their processes. This is accomplished with different tools, including Total Quality Management, Agile, and Six Sigma. These tools provide organizations with the necessary framework to modernise operations and enhance global productivity.

It is worth mentioning that Lean Thinking stands in stark contrast to the traditional concept of mass production and economies of scale. Instead, it prioritizes the needs of the customer and aims to deliver products that precisely meet their requirements. By doing so, Lean Thinking ensures that the customer is able to effortlessly extract value from the product, thereby enhancing their overall satisfaction.

By embracing this approach, organizations can continuously improve, adapt to change, and ultimately deliver products that meet the needs of their customers. In order for Lean Thinking to be effectively implemented, it is crucial to establish a strong connection with strategic objectives. Typically, top management takes the responsibility of disseminating these strategic objectives throughout the organization's processes. This is achieved through the utilization of specialized systems, such as the renowned Balanced Scorecard, which was developed by Kaplan and Norton in the 1990s at Harvard Business School. Another system commonly employed is Hoshin Kanri, a purely Japanese approach that originated in the 1960s at Bridgestone Japan. It is worth noting that Hoshin Kanri was first conceptualized by Miyaji, as documented in his seminal work in 1969, which was later cited by Kondo in 1998.

The Balanced Scorecard, being a classic distribution approach, employs a set of matrices that are structured into four quadrants. These matrices serve as a framework for implementing strategies in a systematic manner. Initially, strategies are translated into actionable tactics or action plans. Subsequently, they are integrated into the organization's processes. Finally, the strategies are evaluated based on the achieved results. This sequential approach ensures that the strategic objectives are effectively cascaded throughout the organization, leading to the desired outcomes. Noteworthy references that further explore this topic include the works of Jackson (2019) and Cudney (2009).

The Lean philosophy, in its essence, is centred around the eradication of any and all forms of waste within various processes. These processes encompass a wide range of areas, including customer relationships, product design, supplier networks, production management, sales, and marketing. The ultimate objective of this philosophy is to minimize human effort, inventory, cycle time, and space requirements in order to efficiently produce and deliver outputs to the customer.

This particular system finds its roots in the well-established practices of Lean Production or Lean Manufacturing (LP/LM). These practices were initially derived from the production techniques implemented by the esteemed Toyota Motor Company as early as the 1950s. The term "Lean Production" or "lean manufacturing" was first introduced to the public in an article penned by John Krafcik (1988) titled "Triumph of the Lean Production System", which was published in the prestigious MIT Sloan Management Review.

The term "lean" gained widespread recognition and popularity following the publication of the highly acclaimed book *The Machine That Changed the World* in 1990. Authored by three esteemed researchers from the Massachusetts Institute of Technology (MIT), namely Jim Womack, Daniel Jones, and Daniel Roos, this best-selling book played a pivotal role in promoting the principles of lean thinking. It is widely acknowledged as the seminal contribution that gave birth to this emerging field of study.

In 1996, Womack and Jones further solidified the concept of lean thinking with their book *Lean Thinking: Banish Waste and Create Wealth in Your Corporation*. In this influential work, they introduced the term "lean thinking" to describe a novel mindset that advocates for the systematic elimination of waste in any given activity. This approach emphasizes the practical implementation of Lean Production (LP), a methodology that aims to optimize efficiency and minimize resource wastage.

While lean production and LP are often used interchangeably, it is worth noting that there may exist a subtle distinction between the two, as highlighted by Barlotti (2013). Lean production specifically pertains to a well-organized form of production that involves coordinated labour and the utilization of tools to achieve tangible outcomes, such as the manufacturing of automobiles.

Overall, the impact of the aforementioned books and the subsequent adoption of lean principles have revolutionized various industries, prompting organizations worldwide to embrace a more streamlined and waste-free approach to their operations.

Lean management sheds light on the birth of such a system, starting from the Toyota Production System (TPS) up to current applications, as well as on the main advantages it offers in terms of reducing production times, waste of raw materials, reduced use of inventory space, and reduction of waste in general (Muda in TPS).

Furthermore, the year 2020 has brought forth a multitude of inquiries regarding the long-standing assumptions upon which both business operations and individuals' lives are predicated. In light of recent events, companies have come to a stark realization regarding the vulnerability of their supply chains, the unreliability of sources of information, and the evolving needs of their customers. In response to this newfound reality, a considerable number of these companies have emerged as pioneers of change.

It has become abundantly clear that the concept of leadership is inextricably linked to technological prowess. The rapid and unprecedented surge in digital advancements has firmly established technology as the bedrock of global leadership.

Moreover, companies have gleaned an invaluable lesson: true leaders do not idly await the arrival of the so-called "new normal", but rather, they actively shape it. The profound transformations that are necessitated in times of great change demand leaders who possess the fortitude to make bold decisions and who assign technology a paramount role in their endeavours.

It is not just about restarting business, but challenging conventions and developing a new vision for the future. It is a unique opportunity to rebuild a better world than before the pandemic. This means expanding the definition of value, including well-being, environmental impact, greater inclusivity, and more. Business and technology strategies are becoming inseparable, if not indistinguishable.

In response to the COVID-19 pandemic, companies around the world are undergoing rapid digital transformations.

These changes and the influx of new technologies have opened a new era for companies, where IT architecture is more important than ever and competition between sectors is a battle between technological stacks.

In the constantly changing environment of today's world, companies find themselves faced with an abundance of technological choices. From the advent of multi-cloud applications, to the emergence of cutting-edge devices, and the continuous evolution of AI models, every element of the technological stack is expanding in new and exciting dimensions (Accenture, 2021).

In this context, we have to set the birth of Lean Thinking 5.0, a highly regarded methodology that places its focus on the elimination of waste and the maximization of value in a systematic and efficient manner. By employing this approach, companies are able to streamline their operations, optimize their resources, and ultimately achieve higher levels of productivity and success. It is a methodology that has proven to be effective in a wide range of industries, providing tangible results and driving sustainable growth.

In essence, Lean Thinking 5.0 offers companies a structured framework through which they can find and remove inadequacies in production processes, enhancing their overall efficiency, reducing costs, and delivering greater value to their customers (Moraes et al., 2023). This methodology has gained significant recognition and adoption across various sectors, as it empowers organizations to endlessly improve and acclimatise to the ever-changing business landscape.

The plethora of technological choices available to companies today presents both opportunities and challenges. However, by embracing methodologies such as Lean Thinking 5.0, organizations can navigate these complexities with ease, ensuring that they remain at the forefront of innovation and achieve sustainable success. It builds upon the principles of Lean Thinking 4.0, with a greater emphasis on continuous improvement and innovation.

Conversely, Lean Thinking 5.0 incorporates advanced technologies to optimize operations and decision-making. It also emphasizes the importance of employee engagement and empowerment, as they are key drivers of success in implementing Lean Thinking 5.0. Overall, Lean Thinking 5.0 is a comprehensive framework that enables organizations to achieve advanced levels of efficiency, quality, and competitiveness.

Birth of Lean Production: Toyota Motor Company and Toyota Production System

Lean Production (or lean manufacturing) is a method of production organization that aims to eliminate waste. Deriving from the Toyota Production System (TPS), it is useful to analyse the history of the Japanese automobile company and its distinctive production system.

The birth of Toyota Motor Company dates back to 1933, as a simple division of the Toyoda Automatic Loom Works, a company that produced the most advanced looms of the time. Following the sale, which took place in 1929, by the founder of the company, entrepreneur Sakichi Toyoda, of the patents to Platts Brothers to allow his son Kiichiro to realize his vision of automotive production (Holweg, 2007). Legend has it that Sakichi, on his deathbed, addressed these words to his son: "I have served our country with the loom, I want you to serve it with the automobile" (Ohno, 1978). However, some analyses by modern scholars seem to refute this version of the story (Wada, 2004).

During that particular period, the Japanese market was predominantly controlled by the local subsidiaries of Ford and General Motors (GM), which had established their presence in the country as early as 1920. Additionally, the fledgling automobile enterprise of Toyoda, following the passing of Sakichi in 1930, encountered significant financial challenges and persistent ownership disputes. However, Kiichiro, with determination and resilience, managed to overcome these obstacles. His success was aided by the enactment of the Japanese law on automobile production in 1930, which provided a favourable environment for his endeavours. Kiichiro embarked on the development of his AA model, skilfully incorporating various components sourced from Ford and GM. These strategic choices played a crucial role in the realization of his vision (Cusumano, 1988).

The company was renamed "Toyota" to simplify pronunciation and give a meaning of good omen in Japanese – to write Toyota in katakana characters (トヨタ), a number of strokes equal to 8 is required, a number considered lucky in Japan. Truck production commenced in the year 1935, followed by the initiation of car production in 1936. Subsequently, in the year 1937, Toyota Motor Company was officially established. Unfortunately, the onset of the Second World War disrupted the production process, causing an unfortunate interruption. Furthermore, the post-war economic challenges exacerbated the situation, resulting in a substantial surplus of unsold cars. Regrettably, this surplus of inventory posed a significant financial burden for Toyota, leading to a period of considerable financial difficulties.

In 1950, the consequences resulted in a forced split of the production and sales divisions of the company, as well as the resignation of Kiichiro. His cousin, Eiji Toyoda, became CEO of the production branch and was sent to the United States in 1950 to study American production methods. Eiji, with unwavering determination, set out to introduce the mass production practices that were prevalent in the United States to Toyota. However, due to certain capital constraints and the relatively low volumes in the Japanese market, the justification for implementing such intensive production methods was not apparent (Holweg, 2007). It is worth noting that the Japanese market, at that time, did not possess the necessary conditions to fully embrace and benefit from the mass production approach.

Therefore, a different production philosophy became necessary, and the TPS was on the horizon. In the year 1930, Kiichiro had successfully obtained a collection of equipment that was both simple and flexible in nature. This particular assortment of tools would prove to be of great significance, as it would serve as the foundation for the implementation of numerous essential concepts for the Toyota Production System (TPS). It was Taiichi Ohno who played a pivotal role in propelling the advancement of a production system that possessed the capability to generate a vast array of products in limited quantities. His contributions were instrumental in revolutionizing the manufacturing industry. Ohno joined Toyoda in 1932 after graduating in Mechanical Engineering, and only in 1943 did he enter the automotive sector. Ohno lacked any experience in automobile production, and some scholars argue that it was precisely his unbiased approach that was fundamental to the development of the just-in-time philosophy (Cusumano, 1988).

In his way of thinking, he saw two major flaws in the Western market. The first concerned the production of components in large batches, which resulted in both a high number of stocks, which tied up considerable funds and large warehouse spaces, and a high number of production defects. The second issue, as described by Hounshell in 1984, pertains to the inability of the company to effectively meet the diverse preferences of consumers. This lack of accommodation and satisfaction has proven to be a significant challenge for the organization.

Beginning in 1948, Ohno, who held the position of manager at the engine processing laboratory, gradually introduced his concept of production in small batches throughout Toyota. This development can be traced in a comprehensive chronology provided by Ohno in 1978. Ohno's primary objective was to reduce costs by eliminating waste. This concept was derived from his prior experience with the automatic loom, where the machine had to halt operations whenever the thread broke, in order to prevent any loss of material or time.

The diligent efforts of the individual in question have facilitated the production of an impressive array of motor vehicles in limited quantities, all while maintaining a commendable level of cost competitiveness. This has resulted in a significant departure from the traditional principles of mass production. Upon reflection, it becomes evident that these alterations were indeed revolutionary in nature. However, it is important to acknowledge that, at the time, they were merely pragmatic adjustments made in response to the prevailing economic circumstances (Holweg, 2007).

Eliminating the 7 Muda

The Japanese term "Muda", which translates to waste, is a fundamental concept in Lean manufacturing. It encompasses actions that are deemed superfluous, thus rendering them unproductive. This concept was first identified by Taiichi Ohno, the esteemed chief engineer of Toyota, as part of the Toyota Production System. In essence, any activity within a process that fails to add value to the customer is classified as "waste". There are seven primary types of waste that can be observed. Firstly, waste within the production system can be discerned through the presence of excessive production resources. These resources are characterized by an overabundance of labour, equipment, and inventory, all of which are deemed unnecessary and therefore surplus to the requirements of the customers. This solely increases costs but does not contribute to value creation. In order to embark upon the journey of attaining a lean organization, as suggested by esteemed scholars Womack and Jones in their seminal work of 1996, it is imperative to meticulously scrutinize and analyse the various processes within the organization. These processes, encompassing both primary and support functions, warrant careful examination in order to identify areas of improvement and potential inefficiencies. By undertaking this comprehensive analysis, organizations can lay the groundwork for streamlining operations, enhancing productivity, and ultimately achieving the coveted state of lean efficiency. To do this, a value stream mapping can be developed to recognise the network of processes and sub-processes. The primary objective of this particular endeavour is to meticulously discern and differentiate between activities that contribute significant value and those that can be promptly and effortlessly eliminated due to their lack of value. This notion is supported by the research conducted by Rother and Shook (1999). By engaging in this process, one can effectively streamline operations and optimize efficiency by focusing solely on activities that yield tangible benefits. In conclusion, constant monitoring, analysis, and improvement of the process are fundamental to ensure its stability and repeatability over time.

The focal point of Lean manufacturing is the customer. Among others, two factors are closely connected to the customer: the delivery time of the product/service and the cost (price). Hunting for waste undoubtedly can impact both these factors by reducing the customer's waiting time (impacting on the "production time") and cutting some production costs that are a significant part of the final product/service price.

For instance, holding disproportionate inventory results in paying financial charges on immobilized capital invested in material that does not add value. The use of excessive resources leads to overproduction, meaning carrying out activities even beyond what is necessary and essential. This situation contributes to creating excess inventory, which, in turn, determines the need for labour, equipment, storage spaces, and resources for movement. The use of excessive resources, overproduction, and excess inventory determines an unnecessary investment of capital in creating warehouses for storing extra inventory, the need to hire operators to move materials from the warehouse, purchase equipment for operators, hire operators for maintenance, with the result of having a very complex management of all processes. These wastes indirectly affect the increase in administrative costs and indirect labour costs without adding value to the customer (Imai, 1986).

Taiichi Ohno identified seven types of waste listed below. The production system promoted by Ohno was based on reducing Muda, i.e., those unnecessary, unproductive, or non-value-adding activities within all business processes.

These Muda are (McBride, 2003):

1. overproduction;
2. downtime;
3. transportation time;
4. inappropriate processing;
5. excess inventory;
6. excessive/unnecessary movements;
7. defects.

The following sections discuss each component in turn.

Overproduction

In brief, the concept of overproduction entails the act of manufacturing an item prior to its actual requirement. This practice, although seemingly innocuous, can prove to be quite burdensome for a production plant. The reason being that overproduction disrupts the seamless flow of materials, thereby impeding the overall efficiency and effectiveness of the production process. Moreover, it is worth noting that overproduction has the potential to compromise the quality of the final product, as well as diminish the productivity levels within the plant. Consequently, it becomes evident that overproduction is a costly endeavour that should be avoided in order to optimize the operations of a production facility. The concept being discussed here is known as Just-in-Time (JIT) production. JIT production is a method where each element is produced precisely when it is needed. In contrast, the alternative approach, known as Just-in-Case, involves producing items in anticipation of future requests. This latter approach often results in excessively long lead times and high storage costs. Additionally, it can make it challenging to identify any defects in the final product. By implementing JIT production, companies can mitigate these issues and streamline their operations. The simplest solution to avoid overproduction lies in "closing the tap"; this requires a lot of courage as it reveals the problems that overproduction hides.

The concept is to only program and produce what can be immediately sold/shipped and increase the machine's set-up capacity.

Downtime

This particular form of waste arises when goods experience a lack of movement or are subjected to processing delays. In the realm of traditional production, it is customary for over 99% of a product's lifespan to be consumed by idle waiting periods, as it awaits its turn for processing. The primary cause of this extended lead time is the inadequate flow of materials, excessively long production cycles, and substantial distances between work centres. Goldratt (1988) has consistently emphasized that an hour lost at a bottleneck equates to an hour lost for the entire factory's output, with no possibility of recovery. By establishing a seamless connection between processes, wherein each operation directly feeds into the next, the waiting times can be significantly reduced.

Transportation time

Transporting a product from one stage to another represents a cost and a non-value-added activity. Excessive movements can cause damage, deteriorating the overall quality of the products. Mechanisms for managing components and/or raw materials are necessary for the transportation of materials, resulting in additional organizational costs that do not add value for the customer. Transportation can be difficult to reduce due to the perceived high costs of moving equipment and processes, as physical distances between them should be reduced. Furthermore, one frequently encounters the challenge of discerning the appropriate sequencing of processes. This can lead to a certain level of ambiguity and uncertainty. However, by employing the technique of mapping process flows, one can effectively alleviate this predicament. The act of mapping process flows provides a visual representation of the various phases involved, thereby facilitating the determination of the logical progression from one phase to the next. This method offers a clear and concise means of establishing the appropriate sequence of processes, ultimately enhancing overall efficiency and productivity.

Inappropriate Processing

It could be defined as "using a sledgehammer to crack a nut". Many organizations use expensive, high-precision equipment even in cases where simpler tools would suffice. This often results in an inefficient plant layout since subsequent operations are far apart from each other. Additionally, it encourages high-speed use (overproduction with minimal changeover), in order to quickly amortize the high cost of equipment. Toyota is instead famous for the use of low-cost, perfectly maintained automation, often represented by older machines than those used by the competition. Investing in small, flexible equipment where possible and creating production cells can greatly reduce the waste of inappropriate processing.

Excess Inventory

Work in progress is a direct result of overproduction and waiting times. Excess inventory tends to hide plant-level problems, which must be identified and resolved in order to improve operational performance. They cause an increase in lead time, consume production space, delay problem identification, and inhibit communication. With the successful attainment of an uninterrupted flow between work centres, numerous manufacturers have experienced notable enhancements in their customer service capabilities, as well as substantial reductions in inventories and the associated costs. This achievement has been made possible through the implementation of streamlined processes and efficient coordination between various work centres. By eliminating unnecessary delays and bottlenecks, manufacturers have been able to ensure a smoother and more efficient workflow, resulting in improved customer satisfaction. Furthermore, the reduction in inventories has not only minimized storage costs but has also allowed manufacturers to respond more promptly to customer demands, thereby enhancing their overall competitiveness in the market. These positive outcomes have been a direct result of the meticulous planning and diligent execution of strategies aimed at optimizing the flow of work within manufacturing facilities.

Excessive/Unnecessary Movements

This waste is associated with the field of ergonomics and can be observed in situations where staff members experience stress as a result of activities such as bending, stretching, moving, and lifting. It is important to note that this waste is not limited to specific cases, but rather occurs universally. This matter also pertains to health and safety concerns, which, in our current litigious society, are increasingly posing challenges for organizations. It is imperative that we thoroughly analyse and subsequently redesign positions that involve excessive movement in order to enhance and optimize the participation of our plant personnel.

Defects

Quality defects, resulting in rework or product scrap, represent a huge cost to organizations. They are also associated with costs for quarantine zones within inventory, re-evaluations, reprogramming, and capacity losses. In numerous corporate entities, it is a common occurrence for the cumulative cost of defects to constitute a substantial proportion of the overall manufacturing expenditure. However, by fostering employee engagement and implementing a culture of continuous process enhancement, it is possible to achieve a noteworthy reduction in defects across various facilities.

The concept of reducing Muda, which refers to non-value-adding activities within a process, is intrinsically linked to the reduction of Mura and Muri. Mura, which can be translated as "non-uniformity", and muri, which signifies "excessiveness", are both detrimental elements that are to be addressed to reach optimal efficiency. By eliminating Muda, we can effectively minimize Mura and muri, thereby streamlining the workload and enhancing productivity.

To successfully reduce Muda, it is imperative to find and remove any activities that do not contribute value to the overall process. These non-value-adding activities often result in wasted time, resources, and effort. By meticulously analysing each phase of the process, we can identify areas where Muda is present and take appropriate measures to eliminate it.

Furthermore, the reduction of Mura is equally crucial in achieving process optimization. Non-uniformity within a process can lead to inconsistencies, bottlenecks, and delays. By striving for uniformity and standardization, we can ensure a smooth and efficient workflow, minimizing disruptions and maximizing output.

Lastly, addressing Muri, or excessiveness, is paramount in achieving a balanced workload. Overburdening individuals or resources can lead to fatigue, errors, and decreased productivity.

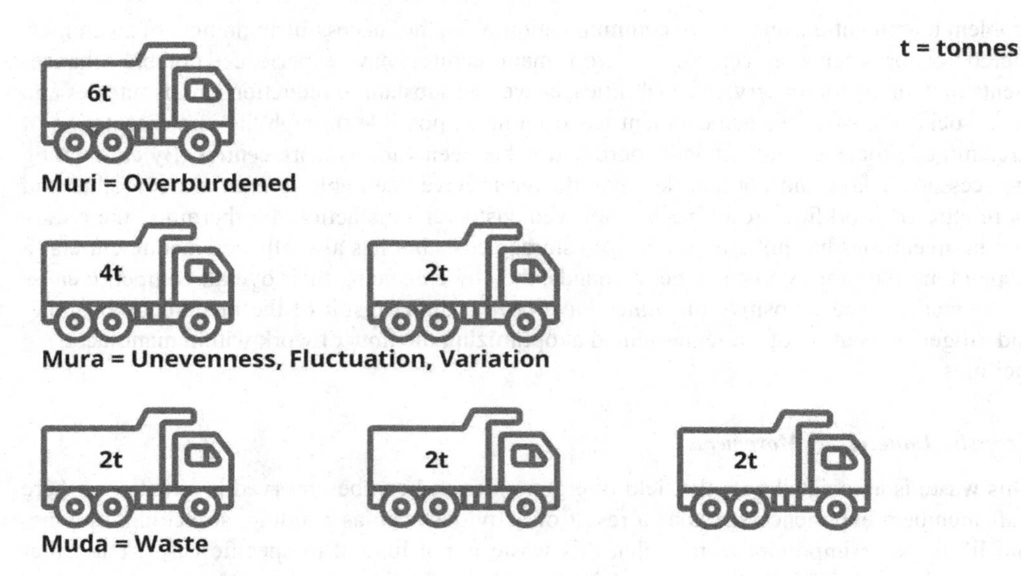

Figure 3.2 Graphic representation of the concepts of Muri, Mura, and Muda.
Source: authors' own elaboration

By carefully assessing the workload and making necessary adjustments, we can ensure that it remains within manageable limits, promoting a healthy and sustainable work environment.

Thus, by focusing on reducing Muda, Mura, and Muri, we can improve the efficiency and effectiveness of our processes. By eliminating non-value-adding activities, striving for uniformity, and avoiding excessiveness, we can optimize our workload and achieve optimal results. It is imperative that we remain vigilant in identifying and addressing these factors, as they directly impact our productivity and overall success.

In this regard, Mura indicates irregularity in the workload, which causes bands of excessive and non-optimal load. This is caused by incorrect management and standardization of demand and results in the emergence or increase of all other internal waste in the production process. Mura can be corrected through the use of Just-in-Time practices (Emiliani et al., 2007).

Muri, on the contrary, pertains to the excessive burden of work, a circumstance that frequently culminates in the occurrence of breakdowns, injuries, and illnesses. These unfortunate events subsequently lead to the unavailability of both machinery and employees, ultimately rendering the company incapable of fulfilling market demand. To remedy this, it is necessary, first and foremost, to monitor the satisfaction and health of workers and then adopt corrective measures such as standardized work methods and machinery that facilitate employees' work and reduce the physical strain required of them (Emiliani et al., 2007).

Furthermore, this list could include the item "unused creativity of employees" (Liker J. K., 2014). This concept includes the loss of time, ideas, skills, and opportunities for improvement and learning because employees are not listened to or interacted with.

In this regard, Toyota was one of the first companies in Japan to use the method of quality circles: these are working groups that include everyone in the company, from executives to workers, whose goal is to contain problems and maintain high production levels and related processes.

Precisely because of the elimination of all possible waste through its TPS, today Toyota is the world's largest automaker in terms of production volume (in June 2012 it produced its 200,000,000th piece) and sales (about 9.98 million vehicles sold in 2013, with a 2% increase over 2012).

Toyota's House of Lean

Fujio Cho, a disciple of Taiichi Ohno, developed a graphical representation of the TPS method: a house, also known as Toyota's House, which synthetically outlines the strategic elements on which the TPS is based.

He used this figure because a house is a structural system and as such, it is a symbol of solidity only if all the elements that make it up, such as the roof, pillars, and foundations, are well-structured and interdependently cooperate with each other. A weak link weakens the entire structure, and in this case the entire system.

Figure 3.3 is a graphical representation developed by Cho, which has become one of the most recognizable symbols of the TPS theory. The roof of the house, which is symbolized by its high quality, low costs, short lead time, and the elimination of wasted time and unnecessary activities, is supported by two outer pillars known as Just-In-Time and Jidoka. Just-In-Time, being the most renowned and widely recognized component of the Toyota Production System (TPS), safeguards that the appropriate portion is delivered punctually and in the correct amount. This is achieved through meticulous planning based on Takt time, which determines the pace of production, as well as the implementation of a continuous flow system, a pull logic system, and an integrated logistics approach. These elements work in harmony to optimize efficiency and minimize waste throughout the production process. Jidoka, the second pillar of the house, is based on the principle that no machine operation presenting a defect or problem should pass

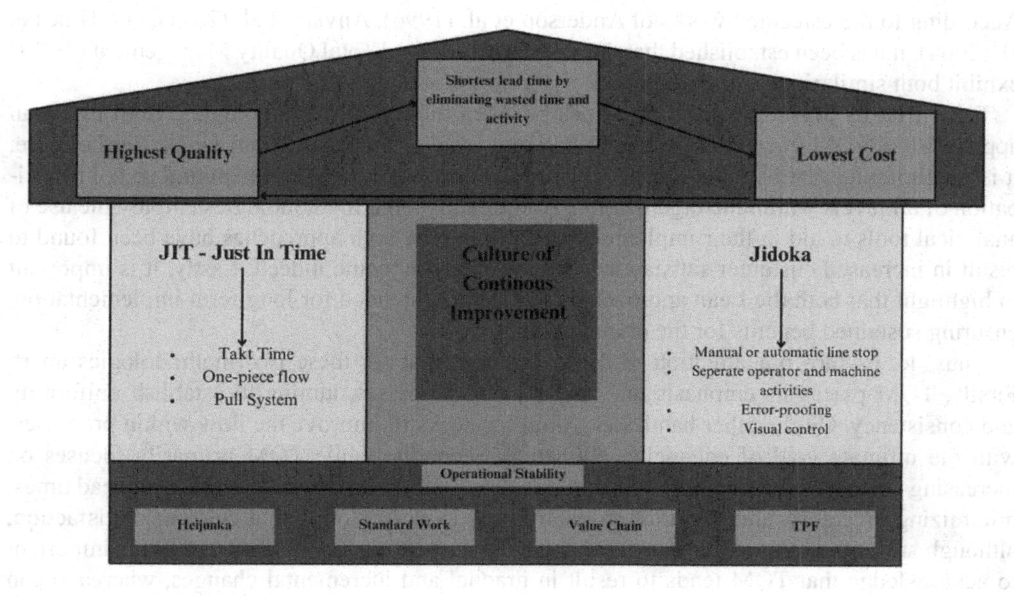

Figure 3.3 Toyota's house.

Source: authors' own elaboration

to the next phase without first resolving it. Between the two pillars, at the centre, lies the culture of continuous improvement, always aimed at satisfying the customer.

Finally, standardized, stable, and reliable processes form the foundation of the Toyota house, in which Heijunka, the proper ratio between production sequences in relation to volume and variety, represents the foundation that allows the system to remain stable, especially by keeping inventory at a minimum.

Lean Production is a management method that has garnered considerable acknowledgment in the field of operations management. According to Womack and Jones (1996), it emphases on the removal of any use of resources that does not directly contribute to creating value for the end customer. It is worth noting that Lean Production builds upon the integration of Just-In-Time and Total Quality Control principles, as highlighted by Crute et al. (2003) and Kojima and Kaplinsky (2004). These principles have been extended to various aspects of production, including processes, logistics, and supply chain systems.

One of the main contributions of the Lean approach is the introduction of the concept of "waste" (W) in production. This concept was initially studied by Ohno (1978) in the context of the Toyota production system, where he referred to it as Muda. The significance of Muda lies in its direct impact on quality, time, and cost variables. By identifying and eliminating Muda, organizations can achieve a synergistic and positive effect on the overall competitiveness of their systems. This can manifest in various ways, such as managing prices more effectively or delivering a higher level of performance in products or services (Rother and Shook, 1999).

To conclude, Lean Production suggests a systematic and comprehensive method to optimizing resource utilization and enhancing overall operational efficiency. By concentrating on the eradication of waste and endlessly enhancing processes, companies can reach significant benefits in terms of quality, time, and cost.

Comparing Lean Production and TQM

According to the esteemed works of Anderson et al. (1994), Anvari et al. (2011), and Hines et al. (2004), it has been established that the Lean approach and Total Quality Management (TQM) exhibit both similarities and differences.

Firstly, let us delve into the similarities between these two methodologies. Both the Lean approach and TQM trace their origins back to the Land of the Rising Sun, Japan. Furthermore, it is worth noting that both approaches require the unwavering commitment and active participation of all levels within the organization. Additionally, both methodologies employ the use of analytical tools to aid in their implementation. Moreover, both approaches have been found to result in increased customer satisfaction, a noteworthy outcome indeed. Lastly, it is important to highlight that both the Lean approach and TQM are intended for long-term implementation, ensuring sustained benefits for the organization.

Thus, let us turn our attention to the differences that set these two methodologies apart. Firstly, TQM places its emphasis on standardizing processes, aiming to establish uniformity and consistency. On the other hand, Lean thinking seeks to improve the flow within processes, with the ultimate goal of enhancing efficiency. Secondly, while TQM primarily focuses on increasing customer satisfaction, Lean thinking directs its attention towards reducing lead times, minimizing inventory, and boosting productivity. It is worth noting that customer satisfaction, although still important, assumes a secondary role in the Lean approach. Lastly, it is important to acknowledge that TQM tends to result in gradual and incremental changes, whereas Lean thinking has the potential to bring about sudden and transformative shifts.

In conclusion, it is evident that the Lean approach and TQM share certain commonalities, such as their Japanese origins, the need for organizational commitment, the utilization of analytical tools, the achievement of heightened customer satisfaction, and their long-term implementation. However, they also exhibit distinct differences, including their respective focuses on standardization versus process flow improvement, the primary effects of increasing satisfaction versus reducing lead times and inventory, and the pace of change they tend to generate. It is crucial for organizations to carefully consider these similarities and differences when determining which approach aligns best with their specific needs and objectives. Therefore, TQM is a business approach in which the entire company must be involved in achieving the objective, in a continuous improvement cycle focused on customer needs, which involves participatory management. This also involves involving and mobilizing employees and reducing waste in an effort to optimize efforts. The key components of this system are employee involvement and training, problem-solving groups, statistical methods application, long-term goals, and the awareness that inefficiencies are produced by the system, not the people.

From a traditional standpoint, it has long been believed that in order to achieve high quality, substantial production costs are necessary. However, it is important to note that this perspective is no longer applicable in today's business landscape. There are two key reasons for this shift in thinking:

- Firstly, the concept of quality must now be defined based on the wants and expectations of the customer. In order to truly satisfy the customer, their needs and desires must be taken into account. This customer-centric approach is the driving force behind the total quality movement, which aims to ensure customer satisfaction above all else.
- Secondly, many companies have come to realize that they are competing not only on price, but also on quality. Customers now expect a certain level of quality at a competitive price, as well as specific supply requirements. This means that companies must strive to meet these expectations in order to remain competitive in the market.

In today's business environment, a company can establish a competitive advantage over its rivals by adopting a comprehensive and consistent approach that impacts all aspects of its operations. A strategy cantered around Total Quality is designed to satisfy all stakeholders involved, including customers, suppliers, social parties, workers, management, and shareholders. It is important to note that the output of a company is no longer limited to just the product or service it offers; rather, the company itself becomes a producer of quality.

In conclusion, the traditional belief that high production costs are necessary for high quality is no longer valid. By embracing a customer-centric approach and recognizing the importance of meeting both price and quality expectations, companies can build a competitive advantage and ensure the satisfaction of all stakeholders involved. In summary, the principles that underlie TQM are as follows:

1. **Customer focus**: TQM places a strong emphasis on understanding and meeting the needs and expectations of customers. This involves actively seeking feedback, conducting market research, and continuously improving products and services to ensure customer satisfaction.
2. **Leadership**: TQM recognizes the crucial role of leadership in driving quality improvement initiatives. Effective leaders establish a clear vision, set goals, and provide guidance and support to employees to achieve these objectives. They additionally cultivate an environment of ongoing enhancement and promote employee engagement.

3. **Involvement of people**: TQM emphasizes the importance of involving all employees in the quality improvement process. This includes empowering employees to contribute their ideas, skills, and knowledge, as well as equipping them with essential training and resources to actively participate in quality initiatives.
4. **Process approach**: TQM adopts a systematic approach to managing quality by focusing on the processes that contribute to the overall value creation. This involves identifying, analysing, and improving key processes to enhance efficiency, reduce waste, and deliver consistent and reliable results.
5. **Systematic approach to management**: TQM advocates for a holistic and integrated approach to managing quality throughout the organization. This entails establishing clear policies, procedures, and performance metrics to ensure that quality objectives are effectively communicated, monitored, and achieved.
6. **Continuous improvement (Kaizen)**: TQM promotes a culture of continuous improvement, where all employees are encouraged to identify opportunities for enhancement and contribute to incremental changes. This involves regularly reviewing processes, gathering data, and implementing improvements to achieve higher levels of quality and efficiency.
7. **Evidence-based decision-making**: TQM emphasizes the importance of making informed decisions based on reliable data and evidence. This involves collecting and analysing relevant information to support decision-making processes, ensuring that choices are objective, rational, and aligned with quality objectives.
8. **Mutually beneficial supplier relationships**: TQM recognizes the significance of building strong and collaborative relationships with suppliers. This involves selecting and working with suppliers who share the same commitment to quality, fostering open communication, and establishing mutually beneficial partnerships to enhance overall quality performance.

It is important to note that while TQM and Lean Manufacturing (LM) both aim to optimize operations, they differ in their approaches and perspectives. TQM generally refers to managing quality at every stage of operations, from planning and design to continuous process monitoring for improvement. On the other hand, LM focuses on removing waste and restructuring processes to achieve operational efficiency. By understanding these distinctions, organizations can choose the most suitable approach to meet their specific quality improvement goals.

In the year 1997, Boaden undertook the task of comparing the definition of TQM to shooting at a moving target. This analogy was employed due to the existence of various definitions that have been put forth over the course of several years. TQM, as it is commonly understood, is a corporate culture that places great emphasis on the enhancement of customer satisfaction through the continuous improvement of processes, as well as the active participation of all employees. Dale, in the year 1999, provided a more comprehensive description of TQM, stating that it is not only a philosophy, but also a set of guiding principles that are employed in the management of an organization. The primary focus of TQM is to exercise control over business processes, while simultaneously ensuring customer satisfaction through a range of activities, including but not limited to improvement initiatives, statistical control, supply control, and quality engineering. It is worth noting that TQM originated within the academic realm and has been significantly influenced by notable contributors such as Feigenbaum, Juran, and Deming.

Total Quality Management (TQM) and Lean Manufacturing (LM) are two management approaches that have garnered significant attention in the field of production optimization.

However, due to the existence of varying interpretations and definitions of these approaches, a spirited debate has ensued regarding their similarities and differences.

Lean Manufacturing, also referred to as Lean Production (LP), is a production practice that strives to eliminate the utilization of any resources that do not directly contribute to the creation of value for the end customer. The term "Lean" is believed to have been initially coined by Womack et al. in their seminal work, *The Machine that Changed the World*. Lean Manufacturing (LM) is a process management philosophy that finds its roots in the esteemed Toyota Production System (TPS). It is worth noting that some esteemed authors have taken the liberty to define Lean in a more comprehensive manner, encompassing not only the people and process components, but also the internal and external components that are intricately linked to the firm, supplier, and customer. One such definition, put forth by Shah and Ward, places great emphasis on the mechanisms that are necessary to achieve the noble objective of waste elimination, which lies at the very heart of Lean.

In its essence, Lean Manufacturing is the art of producing goods with utmost efficiency, minimizing waste, human effort, manufacturing space, investment in tools, and engineering for product development. By meticulously reducing waste, Lean Manufacturing ultimately enhances the overall value that is delivered to the esteemed customer.

Tools of Lean Production

The Japanese term utilized to denote ground-breaking, transformative, or revolutionary alteration is "kaikaku". This methodology was actively employed at Toyota and is accomplished by establishing revolutionary objectives, subsequently working towards their attainment through incremental steps, and actively involving the individuals who possess the most comprehensive knowledge and will be directly impacted by the changes.

In essence, this process is facilitated through "kaizen", which refers to the concept of continuous improvement. By achieving radical change through a series of small, incremental steps, it ensures that all significant matters are thoroughly addressed, interruptions are minimized, and the changes are effectively sustained. This outcome would not be feasible if the changes were implemented hastily and all at once. It is imperative that once a radical or innovative change is successfully accomplished, it is immediately followed by incremental improvements. Failure to do so would result in the stagnation of the S-curve and leave the company vulnerable to being surpassed by its competitors.

For the majority of organizations, the concept of lean production can be seen as a revolutionary and unfamiliar change. It is understandable that some may even find it threatening. However, it is important to approach the transition to lean production with caution and selectivity, rather than attempting to implement it everywhere and quickly. The recommended approach is to embark on a continuous series of small improvements, known as kaizen.

Lean Production, as an integrated system, works in conjunction with other systems such as quality management and economic-financial management. It begins at the top level of the organization, aligning with the objectives outlined in the business plan. From there, it evolves into specific improvement projects, such as Six Sigma, with a primary focus on enhancing production processes and the delivery of products or services. In essence, the implementation of a Lean system is not vastly different from that of a quality management system. Both involve translating business-level goals into actionable steps that target the reduction of the seven categories of waste.

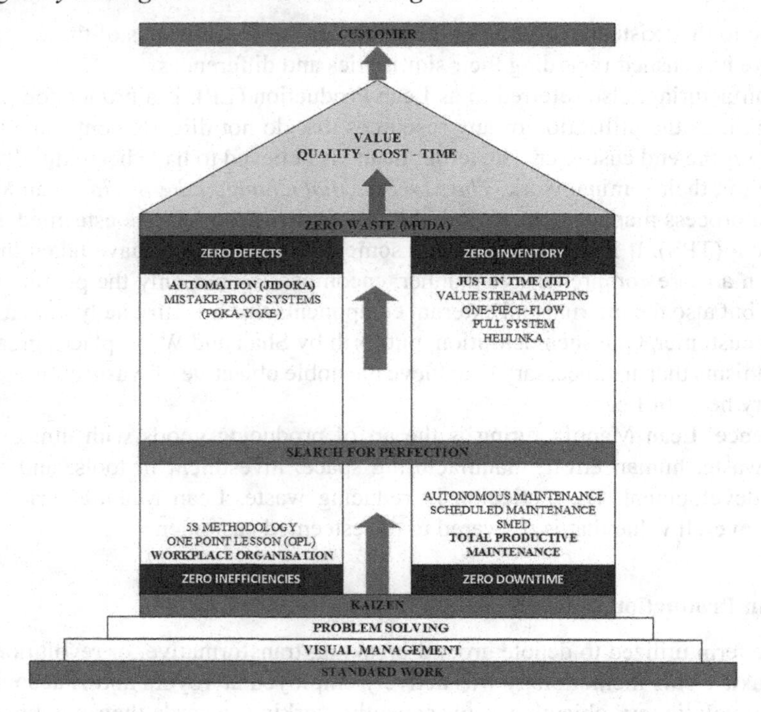

Figure 3.4 Lean House.
Source: authors' own elaboration

The fundamental elements of Lean Production can be succinctly summarized by the concept of the "Lean House". This framework draws inspiration from the Toyota House and incorporates the principles outlined in this chapter. By adhering to these principles, organizations can confidently navigate the path towards lean production, optimizing their operations and achieving greater efficiency.

The four pillars are:

- Just-in-Time (JIT);
- Automation (Jidoka);
- Total Productive Maintenance (TPM);
- Workplace Organization (WO).

There exist two fundamental concepts that serve as the foundation for these pillars:

1. The first concept is Standardization, also known as Standard Work. This concept employs the use of Visual Management, which is a method utilized by both workers and managers to promptly control and visualize any wasteful practices within the workshop. In order to achieve this, it is imperative that all indicators and issues within the workshop are effectively controlled and managed through the implementation of displays, signals, audible alarms, and other real-time systems. By employing these measures, waste can be efficiently identified and addressed in a timely manner, leading to improved productivity and overall efficiency within the workshop.
2. Continuous Improvement (Kaizen), which uses specific problem-solving techniques.

The goal of Lean Production is total waste elimination ("zero target"), not just its reduction. Each pillar has its own zero target:

- JIT: Zero Inventory;
- Jidoka: Zero Defects;
- TPM: Zero Downtime;
- WO: Zero Inefficiencies.

These individual targets are transformed into perceived Customer Value, in terms of Quality, Cost, and Time.

Lean Manufacturing uses various tools (Tozawa, 1995; Monden, 1998; Laraia et al., 1999; Feld et al., 2010; Nahmias, 2001; Galgano, 2008; Bakri and Januddi, 2020). There are numerous tools for process representation and improvement, which can be grouped into the following categories:

(a) Mapping techniques for Just-in-Time.
(b) Techniques used for workplace organization.
(c) Techniques to achieve zero defects in the Jidoka automation process.
(d) Techniques to achieve zero downtime in production.

Mapping Techniques for Just-in-Time

Mapping tools are used to highlight waste, specify value-added activities, and indicate intervention priorities together with the strategic objectives of management.

It should be noted that Just-in-Time (JIT) is an industrial philosophy that is based on a Pull production logic, which means that production is only carried out for what has already been requested or sold, or what is known to be sold. This is in stark contrast to the dominant philosophy in old Western production systems, which were based on Push logic, which involved producing large volumes of inventory prior to ordering. JIT views the origin of the production process in the customer's request, after which all production operations begin. This allows for a drastic reduction in costs and warehouse space, as inventory is minimized, if not completely absent. The primary objective of this initiative is to place a strong emphasis on the reduction of production times. This is achieved through the implementation of a highly efficient raw material supply system that seamlessly integrates with the various stages of the production process. It is worth noting that this system is intricately connected to the initial demand, ensuring a smooth and streamlined workflow. By adopting this approach, we are able to optimize our operations and enhance overall efficiency. In other words, materials, in the exact quantity necessary for production, are transferred from one phase to another in such a way as to prevent waiting times, and their processing has standardized methods and, above all, times that must be strictly adhered to. Therefore, by improving the flow of materials, the overall quality of intermediate and finished products is also improved, as operators will not be overloaded. Finally, JIT allows for a potential cost savings for employees, as in the absence of demand, there will be no production, and therefore the work of the employees will not be necessary. The tools are described in the following sections.

Value Stream Mapping

Value Stream Mapping is a graphical visualization method that originated from Toyota's production philosophy.

This particular methodology gained widespread recognition and acclaim due to the efforts of esteemed individuals, namely Mike Rother and John Shook, who authored the highly regarded publication titled "Learning to See". Additionally, they developed a specialized software, Microsoft's Visio, which greatly facilitated the seamless integration of this methodology into various productivity documents. The essence of this methodology lies in the meticulous mapping of the value chain, which serves to eliminate any form of wastage and significantly enhance overall operational efficiency. In essence, this entails the comprehensive delineation of all processes and activities that actively contribute to the creation of a product, ultimately culminating in the successful delivery of the finished product.

The Value Stream Mapping (VSM) technique serves as the initial tool for conducting internal process analysis. By meticulously mapping the material flows and information flows that govern the material, VSM provides a visual representation that greatly enhances the implementation of Lean principles. This visual aid is instrumental in identifying both value-added and non-value-added activities, thereby streamlining the Lean implementation process.

VSM consists of two distinct maps: the Current State Map and the Future State Map. These maps employ standardized symbols to effectively depict the various processes involved, ensuring a comprehensive understanding of the entire flow of a product, service, or product family from suppliers to customers.

It is important to note that the underlying premise of VSM is rooted in the pursuit of global and continuous optimization. By employing this technique, organizations can systematically analyse and improve their processes, leading to enhanced efficiency and overall performance.

The various types of Value Stream Mapping (VSM) are as follows:

1. The Current State Map provides a comprehensive description of the product's current situation within the value flow. By examining the existing processes, this map allows us to identify areas of improvement and potential waste.
2. The Future State Map, on the other hand, serves as a guide to envision how the product should ideally be positioned within the value flow. It outlines the desired state of the processes, highlighting the improvements that need to be made to achieve optimal efficiency.

The primary objectives of Value Stream Mapping are not limited to individual processes, but rather focus on the overall flow of value. By analysing the entire value stream, we can identify and eliminate any sources of waste that hinder productivity. This approach empowers the entire organization by equipping them with the necessary tools to comprehend and interpret the flow of value.

Furthermore, Value Stream Mapping enables us to visualize the aspects that contribute to process efficiency. By identifying and understanding these factors, we can implement strategies to enhance productivity and streamline operations.

In summary, Value Stream Mapping is a tool that consents companies to gain a comprehensive understanding of their value flow. By utilizing this technique, we can recognise areas for enhancement, abolish waste, and ultimately enhance our processes for maximum efficiency.

Cellular Manufacturing and One Piece Flow

Cellular Manufacturing is a highly efficient method of organizing the production process for a specific product or a group of similar products. It involves the creation of a dedicated group, known as a cell, which encompasses all the necessary machines, equipment, and operators. The resources within these cells are meticulously arranged in a straightforward manner, ensuring

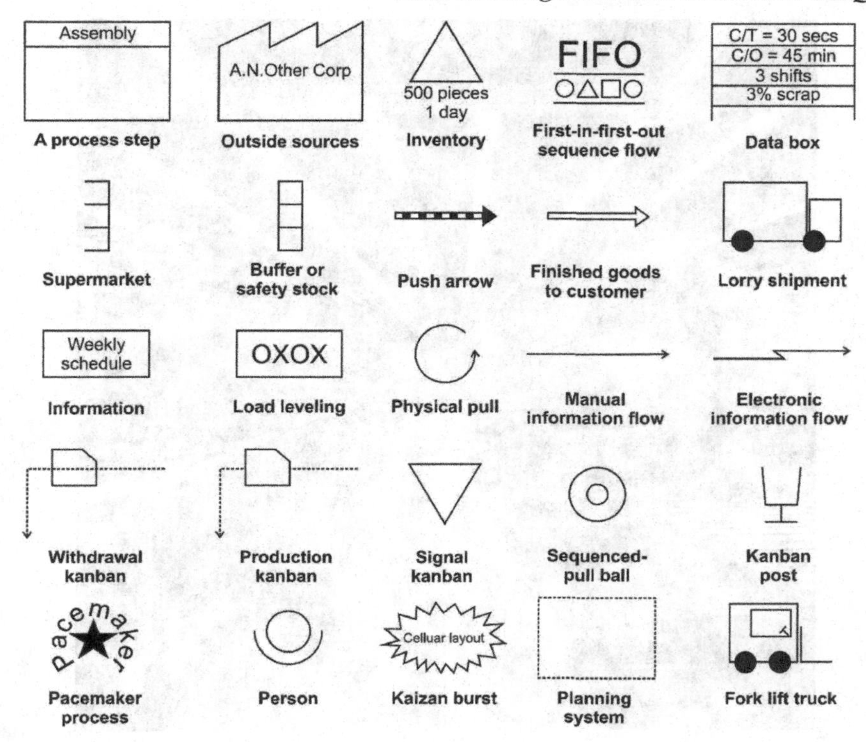

Figure 3.5 VSM's symbols.

Source: authors' own elaboration

seamless operations. This approach can be likened to having multiple "plants within the plant", where each cell is responsible for executing all the essential operations required to convert raw materials into a finished product.

By implementing Cellular Manufacturing, companies can reach remarkable improvements in productivity and efficiency. The streamlined organization of resources within each cell eliminates unnecessary movement and minimizes the time and effort mandatory to conclude each task. This results in a noteworthy reduction in production lead times and increased throughput.

Furthermore, the simplicity and focus of Cellular Manufacturing enable companies to enhance quality control measures. With all operations consolidated within a single cell, it becomes easier to monitor and maintain consistent quality standards throughout the production process. This ensures that the finished products meet or exceed customer expectations, leading to higher levels of customer satisfaction and loyalty.

In conclusion, adopting Cellular Manufacturing can revolutionize the way companies approach production. By grouping resources into dedicated cells and empowering them to perform all necessary operations, businesses can achieve remarkable gains in productivity, efficiency, and quality control. Embracing this methodology will undoubtedly position companies at the forefront of their industries, enabling them to meet the demands of a competitive market with confidence and success. Machines are located close to each other in sequence to minimize

Figure 3.6 A robotized production cell.
Source: authors' own elaboration

product movement and maintain a continuous flow, eliminating storage points between operations and drastically reducing processing times. The cell is monitored by a team of operators with diverse skills who have complete responsibility for the quality and output of their cell, which certainly contributes to increasing the involvement and accountability of the operators themselves.

This type of production allows for a reduction in lead-time through a reduction in material lead times across all stages of the process, as well as reduced work organization times, as each cell has all the necessary tools for production, and less stress for operators, who are subject to significantly lower efforts.

To avoid overproduction, it is imperative to obtain a delicate equilibrium between capacity and workload. One of the most widely recognized methods for streamlining production layout is known as "cellular manufacturing". This method not only aids in lessening various forms of waste but also propels the organization towards a pull system. In stark contrast to traditional work practices, a cell operates in a manner that is entirely antithetical. In the traditional set-up, a product would have to traverse several hundred meters within the production process, resulting in wastage of time and a substantial accumulation of work in process (WIP) between different areas. Conversely, in a cell, machines and equipment are strategically positioned to eliminate the need for movement or product transport.

The numerous advantages of implementing cellular manufacturing include the following:

1. Waste reduction in terms of worker transport and movement.
2. Decreased delivery times.
3. Optimal utilization of space.
4. Activity balancing and reduction in production batches.
5. Minimized internal set-up times.
6. Elimination of causes of waste.

By adhering to these principles, organizations can effectively mitigate overproduction and optimize their production processes. Cellular manufacturing is an absolutely essential practice when it comes to maintaining the desired level of flexibility in order to meet the ever-changing demands of customers. One highly effective tool that can be employed in this regard is known as "one-piece flow". This tool involves the meticulous process of working with one product at a time, skilfully mixing similar products with different codes within the same process.

In the contemporary rapid-paced business landscape, where product and service life cycles are becoming increasingly shorter and the demand for customization is on the rise, it has become quite challenging to produce goods using traditional production methods that are primarily designed for large quantities. This is where the concept of "one-piece flow" comes into play. By replacing the conventional assembly lines with a U-shaped cell, which encompasses all the necessary activities and equipment required for the production of a specific product or service, manufacturers can effectively address these challenges.

By adopting this approach, manufacturers can achieve a higher level of efficiency and productivity, as well as reduce waste and minimize the risk of errors. The U-shaped cell layout allows for a smooth and continuous flow of materials and information, eliminating the need for excessive transportation and unnecessary waiting times. This not only conserves valuable time but also enhances resource utilization, leading to cost reductions and improved performances.

Therefore, the implementation of cellular manufacturing, particularly through the utilization of "one-piece flow", is a highly recommended strategy for organizations seeking to enhance their operational flexibility and meet the ever-evolving demands of their customers. By embracing this approach, manufacturers can become leaders in their businesses, ensuring their long-term success and profitability. In order to optimize production efficiency, it is imperative to allocate cells either exclusively to a singular product in cases of large quantities, or to multiple products through a mixed model approach. When employing this particular methodology, it becomes absolutely essential to swiftly transition from one product, identified by its unique code, to the subsequent product. This practice ensures a seamless workflow and maximizes productivity. This method refers to a type of production or assembly in which manufacturing occurs on the basis of single pieces that "flow" through the entire production system without intermediate stocks or inventory. The assembly line, as organized in this particular manner, possesses the remarkable capability of swift conversion. This means that it can be readily adapted to accommodate the assembly of various products, regardless of their quantity, be it small or even large in numerical terms. The line's flexibility allows for a seamless transition from one product to another, ensuring a smooth and efficient production process. This feature is particularly advantageous for manufacturers who require the ability to assemble different products in succession. By implementing this assembly line configuration, manufacturers can optimize their production capabilities and enhance their overall operational efficiency. With the one-piece flow method, processing times and set-up times are reduced, production inventories are reduced, and material flow follow-up is guaranteed.

Heijunka

Heijunka is a production levelling technique that serves the noble purpose of balancing the workload within a production cell, thereby minimizing those pesky supply fluctuations that can cause such distress. First and foremost, we have production volume levelling. This magnificent concept involves the distribution of production in a uniform manner over a specific period of time. By achieving this harmonious balance, we can ensure a steady and consistent production volume. However, do not forget that the levelled production volume is heavily reliant on another crucial element, known as levelled production variety. This entails the uniform distribution of the production mix over the same period of time. It is through the combination of these two elements that the true magic of Heijunka production is realized. Let us delve into the realm of Heijunka production control. This remarkable tool guarantees the uniform distribution of labour, materials, and movements. How, you may ask? Well, it all begins with the preparation of a specific board. This board serves as the guiding light, providing a clear and concise visual representation of the production process. With this invaluable tool in hand, one can confidently navigate the intricate web of production, ensuring that each and every aspect is meticulously balanced and harmonized. Heijunka is a force to be reckoned with in the realm of production levelling. By implementing the principles of production volume levelling, production mix levelling and Heijunka production control, one can achieve a state of equilibrium that will leave supply fluctuations trembling in their wake. The Heijunka card (or box) was developed by Toyota in the 1960s to level production quantities and mix within a cell. The concept of levelling can also be applied in traditional companies, but only if set-up times have been reduced, allowing workers to quickly switch from one code to another. The advantages include:

- reduction in the quantity of lots processed;
- reduction in delivery times;
- reduction of frozen capital;
- better organization of the value stream.

This particular method, known as the Heijunka board, is not widely utilized in the Western world. The reason for this is that its implementation requires a prior redesign of the cell, which involves the introduction of both Kanban and quick change concepts. Traditionally, the Heijunka board is positioned near the "pacemaker" operation of the cell. The pacemaker, in this context, refers to a production program that is strategically placed at a specific point within the cell. Its purpose is to establish the pace of production and to pull the various processes upstream. For instance, it may be positioned at the final operation, from which the finished products are subsequently dispatched to the customer. On a daily basis, the heijunka board is used to determine the quantities and codes that need to be produced in accordance with the takt time. This information is then utilized by the Kanban tags to initiate the ordering of products from the upstream processes. Despite its potential benefits, the heijunka board has not gained widespread popularity in the Western world due to the aforementioned requirements and the need for a comprehensive redesign of the cell.

Makigami

Makigami, unlike Value Stream Mapping (VSM), is a tool that is better suited for capturing the essence and direction of processes in the realm of services. It is important to note that Makigami is not a tool that originated within the traditional Lean context. On the other hand, Value Stream Mapping, being a pure Lean matrix, tends to lose its significance when applied to pure service processes. This is due to the fact that Work in Progress (WIP) often holds less value in terms of immobilized capital and can be transformed into cycle time.

Since the 1980s, when Business Process Reengineering (BPR) principles were introduced to enhance both transactional and non-transactional processes, various mapping tools have been employed. Examples of these tools include the Interfunctional Flow Diagram and the more advanced Metrics-Based Flow Diagram (MBF), also known as Metrics-Based Process Map (MBPM). These tools have been utilized to provide a comprehensive understanding of the intricacies involved in improving processes. They have long been recognized for their exceptional mapping capabilities and waste reduction potential, much like the renowned Makigami. In fact, when combined with the powerful VSM, they form a formidable arsenal of tools for waste reduction. Their effectiveness in identifying and analysing waste is unparalleled. By utilizing these tools, one can confidently navigate the complex landscape of waste reduction, armed with the knowledge and insights necessary to make informed decisions. Makigami, born in Japan in the early 1990s and clearly derived from BPR mapping techniques, has found application in the service processes of manufacturing companies (e.g. marketing, after-sales support, administration, acquisitions, etc.) as well as in the processes of service organizations, including the Public Administration. The utilization of this particular method is commonly observed in conjunction with the Value Stream Model (VSM), especially in cases where the VSM exhibits macro activities or processes that encompass a sequence of activities or transactions that are overseen by multiple individuals. In the United States, the application of Makigami is known as MBPM (Metrics-Based Process Map) or MBF (Metrics-Based Flow). The basic applicative principles of Makigami remain largely unchanged in this context. It is typically developed within the framework of a Kaizen workshop, where a team conducts a thorough analysis of activities. This analysis may involve further division of activities and identification of key metrics for each of them. The primary objective is to swiftly reduce waste in a specific area.

Makigami is commonly compiled using mobile sheet boards that are affixed to the walls of the room and interconnected. This compilation method is reminiscent of the process employed in Value Stream Mapping. Similar to the latter tool, Makigami is compiled in two distinct forms:

1. Current State: This represents the process in its existing state, commonly referred to as "as is".
2. Future State: This depicts the desired appearance of the process after undergoing re-engineering.

These two forms provide a comprehensive view of the current and desired states of the process, facilitating effective analysis and improvement.

Spaghetti Chart

The Spaghetti Chart is an emerging mapping method that proves to be highly valuable in various workplace settings. Unlike traditional mapping techniques that focus on documenting the paths of documents or people in service, the Spaghetti Chart specifically maps the paths of product codes and families. This unique approach allows teams to identify and reduce waste in terms of movement, leading to increased efficiency and productivity.

One of the main advantages of Spaghetti Chart is its versatility in gathering information. In addition to the traditional method of charting paths, teams can also utilize photos or videos to further enhance their understanding of the area. This multi-faceted approach ensures that all relevant details are captured, enabling teams to make informed decisions and implement effective strategies.

Moreover, the Spaghetti Chart is not limited to a specific industry or sector. It can be applied both in manufacturing and office settings, making it a valuable tool for a wide range of professionals. By visualizing the physical flows of materials, people, or documents, the Spaghetti Chart provides a comprehensive overview of the workflow, allowing teams to identify bottlenecks, streamline processes, and optimize resource allocation.

Hence, the Spaghetti Chart is a highly useful mapping method that provides myriad advantages in terms of minimizing waste and process optimization. Its flexibility, combined with the ability to gather information through various methods, makes it an invaluable tool for teams seeking to improve efficiency and productivity in both manufacturing and office environments.

In a production context, for example, a reference product or product family is taken into consideration (important for sales volumes, revenue, margin, etc.) and the entire production flow is traced, or only the part that needs to be analysed, on a layout of the plant. This can be done using coloured pens and paper. The typical trajectory that the product follows within the company is meticulously delineated on the map, effectively illustrating the various stages of transformation, designated storage locations, rigorous quality control measures, and other pertinent details. This comprehensive mapping serves as a valuable tool in ensuring the smooth and efficient progression of the product throughout its journey within the organization. This mapping allows highlighting all the movements (Muda) made, all the intersections resulting from a non-optimal layout, the metres – or sometimes kilometres – travelled during the production cycle, and numerous other useful information.

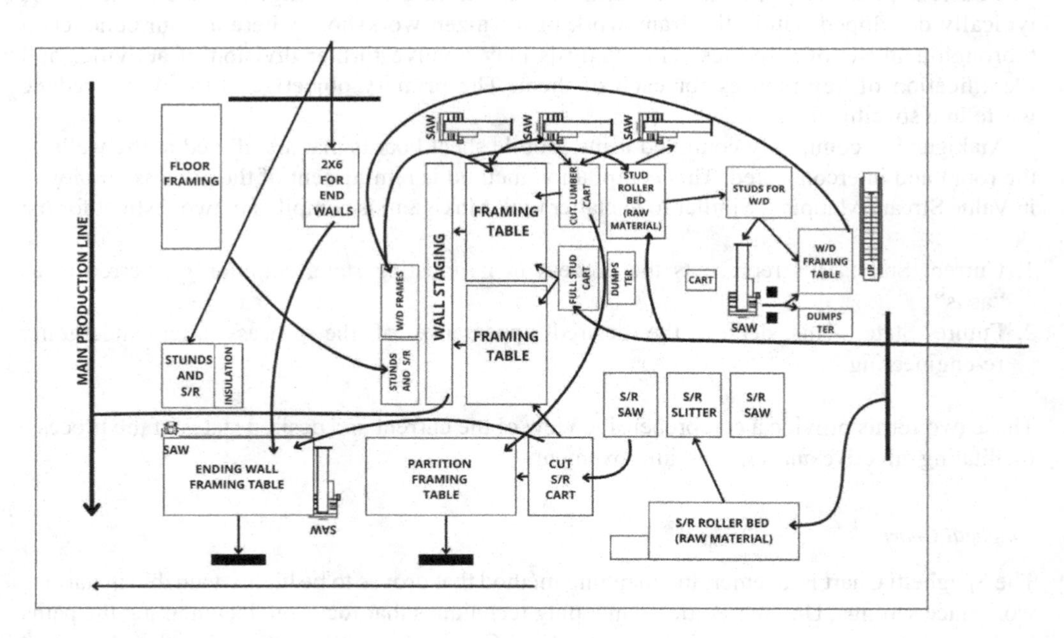

Figure 3.7 Spaghetti Chart.
Source: authors' own elaboration

Kanban

In the traditional push system, products move through the Value Stream through production planning and scheduling based only on demand forecasts, resulting in inevitable inventory. The pull system opposes the push system as it "pulls" the flow of the Value Stream starting from customer orders, connecting upstream and downstream processes through Kanban.

Kanban is a highly effective tool that enables us to shift our perspective and view the production process as a seamless operation that progresses from downstream to upstream, focusing solely on the necessary components when they are required. This concept, aptly referred to as "thinking backward" by Ohno, is characterized by the utilization of cards to manage the prioritization of progress between cells. These cards facilitate the self-regulation of cell work, allowing for swift adjustments in response to any changes in the production rate. It is worth noting that each cell is equipped with two bins: one for withdrawal Kanban and another for production Kanban. These bins play a decisive role in providing the operator with a clear understanding of the precise quantity and type of products that need to be produced or supplied, based on the contents of the respective bin. Each card or Kanban serves as a valuable tool in the realm of production management. These cards play a crucial role in identifying specific products or components, while also providing information regarding their origin and intended destination. By acting as an efficient information system, the Kanban seamlessly integrates various production processes, thereby reducing the occurrence of overproduction. This reduction is achieved by aligning all processes with the ever-important customer demand. Furthermore, the Kanban system ensures that production adheres to the Takt Time, which is essentially the rate at which production should occur. This term, derived from the German "Produktionstak" system implemented at Focke-Wulff aircraft plants, holds significant influence over all stages of the production process. From sales to suppliers, the Takt Time dictates the pace at which the product and its components must be manufactured. It is important to note that any deviation from the Takt Time could result in undesirable consequences. For instance, faster production may lead to an accumulation of excess inventory, while slower production could potentially cause delays in product delivery. Therefore, it is imperative to maintain a harmonious balance in order to ensure optimal efficiency and customer satisfaction.

Takt Time is the time required to produce a single component or the entire product, also known as the Sales Rate. Indeed, when the production stations are meticulously synchronized with one another, a remarkable outcome is achieved: a continuous and harmonious flow of production. This synchronization ensures that each station operates in perfect harmony, resulting in a balanced and efficient production process. To calculate this synchronization, a formula is employed, which is as follows:

Takt Time = Total Available Time per day ÷ Total Customer Demand per day

By dividing the total available time in a day by the customer demand within that same timeframe, we are able to determine the optimal Takt Time. This Takt Time serves as a guiding principle, allowing us to allocate our resources and time effectively, ensuring that our production process remains in sync with the demands of our valued customers.

The set-up time, conversely, encompasses a sequence of operations that are indispensable in order to adequately prepare a work station for the commencement of a new activity. It is precisely defined as the duration between the completion of the last unit produced in the preceding batch and the initiation of the first unit in the subsequent batch that satisfies the established production standards. This period encompasses all operations that are directly associated with the aforementioned process:

- finding the necessary information to carry out the next batch;
- preparing the necessary equipment on board the machine;
- removing equipment related to the previous batch and installing new ones;
- any modifications and adjustments;
- tests and checks on the first unit produced.

In order to safeguard the effective application of an LM system, it is of utmost importance to consistently strive towards minimizing machine set-up times. The reduction of these time intervals offers several advantages. Firstly, it allows for increased efficiency in the overall production process. By minimizing the time required for machine set-up, valuable resources can be allocated more effectively, resulting in improved productivity. Additionally, reducing set-up times enables a higher level of flexibility in the manufacturing process. This allows for quicker adaptation to changing production demands and facilitates the implementation of just-in-time manufacturing strategies. Furthermore, shorter set-up times contribute to cost savings by minimizing downtime and maximizing machine utilization. By striving to minimize the duration of machine set-up times, organizations have the opportunity to significantly improve their operational performance and attain heightened levels of success in the implementation of Lean Manufacturing (LM) systems. This pursuit of efficiency yields several notable benefits, including the reduction of downtime, resulting in lower operating and non-production costs. Additionally, the increased standardization of operations leads to enhanced product quality, thereby bolstering overall organizational performance. Furthermore, the reduction in batch size during production becomes a viable option, further optimizing the manufacturing process. There are two types of Kanban:

- Production Kanban;
- Transfer Kanban.

The production Kanban is used in the station where the production of a particular code is carried out and specifies the quantity to be produced.

The transfer Kanban is used to trace consumption between various processing phases and is used in the department that uses a certain component and reports the value of the quantities to be withdrawn from the upstream station's supermarket.

In other words, the **Kanban supermarket** is an area that resides between two production processes. The downstream process draws what it has consumed and needs from the supermarket, and such withdrawal generates input to the upstream process, which supplies materials without attempting to program or forecast consumption. The upstream process can only produce if authorized by the downstream process.

It is an effective control system that limits overproduction, in fact:

- indicates the quantity to be produced;
- maintains discipline;
- the production Kanban authorizes production;
- the transfer Kanban authorizes material movement.

The concept of the kanban supermarket originates from the first self-service store in history, Piggly Wiggly, which was established in Memphis, Tennessee in 1916. The innovative idea was that customers could serve themselves directly from the shelves without intermediaries, automatically triggering the restocking of consumed goods based on the simple depletion of the shelves.

Article	**VIT0027**
Description	**BOLT**
Supplier	F001 DEWAL SPA
Customer	R006 MAG. CENTR. MP
Lead Time	10 WL (Working Days)
Container	B008 MEDIUM YELLOW BOX
Quantity	**300 pcs**

Figure 3.8 Example of Kanban.
Source: authors' own elaboration

These supermarkets unintentionally inspired the direction of Toyota Motor Company in the early 1950s while visiting the United States to draw inspiration from their competitors in the automotive industry, Ford and General Motors.

According to this approach, every material required for production is moved in standard-sized containers, each of which is associated with a card that displays information such as the identification code of the piece and the quantity of pieces to be produced.

In essence, the kanban represents the fundamental signalling system for the implementation of JIT. In Japanese, it literally means "card", and that is what the kanban is: a set of cards designed to communicate the need for raw materials by a production stage. The card describes the type of material used for the job and is attached to the respective container. When the container is emptied, the kanban is separated and placed on a board. The need for material is thus communicated, and the container promptly refilled. The real-time flow of procurement avoids warehouse stocks and associated costs. Of course, for this system to be applied, production must be standardized, as must the containers necessary for material movement.

CONWIP, an alternative to Kanban, is a variant that operates on the basis of a specific parameter: the total number of cards in the production process. This parameter directly influences the maximum level of Work in Progress (WIP) within a production system. In other words, it determines the number of pieces or batches that can be worked on simultaneously.

CONWIP

It stands for "constant work in progress" and was developed by Mark Spearman and Wallace Hopp in 1990. It is almost identical to Kanban, but with one major difference. A Kanban is permanently assigned to a part of the supply chain, whereas CONWIP cards change parts at each iteration in the cycle. This makes CONWIP suitable for make-to-order production. CONWIP is the make-to-order equivalent of make-to-stock Kanban.

Techniques Used for Workplace Organization

5S and Standard Work

This methodology focuses on an effective workplace organization and standardized work procedures. The 5S are Seiri, Seiton, Seiso, Seiketsu, and Shitsuke, which are translated from Japanese to English as Sort, Straighten, Shine, Standardize, and Sustain.

Seiri refers to the act of discarding all unwanted, unnecessary, and non-work-related materials. Those involved should have no hesitation in disposing of anything that is not strictly necessary, as the idea is to ensure that everything left in the workplace is relevant to it. The number of necessary items in the workplace should also be kept to a minimum. Following the implementation of seiri, tasks will be simplified, space will be used more effectively and efficiently, and materials will be used more carefully.

Seiton, which concerns efficiency, involves placing each necessary element for production in a designated position, ensuring that it can be quickly retrieved by anyone. Additionally, this makes it easier to store tools once they have been used. When every individual has rapid access to an object or material, the workflow becomes more efficient, and the worker more productive. The appropriate place, position, and/or support for all tools, elements, or materials must be carefully chosen in relation to how the work will be executed and who will be using them. Each individual item must be assigned its own storage location, and each position must be labeled for easier identification.

According to Seiso, it is the responsibility of everyone to keep the workplace clean and "make it shine". The task of cleanliness should be carried out by all members of the organization, from operators to managers. It would be a good idea to assign each area of the workplace to a person or group of people responsible for cleaning. No area should be neglected, and everyone should view the workplace through the eyes of a visitor.

Seiketsu, which can be translated as "standardized cleanliness", aims to establish the standards by which personnel should measure and maintain cleanliness, both personal and environmental. It is expected that employees practice Seiketsu starting with their own personal cleanliness. Visual management is an important component, as the color coding and standardized coloring of the surroundings allow for easier visual identification of anomalies in a specific environment. Personnel must be trained to detect anomalies through all five senses and to immediately correct such anomalies.

Shitsuke, which translates to "discipline", signifies the commitment to maintaining order and practicing the first 4 S's as a way of life. It focuses on the elimination of bad habits and the consistent practice of good ones. This desired outcome can be considered achieved when the staff willingly adheres to cleanliness and order rules at all times, without requiring any reminders from management. By embodying the principles of Shitsuke, individuals demonstrate their dedication to upholding a high standard of cleanliness and organization, contributing to a harmonious and efficient work environment.

To sum up, this methodology consists of five phases:

1. Seiri – Sort and Separate. Eliminate anything that is not needed in the workplace.
2. Seiton – Set in Order and Organize. Efficiently arrange tools, equipment, materials, etc.
3. Seison – Shine and Clean. Control the order and cleanliness created.
4. Seiketsu – Standardize and Improve. Maintain the order and cleanliness created, trying to improve by repeating the phases continuously: Seiri, Seiton, Seison.
5. Shitsuke – Sustain. Impose discipline and rigor for the continuation.

The 5S is a tool used to achieve cleanliness and order in the workplace, and thereby increase productivity, quality, and safety. Hiroyuki Hirano, who made it famous, defines it as the pillar of "visual work", the visual control that allows, with a single glance, to have under control at all times the situation of a production department.

Subsequently, the new working method is standardized through Standard Work, which is subject to continuous improvement. The typical Japanese Standard Work includes visual and simple instructions for the personnel.

Standardizing work means:

- Organizing individual activities by documenting them in written form.
- Allowing them to be continually reproduced under controlled conditions (machinery with controlled maintenance, adjustments for the best possible performance, etc.).
- Achieving maximum effectiveness and efficiency.

To document and standardize an activity, it is necessary to:

- ensure a complete view of the process;
- simplify the work as much as possible;
- analyse the work (write down activities in detail, including all necessary steps to ensure quality, effectiveness, efficiency, safety, compliance with requirements, etc. If there are different ways of doing something, they should all be examined);

- ensure that the work can be continuously monitored through visual management;
- document the process;
- document the work of each operator;
- document individual activities (sequence, reference regulations, methodologies, flows, materials, number of parts, locations, tools, quality requirements, timing, etc.);
- consider training;
- provide for periodic improvements (ask yourself, after some time, if it is possible to perform an activity in a simpler or faster way and how to reduce the percentage of errors made);
- test the new way of working to verify that the reference standards are respected and correct any errors.

One Point Lesson

The One Point Lesson is also used when a problem arises: it instructs employees on what to do to prevent the problem from reoccurring or to avoid accidents and errors, or it describes the behaviour to adopt if the problem reappears. The One Point Lesson, or "brief lesson", is a short lesson on a specific topic, developed in all key points that summarize a topic addressed by a team or supervisor within the context of continuous improvement worksites. As all relevant information is collected on a single sheet, it can easily be used to educate and inform personnel on the topics discussed. It is a synthesis developed following a study on a specific topic, developed on an A4 or A3 format module. Topics that are addressed can include, for example: modifications made to a device that improve its performance (before/after), specific training on a procedural sequence during a set-up, how to perform adjustments, oil refills, safety instructions, how to inspect a device, and how to respond in case of restoration interventions, descriptions of the workplace, and so on. Additionally, it is:

- A tool to transmit knowledge and technical skills, to provide examples of issues, and to present concrete cases of improvement.
- A way to deepen theoretical competencies and improve practical skills, when necessary and at the appropriate time, in a simple and short amount of time.
- A means to raise the overall competency level of the entire group.

SOP (Standard Operating Procedure)

It is generally accepted that requiring employees to perform their tasks according to Standard Operating Procedures (SOPs) can improve production results in the context of repetitive production. Attempts to link the use of SOPs to intrinsic motivation – a requirement for creativity – have, however, led to debates (De Treville et al., 2005). An SOP, which stands for Standard Operating Procedure, is essentially a comprehensive set of instructions that meticulously outline all the pertinent phases of a given process or procedure. The purpose of defining an SOP is to clearly establish the specific activities that are to be included within it. By doing so, administrators are able to effectively organize personnel, information, and activities in order to respond to events and incidents, ultimately achieving a comprehensive level of control over the operation at hand. An SOP consists of several key components that work together harmoniously to ensure its effectiveness.

First and foremost, we have the **SOP definition**. This serves as the template that is utilized when creating an SOP instance in response to a particular event. The SOP definition is meticulously composed of various activities, each of which is described in detail by the corresponding activity definitions.

Speaking of **activity definitions**, these are an integral part of any SOP. An SOP typically contains one or more activity definitions, which serve to establish the individual instructions that must be diligently followed as part of the overall SOP. These activity definitions provide the necessary guidance and clarity to ensure that each step is executed precisely and accurately.

Lastly, we have the **SOP instance**. This refers to a single occurrence or event that prompts the creation of an SOP instance. It is worth noting that an SOP definition can be utilized for multiple SOP instances, depending on the nature of the event or occurrence. An SOP instance can exist in one of these states: active, started, stopped, completed or cancelled. These states serve to indicate the progress and status of the SOP instance, providing administrators with valuable insights into its current stage of implementation.

In summary, an SOP is a meticulously crafted set of instructions that serves as a guiding framework for various processes and procedures. By defining the SOP, administrators are able to effectively organize personnel, information, and activities, thereby achieving a comprehensive level of control over the operation. The SOP definition, activity definitions, and SOP instances all work together harmoniously to ensure the successful execution of the SOP.

A **task instance**, in the context of this system, refers to an individual occurrence of a specific task definition. It is worth noting that a single task definition can be utilized to generate multiple task instances. These task instances can exist in various states, each representing a different stage in the task's lifecycle. The states include: Active, Waiting, Started, Ignored, and Completed. Each state serves a distinct purpose and contributes to the overall management and tracking of tasks within the system.

References, within the scope of this system, are additional pieces of pertinent information that can be associated with a Standard Operating Procedure (SOP) or a specific task. These references play a crucial role in providing supplementary details and context to the SOP or task at hand. One notable application of references is their ability to define email templates, which can greatly enhance communication efficiency and consistency within the system.

Within this system, there are two primary **roles** that users can assume: Owners and Readers. These roles serve different functions and are designed to cater to both administrative and user-related responsibilities.

A Reader, as the name suggests, is granted the ability to monitor the activities associated with a standard operating procedure. This role allows users to stay informed about the progress and status of tasks, ensuring that they are up to date with any developments or changes.

On the other hand, an Owner possesses a broader set of privileges and responsibilities. In addition to monitoring the activities associated with a standard operating procedure, an Owner is also empowered to actively participate in and complete the tasks themselves. This elevated level of access and control enables Owners to effectively oversee and manage the entire lifecycle of a task, ensuring its successful completion within the established parameters.

By implementing these distinct roles, the system provides a clear and structured framework for users to engage with tasks and SOPs, promoting efficient collaboration and accountability throughout the organization.

The **task type** is a crucial element in describing the response to a given task. It is important to note that tasks can vary in terms of their types and execution patterns. In the context of Standard Operating Procedures (SOPs), it is possible to combine different types of tasks to achieve the desired outcome.

One type of task is the manual task. As the name suggests, this type of task requires manual execution by the SOP Owner. It is essential for the SOP Owner to personally carry out the necessary steps to complete the task.

Another type of task is the If-Then-Else Task. This particular task allows for conditional branching based on a specific criterion. The user has the flexibility to choose which SOP definition to utilize when initiating the task. It is crucial to enter the proper values for the Then and Else conditions to ensure the desired outcome.

The Alert Task is yet another type of task. This task involves displaying an email template that the SOP owner must complete. Additionally, it sends a notification via email to the predefined personnel. This ensures that the relevant individuals are promptly informed about the task and can take appropriate action.

The REST Task is a task that involves making a call to a REST service. When initiating the task, the user has the ability to specify the URL of the service and provide any necessary authentication information. This ensures that the task can successfully interact with the designated REST service.

Lastly, the SOP Task is a task that initiates another SOP. This means that the task itself serves as a starting point for another set of procedures. This allows for a seamless flow of tasks and ensures that the desired outcomes are achieved in a systematic manner.

Thus, understanding the different task types and their execution patterns is crucial in effectively managing and executing tasks within the context of SOPs. By utilizing the appropriate task types, SOP Owners can ensure that tasks are carried out efficiently and in accordance with the desired outcomes.

In order to effectively manage Standard Operating Procedures (SOPs), it is important to understand the various roles and their corresponding capabilities. The following is a comprehensive list of the capabilities associated with each **SOP role**:

- SOP Administrator Roles:
  - The SOP Administrator has the ability to view and delete an SOP definition.
  - They can also start, view, and modify an SOP instance.
  - Additionally, they have the authority to start and complete tasks within an SOP instance.
- SOP Author Roles:
  - SOP Authors possess the necessary privileges to create, modify, view, and delete an SOP definition.
  - They are also able to create an SOP draft.
  - Furthermore, SOP Authors have the ability to view, modify, and delete an SOP task.
  - They can forward an SOP draft for approval and approve an SOP draft.
- Reference Library Manager Role:
  - The Reference Library Manager is responsible for creating shared references, which are essential for the smooth functioning of SOPs.
- SOP Owner Roles (Definition):
  - SOP Owners, in their capacity as definition owners, have the authority to create an SOP draft.
  - They can view, modify, and delete an SOP definition.
  - They are also able to modify and delete an SOP task.
  - Additionally, they have the ability to forward an SOP draft for approval and approve an SOP draft.
  - Lastly, SOP Owners can start, view, and modify an SOP instance.

- SOP Reader Roles (Definition)

  SOP Readers can view the definition of an SOP.
  - They can view an instance of an SOP from their Personal Tasks.
  - They can view a specific task within an SOP, provided they have the Reader role assigned to them for that particular task.

- SOP Owner Roles (Tasks)

  - SOP Owners can view an instance of an SOP from their Personal Tasks.
  - They have the authority to initiate and complete tasks within an SOP instance, specifically for the tasks assigned to them from their Personal Tasks.

- SOP Reader Roles (Task)

  - Users with the Reader role are able to view an instance of an SOP from their Personal Tasks.

By clearly delineating the capabilities associated with each role, SOP management becomes more efficient and streamlined.

The Approval Lifecycle for a Standard Operating Procedure (SOP) encompasses various states that the SOP definition can undergo. Firstly, there is the Draft state, wherein an initial draft version of the SOP is saved. Additionally, it is possible to create another draft version from an approved SOP, should the need arise to modify the SOP definition using the approved version as a foundation. Drafts can be modified, forwarded for approval, or deleted as necessary.

Next, we have the Awaiting Approval state, wherein an SOP draft definition is forwarded for approval and is ready to be either approved or disapproved. It is in this state that the version name is determined, and if the SOP is approved, the version name becomes the official name of the definition. However, if this particular version fails to receive approval, the SOP definition reverts back to the draft version status.

In summary, the Approval Lifecycle for an SOP involves the Draft state, where initial drafts are saved and modifications can be made, and the Awaiting Approval state, where SOP drafts are forwarded for approval and either become approved versions or revert back to the draft status.

The status of "Approved" is assigned to an SOP definition when it has successfully undergone the approval process and is deemed ready for implementation. This signifies that all necessary stakeholders have reviewed and given their consent to the SOP definition. At this stage, the SOP definition is considered finalized and can be launched without any further modifications.

The One Point Lesson, also known as the "lesson in one point", is a valuable tool that aims to concentrate on a singular aspect of training. This tool serves as an illustration of a concise Standard Operating Procedure (SOP) and effectively communicates new or revised standards and work methods to operators and staff. The beauty of this tool lies in its simplicity, as it does not require a complete SOP.

Typically, the entire training subject is condensed onto an A4 sheet, which includes succinct concepts that can be immediately grasped. Additionally, this sheet may contain visual aids such as photographs, sketches, and drawings. The purpose of this approach is to provide swift instruction to individuals interested in various concepts, such as machinery usage, the functionality of plant components, or even process management techniques.

Techniques to Achieve Zero Defects in the Jidoka Automation Process

Jidoka is an automated system that diligently examines the specific attributes of the machinery or product at hand, and promptly halts production in the event of any non-conformity. It is worth noting that this system operates independently, without necessitating any form of worker intervention. Jidoka operates on the same fundamental principle as Poka-Yoke or Mistake Proofing tools, which are designed to prevent, albeit not necessarily in an automated manner, any human errors that may occur during the various processes involved, thereby effectively reducing the occurrence of defects.

Poka-Yoke

The poka-yoke, also known as the foolproof approach, is a method aimed at preventing errors. It involves determining operational conditions in such a way that the operator is unable to perform an incorrect manoeuvre. Shigeo Shingo, a highly regarded engineer at Toyota, was one of the major proponents of Zero Quality Control. This approach extensively utilizes poka-yoke principles. These mechanisms serve two purposes: preventing specific causes of errors and ensuring that every item produced is defect-free, all while keeping costs low.

A poka-yoke method refers to any mechanism that can prevent an error from occurring or make the error immediately obvious. The ability to quickly identify errors at a glance is crucial. As Shingo asserts, product defects are a result of worker errors. Therefore, it is imperative to carefully identify and analyse these shortcomings. By doing so, worker errors can be eliminated in advance, preventing them from converting into defects.

Hence, the poka-yoke approach is a valuable tool in error prevention. By implementing mechanisms that prevent errors and ensuring that workers are able to identify and eliminate their own mistakes, the production of defect-free items can be achieved.

An example cited by Shingo shows how "spotting" errors at a glance allows defects to be overcome. Suppose a worker has to assemble a piece consisting of two buttons. Each button must be accompanied by a properly positioned spring to ensure optimal functionality. However, it is not uncommon for a worker to inadvertently overlook this crucial step, resulting in an error. To address this issue, a straightforward poka-yoke application has been devised. The worker must extract from the container exactly the number of springs needed for the operation and arrange them on a disc. If, once the assembly is completed, a spring remains on the disc, an error has been made. The operator immediately realizes that a spring has been omitted and corrects the error immediately. The cost of the check is minimal, as is the cost to eliminate imperfections. This example demonstrates that the poka-yoke approach is particularly suitable for eliminating mistakes and omissions. Poka-yoke aims to:

- Embed quality in design and processes.
- Ensure it is impossible to produce defective products.
- Use poka-yoke protections incorporated into products and processes.
- All unintentional errors and defects can be eliminated. One must firmly believe that errors can be eliminated.
- Errors and defects can be eliminated when everyone works together.
- Seek the root causes: one must understand the actual cause of the problem and apply countermeasures.

Shingo identified three different types of inspections:

- **Judgment inspection** involves separating defective products from those deemed acceptable, also known as "quality control".
- **Informative inspection** involves using data obtained from inspections to control processes and prevent defects. The more frequent the feedback, the faster the improvements that can be

achieved in terms of quality. It is important to conduct inspections throughout all phases of the process to intervene as quickly as possible. If the operator is also responsible for checking their own work, the benefits will be even greater. In fact, since workers check each individual unit produced, they are able to immediately identify the cause of the defect. Such inspections and controls allow for information to be obtained "after the fact".

- **Source inspection** involves determining "before the fact" whether the necessary conditions for a high level of quality exist.

Poka-yoke mechanisms are used in this type of inspection, for example, to prevent production until adequate operating conditions are met.

Techniques to Achieve Zero Downtime in Production

Total Productive Maintenance (TPM)

The approach to equipment maintenance being discussed here is truly innovative. The primary objective of TPM, or Total Productive Maintenance, is to not only enhance production levels, but also to boost employee morale and overall satisfaction. TPM achieves this by implementing a regular maintenance schedule for equipment, which allows for the early detection of any anomalies or the need for maintenance interventions, such as revisions, replacements, or repairs, before a component experiences a breakdown. It is worth noting that the operators themselves are entrusted with the initial maintenance and monitoring tasks, and are required to promptly report any malfunctions they encounter. This approach instils a sense of responsibility and heightened awareness among the operators, as they become intimately familiar with the equipment they utilize. Furthermore, this method effectively curtails instances of equipment abuse and misuse, as the operators are held accountable for the maintenance of said equipment, thereby encouraging them to exercise greater caution in its usage. The aim is to strike a balance between preventive and repair maintenance, minimizing the company's overall maintenance costs. TPM coordinators must balance these two factors to minimize the overall cost. TPM allows for the rapid and economical identification of malfunctions and the implementation of appropriate corrective actions, reducing waste due to production losses caused by machine breakdowns.

The TPM technique shares several key aspects with TQM. One of these aspects is the requirement for the commitment and support of top management. In order for TPM to be successfully implemented, it is crucial that top management demonstrates their dedication to the process.

Furthermore, TPM emphasizes the importance of empowering employees. By empowering employees, they are given the authority to take corrective actions when necessary. This not only increases their sense of ownership and responsibility, but also motivates them to actively contribute to the success of the TPM initiative.

It is worth noting that TPM is not a short-term endeavour. On the contrary, it requires a long-term perspective. Implementing TPM is a process that typically takes more than a year to complete. This extended timeframe allows for a thorough and comprehensive integration of TPM principles and practices into the organization.

One of the fundamental principles of TPM is the recognition of maintenance as a vital division of the company. It is not merely a single operational task or an occasional intervention. Instead, TPM views maintenance as an ongoing effort to minimize emergencies and unscheduled maintenance interventions. By prioritizing proactive maintenance measures, TPM aims to prevent breakdowns and optimize the overall efficiency of the company's operations.

In summary, TPM and TQM share common elements such as the need for top management commitment, employee empowerment, and a long-term perspective. Additionally, TPM places great importance on maintenance as a crucial division of the company, with the goal of minimizing emergencies and unscheduled maintenance interventions. To implement TPM, the entire workforce must be motivated by the fact that top management supports the project. The first step is to designate a TPM coordinator to educate employees and introduce them to the principles of TPM. This may take more than a year.

Next, autonomous teams are created, including operators, maintenance personnel, department supervisors, and managers. Every person involved in the process is encouraged to contribute to the team's success. The TPM coordinator leads the team until the members become familiar with the process and a team leader emerges spontaneously.

The team's responsibilities include:

- defining problems accurately;
- detailing the corrective action list;
- executing the corrective process.

A proper implementation of the project involves visits to other facilities to observe TPM methods. This comparative process is part of the benchmarking practice. Once experience is gained on minor problems, more significant issues can be addressed.

Operators are required to actively participate in machine maintenance, which is one of TPM's main innovations. The operator's responsibility extends to performing daily maintenance checks to reduce unscheduled repair interventions, such as lubrication or component replacement. More thorough checks and more severe failures are handled by specialized maintenance personnel, always assisted by the operator.

Therefore, the application of TPM within the organization occurs through five fundamental steps:

- introduction of improvement activities to increase the efficiency of plants and equipment;
- implementation of an autonomous management system (connected with the organization's objectives), maintenance carried out by trained and aware operators;
- implementation of a scheduled maintenance system with data collection on component reliability (predictive maintenance); continuous updating of intervention programming based on collected data;
- implementation of a design and development system for equipment and plant parts that require less maintenance and are quicker;
- continuous training, emphasis, and dissemination of results obtained.

The advantages of TPM are indeed numerous and worth considering. Firstly, TPM leads to a more efficient use of plants and equipment, resulting in an overall improvement in efficiency. This is achieved through the implementation of a methodology of maintenance that is spread throughout the organization. This methodology is based on preventive-predictive maintenance, which relies on statistical data to identify potential issues before they become major problems.

Furthermore, TPM requires the participation of various departments within the organization, including design and development, production, and maintenance. This ensures that all aspects of the organization are involved in the maintenance process, leading to a more comprehensive and effective approach.

In addition, TPM involves both management and operators, recognizing the importance of collaboration and cooperation between these two groups. By involving both parties, TPM promotes a sense of shared responsibility and accountability for the maintenance activities.

Lastly, TPM promotes and improves maintenance activities through the establishment of specific autonomous teams, such as Six Sigma teams. These teams are dedicated to enhancing maintenance practices and are empowered to make decisions and implement improvements.

In conclusion, the advantages of TPM are clear. It not only leads to a more efficient use of plants and equipment but also introduces a methodology of maintenance that is based on preventive-predictive maintenance. With the participation of various departments, management, and operators, TPM ensures a comprehensive approach to maintenance. Furthermore, the promotion of specific autonomous teams further enhances maintenance activities. In addition, it is worth noting that the TPM technique is an indispensable tool for organizations seeking to gain and sustain a competitive edge in the market. This technique encompasses a range of activities, each serving a specific purpose. On one hand, there are activities that primarily focus on the psychological aspect, aligning with the traditional Eastern approach. These activities aim to cultivate a mindset of continuous improvement and foster a culture of excellence within the organization. On the other hand, there are activities that are purely practical in nature. These activities involve implementing efficient maintenance practices, optimizing equipment performance, and minimizing downtime. By combining both psychological and practical elements, the TPM technique offers a comprehensive approach to achieving long-lasting competitive advantages.

SMED

The acronym "Single Minute Exchange of Die" (SMED) represents an organizational method that aims to efficiently minimize set-up time. This method is a specific tool designed to eliminate downtime and streamline set-up operations. It was initially developed by Shigeo Shingo within the Toyota organization in the year 1955 (Shingo, 1986). By reducing the time required for set-up, workers are able to swiftly transition between different codes that are more frequently processed by the machine. Consequently, this reduction in set-up time leads to a decrease in work-in-progress (WIP) inventory. In diversified production with small lot sizes, as soon as an operation begins to gain momentum, production must switch to a new different lot and a new set-up. This method distinguishes between:

- External exchange of die activities: operations that can be carried out with the machine running (preparation of tools, positioning, etc.).
- Internal exchange of die activities: operations that require the machine to stop.

The primary objective at hand is to effectively reduce the amount of time that machines are not in operation, thus leading to a decrease in internal set-up procedures. This is of utmost importance as it allows for a more efficient workflow and maximizes productivity within the organization. By minimizing machine downtime, we are able to optimize the utilization of our resources and ensure that our operations run smoothly. Consequently, this results in decreased expenses and heightened overall efficiency. Therefore, it is imperative that we focus our efforts on minimizing machine downtime and streamlining internal set-up operations.

The implementation of the SMED technique can be achieved by following four essential steps. Firstly, it is imperative to eliminate any unnecessary operations and transform internal activities into external ones. During this phase, it becomes crucial to differentiate between tasks that must be performed with the machine stopped (internal set-up activities) and those that can be executed while the machine is running, specifically before the tool change (external set-up activities). By questioning whether the activities performed with the machine stopped can also be accomplished while the machine is running, one can effectively convert internal set-up activities into external set-up activities.

Secondly, simplifying clamping procedures is of utmost importance in order to standardize tools and minimize clamping times. This entails streamlining the clamping process to ensure that it adheres to a standardized approach, thereby reducing the time required for clamping.

Thirdly, working in groups and dividing tasks among team members can significantly contribute to time-saving efforts. By performing tasks in parallel, the overall efficiency of the process can be enhanced, resulting in considerable time savings.

Lastly, the elimination of adjustments and controls is a crucial step in the implementation of the SMED technique. By establishing well-defined procedures and standardizing them, the need for frequent checks and adjustments can be minimized. This not only conserves time but also guarantees a more streamlined and efficient workflow.

In conclusion, by diligently following these four steps, one can successfully implement the SMED technique and achieve improved operational efficiency.

UNI 11063:2017 and UNI 10147:2003

Both UNI 11063:2017 and UNI 10147:2003 contribute to lean and quality by helping to ensure that goods and assets are maintained in a way that maximizes their uptime and minimizes their downtime.

UNI 11063:2017 helps to ensure that goods and assets are maintained in a way that maximizes their uptime by providing a framework for classifying maintenance activities and ranking them based on their importance. This can help to ensure that the most critical maintenance activities are carried out first, which can help to prevent downtime and disruptions. UNI 11063:2017 also helps to reduce costs by providing guidance on how to plan and schedule maintenance activities. This can help to ensure that maintenance activities are carried out efficiently and that resources are not wasted.

UNI 10147:2003 helps to improve efficiency by providing a template for a maintenance plan. This template can help to guarantee that all of the necessary information is collected and documented, which can help to prevent mistakes and rework. UNI 10147:2003 also helps to reduce costs by providing a framework for estimating the cost of maintenance activities. This can help to ensure that budgets are set realistically and that resources are allocated appropriately.

In addition to the specific benefits outlined above, UNI 11063:2017 and UNI 10147:2003 can also contribute to lean and quality by helping to:

- improve communication and collaboration between different departments;
- increase employee engagement and morale;
- reduce the risk of accidents and injuries;
- improve customer satisfaction.

UNI 11063:2017 and UNI 10147:2003 are undeniably valuable tools for organizations that are deeply committed to the principles of lean and quality. These standards, when diligently adhered to, have the potential to bring about significant improvements in the efficiency, effectiveness, and safety of maintenance activities within organizations.

In the year 2017, the introduction of UNI 11063 as a standard for Lean Thinking 5.0 marked a pivotal moment in the pursuit of enhanced efficiency and the eradication of waste across various industries. This standard, meticulously crafted and thoughtfully designed, aimed to provide organizations with a complete framework to update their operations and reduce resource consumption. By embracing the principles outlined in this standard, organizations can expect to witness a remarkable boost in their operational efficiency, leading to considerable cost savings and better productivity.

Furthermore, it is worth noting that in the year 2003, UNI 10147 was established as a standard for TQM. This standard, meticulously developed and rigorously tested, serves as a guiding light for organizations seeking to achieve excellence in their operations. By embracing the principles of TQM, organizations can foster a culture of continuous improvement, ensuring that every aspect of their operations is aligned with the highest standards of quality. The ultimate goal of both Lean Thinking 5.0 and TQM is to enhance organizational performance and customer satisfaction, and by adhering to these standards, organizations can confidently embark on a journey towards achieving these noble objectives.

In conclusion, the adoption of UNI 11063:2017 and UNI 10147:2003 can prove to be a game-changer for organizations that are truly committed to the principles of lean and quality. These standards, meticulously crafted and supported by extensive research, provide organizations with a robust framework to enhance their operational efficiency, effectiveness, and safety. By embracing these standards, organizations can confidently navigate the complex landscape of maintenance activities, ensuring that every action taken is aligned with the highest standards of excellence.

By implementing Lean Thinking 5.0, companies can find and remove non-value-added activities, resulting in streamlined processes and increased productivity. TQM, on the other hand, focuses on continuous improvement and customer-centricity, aiming to exceed customer expectations and deliver high-quality products and services. The connection between Lean Thinking 5.0 and TQM lies in their shared principles of waste reduction, process optimization, and customer focus. By integrating Lean Thinking 5.0 and TQM, organizations can reach operational excellence and create a culture of continuous improvement.

The Five Lean Principles

Lean Production and Lean Thinking, more generally, are based on five basic principles that are as simple as they are effective. When reorganizing operational activities, and production in particular, there is a natural tendency to "jump right into action" without considering what should be the guidelines for the reorganization. Fortunately, there is a reference that is so effective in its simplicity that it is a true "milestone" of organizational discipline. We are talking about the 5 Principles of Lean Production (and Lean Thinking in general).

The primary principle of Lean Production revolves around the crucial task of identifying value for the esteemed customer. Value, in this context, is the perception of the customer and, as such, it is the customer who defines it, not the supplier. It is important to note that value is always contingent upon specific factors such as the moment, price, and place. Therefore, it becomes imperative to comprehend which attributes of the product or service generate perceived value for the customer and subsequently strive to enhance those attributes. It is worth mentioning that it would be futile to invest efforts in improving an attribute that the customer deems of little value.

The second principle of Lean Production is the identification of the flow of value and the elimination of activities that do not generate value. The most effective technique for identifying and mapping the flow of value is Value Stream Mapping.

The third principle of Lean Production is to create the flow of value-creating activities so that they run without interruption.

The fourth principle is to ensure that value is pulled by the customer. In this case too, Lean Production responds with specific techniques, in particular pull systems (Kanban and Supermarket) and controlled management of product inventories.

The fifth principle is to pursue perfection through continuous improvement. It is important to note that the ideal perfection is the complete elimination of waste, so that all activities create value for the end customer. This tension is the reference point for maintaining a systematic improvement process; perfection is not a static concept, but dynamic, as the value for the customer changes over time.

There are several widely used application techniques that are worth mentioning. One such technique is Failure Mode Effect Analysis (FMEA). FMEA is a systematic method that is employed to identify and evaluate potential failures in a process or system. By analysing the effects of these failures, organizations can implement proactive measures to prevent their occurrence.

Another notable technique is KANBAN, that is a visual system through which companies manage their inventory and production levels. By using visual cues, such as cards or boards, KANBAN enables teams to easily track the status of tasks and materials, ensuring a smooth workflow.

SIX SIGMA is yet another popular application technique. It is a data-driven approach that aims to improve the quality of processes by reducing defects and variations. By utilizing statistical analysis and rigorous problem-solving methodologies, organizations can achieve significant improvements in their operations.

TPM, which stands for Total Productive Maintenance, is a technique that aims at improving the efficiency and effectiveness of equipment and machinery. In this way, companies can reduce downtime, improve equipment reliability, and enhance productivity.

Another technique worth mentioning is Value Stream Mapping (VSM), a visual tool that helps organizations recognise and remove waste in their production processes. By mapping out the flow of materials and information, organizations can identify areas of improvement and implement strategies to streamline their operations.

Lastly, Visual Management is a technique that utilizes visual cues and displays to communicate information effectively. By using visual aids, such as charts, graphs, and signs, organizations can enhance communication, improve understanding, and promote a more efficient work environment.

In conclusion, these application techniques, including FMEA, KANBAN, SIX SIGMA, TPM, Value Stream Mapping, and Visual Management, are extensively adopted in various industries to enhance productivity and achieve organizational goals.

Lean Approach: Push versus Pull

In the lean approach, it is imperative to ensure that demand is met precisely when required, without any delay or anticipation, and always maintaining the highest standards of quality. This task becomes notably simpler when demand is predictable and, ideally, remains relatively stable over time. Consequently, the implementation of what is commonly referred to as lean principles is greatly facilitated when an organization possesses a comprehensive understanding of its resource demands, and is able to exert some degree of control over them.

The most prevalent method employed to achieve customer-focused management is known as the pull approach, which stands in contrast to the push approach. The push approach occurs when items are immediately transferred to the subsequent phase of processing, regardless of whether the customer has actually requested them at that particular moment. On the other hand,

Table 3.1 Objectives of a Lean Production process.

↑	Intelligent operational processes	↓	Waste
↑	Agile introduction of innovations	↓	Direct costs
↑	Development of new products and services	↓	Process rigidity
↑	Sustainable measures	↓	Lead time
↑	Environmental respect	↓	Non-value-added activities

Source: authors' own elaboration

the pull system operates by "pulling" items forward only in response to a specific customer request. This ensures that resources are allocated efficiently and effectively, in direct response to the needs and desires of the customer.

Push and Pull are two distinct production methods that differ in their approach and execution. The Push approach is primarily driven by sales forecasts, where the organization proactively produces goods and pushes them into warehouses, following the Make To Stock strategy. In contrast, the Pull method is initiated when production is triggered by specific orders, adhering to the Make To Order principle.

The Pull system is commonly employed in mass production scenarios, particularly when the demand for the product remains stable and predictable. This method is particularly suitable when there is minimal customization required for the products. Additionally, the cost of inventory associated with the Pull system is relatively low.

It is important to note that the Push and Pull methods represent two distinct approaches to production, each with its own set of advantages and considerations. By understanding the characteristics and requirements of each method, organizations can make informed decisions regarding which approach best aligns with their specific needs and objectives.

Embracing Digital Lean 4.0

The Lean approach can help to outline the model of the Factory of the Future. The new proposal is based on the integration between Lean principles interpreted with Industry 4.0 technologies, with a production process that starts from the customer rather than the raw materials. The innovation being sought involves not only a modification of the procedures used in production, but also the industrial accounting systems and KPIs used to evaluate manufacturing performance. Manufacturing production will have to be characterized by:

- low volumes;
- high variety (customization);
- intermittent flows;
- complex and variable processes.

In order to gain a competitive advantage, it is essential for a company to implement innovative technologies. Woodward classified companies based on the degree of mechanization of the manufacturing process back in the 1950s:

- small batch production where labour prevails over automation;
- mass production where automation is high;
- continuous process production where the process is entirely controlled by machines.

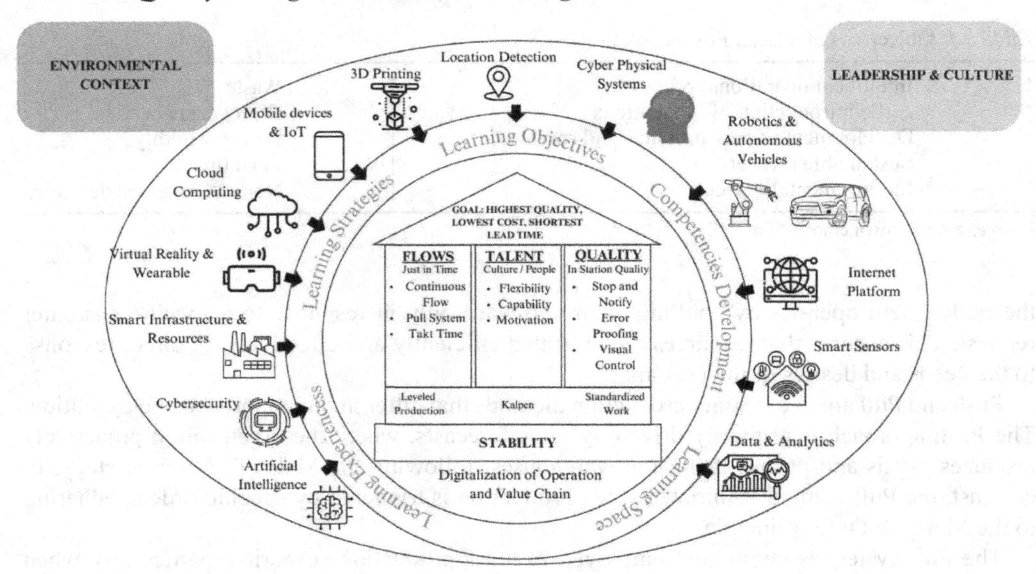

Figure 3.9 Digital Lean 4.0.
Source: authors' own elaboration

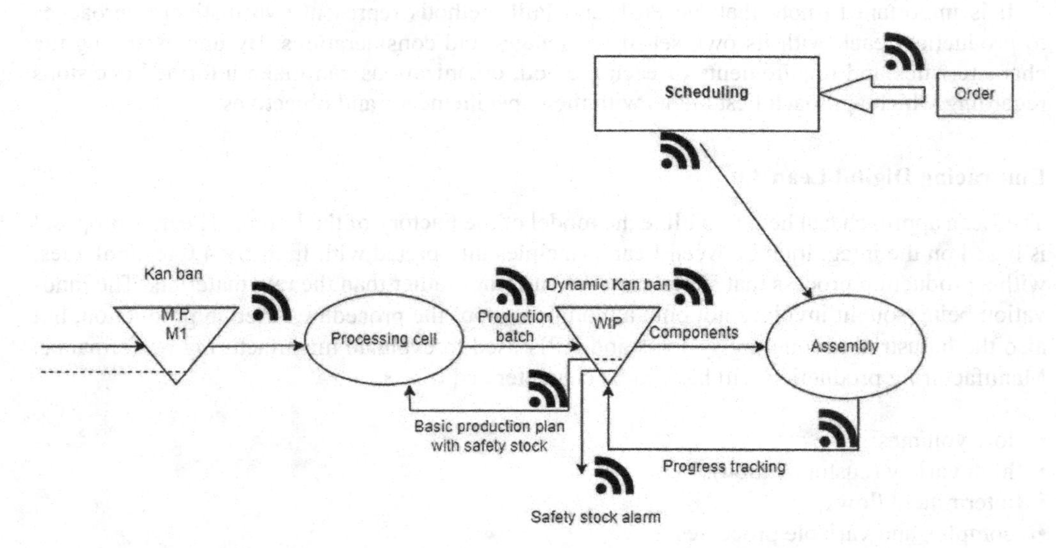

Figure 3.10 Material requirements planning: master plan and production data management.
Source: authors' own elaboration

Woodward also highlighted that when technologies, structures, and strategies are not aligned and complementary, poor performance will result, leading to losing the competitive advantage over other companies in the long run. Today, the new global challenge is the creation of digital factories, automated supply chains, and the integration of emerging technologies into Lean manufacturing processes, ranging from Artificial Intelligence to Predictive Analytics, Digital

Twin, and Cobots. Thus, a manufacturing cycle will be based on a Pull logic with flows where each area operates "pulled" by downstream activities, but in a synchronous manner thanks to the technologies that keep the operation under real-time control.

Figure 3.10 describes a manufacturing cycle based on a Pull flow model where each area operates "pulled" by downstream activities, but synchronously thanks to technologies that monitor operations in real-time.

In this way, customer orders will be entered into the MRP (Material Requirements Planning) for control and aggregation of the base bill of materials. Short-term scheduling (2/3 days) of the final operation (e.g. assembly) that produces the desired product will be carried out. Based on the plan, the base bill of materials and production progress, the necessary components are fed in order to guarantee a taut and synchronized flow between the WIP warehouse and assembly. The component warehouse (WIP) will be managed with a dynamic two-level Kanban logic: undersupply and safety stock. The first will send a warning signal to the upstream processing so that it can produce the required quantity, while the second will generate an alarm for reaching the minimum safety level, indicating a priority. The undersupply level is dynamically calculated based on the production trend. In this case, upstream processing also operates short-term planning based on the warning signals received, and its objective is to produce the expected component in a quantity defined by consumption, but also with economic batch logic closer to the expected value, which, using the flexibility of new technologies, must be as low as possible to make the system flexible. The WIP warehouse will measure the actual lead times between the "warning" signal and the actual load of the new batch. The process will be repeated in the upstream operative cells according to the number of necessary operations. The cycle can also involve external suppliers instead of internal operations using the same logic. Such a model can only be realized with widespread digitization of the factory.

The Six Sigma Model

The implementation of a quality management and improvement program can exhibit significant variations in methodology. It is worth noting that such programs may be referred to by different names, including TQM (Total Quality Management), Six Sigma, Lean Six Sigma, BPR (Business Process Re-engineering), or Operational Excellence. Despite the nomenclature differences, each program shares a common characteristic: the utilization of a specific set of tools and techniques during the implementation process.

Six Sigma, for instance, serves as a measurement standard for product variation. Its origins can be traced back to the 1930s, when Walter Shewhart (1939) conducted a study that demonstrated a correlation between sigma levels and defects in a process. By establishing an interval around a defined target, statistical analysis reveals that the higher the number of sigma levels that remain within the interval, the lower the probability of failure. In this context, failure refers to results falling outside the established interval, which consequently renders the products or services defective.

Sigma (σ) is a letter of the Greek alphabet that has been adopted as the statistical symbol and metric for measuring process variation. It serves as an indicator of the standard deviation of a process, providing valuable insights into its performance. The Sigma measurement scale exhibits a perfect correlation with key characteristics such as defects per unit, parts per million defects, and the probability of failure.

In the realm of process improvement, the concept of Six Sigma holds great significance. It represents the level of quality achieved when a process is measured at six Sigma. To put it into perspective, a Six Sigma level of quality corresponds to a mere 3.4 defects per million opportunities. This remarkable level of precision has been estimated to result in a long-term shift of the process mean to a maximum of 1.5 standard deviations.

By embracing the principles of Six Sigma and striving for this exceptional level of quality, organizations can significantly enhance their operational efficiency and minimize defects. The meticulous measurement and analysis of process variation, as symbolized by Sigma, provide a solid foundation for continuous improvement and the pursuit of excellence.

Six Sigma is a method focused on measuring the extent to which a given process deviates from perfection. It derives its name from the fact that it encompasses six standard deviations on either side of the specification window. This disciplined and data-driven approach is designed to eliminate defects and improve overall quality.

The fundamental principle underlying Six Sigma is the notion that by quantifying the number of "defects" present in a process, one can systematically identify and eliminate them, thereby moving closer to achieving "zero defects". This is achieved through the application of statistical methods, which enable the translation of customer information into specific product or service specifications that need to be developed or produced.

In essence, Six Sigma serves as a powerful tool for enhancing capacity and reducing defects in any given process. By meticulously analysing data and employing statistical techniques, enterprises can acquire valuable insights into their operations, enabling them to make informed decisions to fuel ongoing improvement.

It is crucial to note that Six Sigma is not a one-size-fits-all solution. Rather, it is a highly adaptable methodology that can be tailored to suit the unique needs and requirements of different industries and organizations. By embracing Six Sigma, companies can foster a culture of quality and efficiency, ultimately leading to enhanced customer satisfaction and improved business performance.

To conclude, Six Sigma offers a comprehensive and systematic approach to process improvement. By leveraging statistical methods and diligently addressing defects, organizations can strive towards achieving excellence and delivering products and services of the highest quality.

Numerous measurement standards have made their way into the realm of scientific inquiry and managerial literature over the years. However, it was a distinguished engineer by the name of Bill Smith, employed by the esteemed company Motorola, who first coined the term "Six Sigma". It was during the early to mid-1980s, under the leadership of President Bob Galvin, that the engineers at Motorola came to a profound realization. They recognized that the conventional quality levels, which measured defects in thousands of "opportunities", were simply inadequate in providing the desired quality outcomes. Their ambition was to establish a more refined metric that would measure defects per million opportunities (DPMO). Thus, Motorola embarked on a journey to develop a novel standard known as Six Sigma. This groundbreaking standard not only entailed the creation of a comprehensive methodology but also necessitated a substantial cultural shift.

Motorola, a renowned organization, experienced remarkable financial success across its entire operation, thanks to the implementation of the Six Sigma methodology. Astonishingly, Motorola meticulously documented savings exceeding a staggering $16 billion, all attributed to the effective utilization of Six Sigma. This monumental achievement did not go unnoticed, as numerous companies worldwide eagerly embraced Six Sigma as their preferred approach to conducting business. The widespread adoption of Six Sigma can be directly attributed to the influential endorsements of esteemed leaders within the United States. Distinguished figures such as Larry Bossidy, the former CEO of Allied Signal (now Honeywell), and Jack Welch, the esteemed former CEO of General Electric Company, have openly extolled the myriad benefits of Six Sigma (Harry and Schroeder, 2000). Six Sigma is a highly regarded management system that shares similarities with other well-known methodologies such as TQM, BPR, and Lean Thinking. It emphasizes the importance of reducing defects and improving process efficiency through data-driven decision making and rigorous statistical analysis. Like TQM, Six Sigma focuses on customer satisfaction and continuous improvement. Similar to BPR, it aims to streamline and optimize business processes. Additionally, Six Sigma incorporates principles from Lean Thinking, such as eliminating waste and enhancing value for customers. Overall, Six Sigma offers a structured approach to quality management that draws on the strengths of various methodologies to drive organizational excellence.

It is widely recognized as a system that enables organizations to achieve business excellence, as supported by the research conducted by Klefsjo et al. (2001) and Adebanjo (2001).

At its core, Six Sigma focuses on a precise and structured application scheme known as DMAIC, which stands for Define, Measure, Analyse, Improve, and Control. This methodology provides a systematic approach to problem-solving and process improvement, ensuring that organizations can identify and address areas of inefficiency or defects in their operations.

By following the DMAIC framework, organizations can define the problem at hand, measure the current performance metrics, analyse the root causes of any issues, implement improvements based on data-driven insights, and establish control mechanisms to sustain the improvements over time. This rigorous approach allows organizations to achieve significant enhancements in quality, efficiency, and customer satisfaction.

Furthermore, Six Sigma emphasizes the importance of data-driven decision making and the use of statistical tools and techniques to drive process improvements. This evidence-based approach ensures that organizations can make informed decisions and implement changes that are grounded in objective analysis rather than subjective opinions.

In conclusion, Six Sigma is a comprehensive management system that offers organizations a structured and disciplined approach to achieving business excellence. By utilizing the DMAIC methodology and embracing data-driven decision making, organizations can drive significant improvements in their processes, leading to enhanced quality, efficiency, and overall performance.

This particular table holds a significant role within the Six Sigma model, as it serves to verify the attainment of a specific sigma level for a given project or process, extending its influence to the entire organization. Through the continuous pursuit of higher sigma levels, the organization is able to tangibly showcase the reduction of critical process quality issues (CPQs) and achieve a notable gain. It is worth noting that the implementation of Six Sigma not only enhances customer satisfaction, but also contributes to the overall performance of the organization (Przekop, 2003). It is of utmost importance to acknowledge that numerous major global companies, particularly those that are prominently featured on Wall Street, have embraced the principles of Six Sigma (Pande et al., 2018; Senapati, 2004). After conducting extensive research and analysis, it has been observed that several prominent companies, including Motorola, GE,

Table 3.2 Correlation between process sigma, DPMO, and CPQ.

Sigma level	*DPMO (Defects per Million Opportunities)*	*CPQ (Cost of Poor Quality)*
2	308,537	Not applicable
3	66,807	25–40% of turnover
4	6210 (typical company)	15–25% of turnover
5	233	5–15% of turnover
6	3.4	<1% of turnover

Allied Signal, and Caterpillar, have made a strategic decision to adopt the Six Sigma methodology. This approach has proven to be highly effective in achieving substantial cost savings in terms of the "Cost of Poor Quality" (COPQ). The implementation of Six Sigma has enabled these organizations to identify and rectify inefficiencies, thereby enhancing overall operational performance.

In the realm of organizational theory, scholars such as McAuley et al. (2007) have delved into the neo-modernist perspective, introducing the concept of "new wave management". This theoretical framework offers a critical viewpoint on management systems like Six Sigma. According to these authors, Six Sigma may potentially exert a heightened level of control over organizations, which could potentially impede the professional autonomy and independence of employees.

It is important to note that the critical perspective presented by McAuley et al. (2007) should be considered alongside the numerous success stories of companies that have implemented Six Sigma. These success stories highlight the positive impact of Six Sigma on organizational performance, particularly in terms of cost reduction and quality improvement. However, it is crucial to strike a balance between the benefits of Six Sigma and the preservation of professional autonomy within organizations.

The Six Sigma Model and Lean Thinking

The Six Sigma model differs from Lean Thinking in some significant traits, which are summarized below.

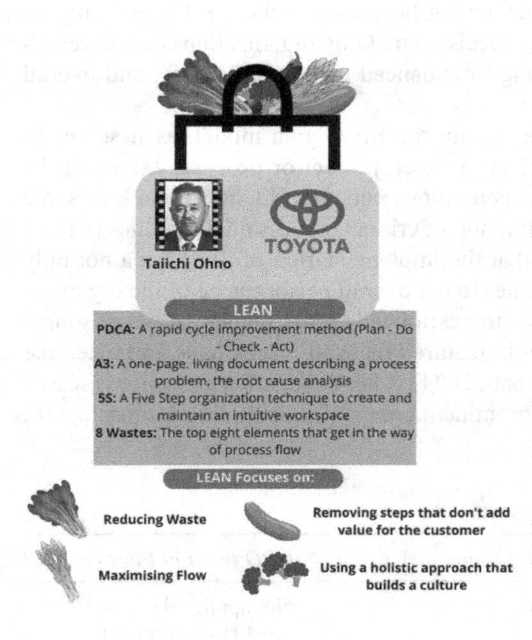

Figure 3.11 Lean vs Six Sigma.

Source: authors' own elaboration

Lean production, also known as Lean Manufacturing (LM), is a methodology that aims to eliminate waste and non-value-added activities in a process. It emphasizes the efficient use of resources and the continuous improvement of processes to increase customer value. LM focuses on optimizing the flow of materials and information to reduce lead times, improve quality, and lower costs. It involves techniques such as value stream mapping, 5S, and Kaizen.

Since the late 1990s, a multitude of articles have been meticulously crafted and published, delving into the intricacies of Six Sigma. It is worth noting that the model proposed by Harry and Schroeder in the year 2000 has now become the quintessential and revered Six Sigma model within the manufacturing sector. In their highly esteemed literary work, aptly titled "Six Sigma: The Breakthrough Management Strategy Revolutionizing the World's Top Corporations", Harry and Schroeder meticulously outline the fundamental model that is widely employed in contemporary times. It is of utmost importance to highlight that they establish a direct correlation between the sigma level of a given process, be it marketing or customer service, and the number of defects that may arise, as well as the consequential Costs of Poor Quality (COPQ) that an organization may incur. Conversely, Six Sigma, as a methodology, is laser-focused on the reduction of process variation and the mitigation of defects. It aims to achieve a process performance level of 3.4 defects per million opportunities (DPMO) or a 99.99966% defect-free rate. Six Sigma uses statistical tools and techniques to identify and quantify sources of process variation and implement solutions to reduce or eliminate them. It involves a structured problem-solving approach called DMAIC (Define, Measure, Analyse, Improve, Control) to achieve process improvement.

In summary, Lean production highlights on the eradication of waste and improvement of efficiency whereas Six Sigma aims at dropping down process variation and defects. Both methodologies can be used together to achieve process improvement and optimize overall performance.

BOPI (Business, Operational, Process, and Improvement) goals are used to set specific targets for a company or organization in various areas of operation.

Business goals focus on achieving financial objectives, such as increasing revenue or reducing costs, to improve the overall profitability of the company. Operational goals are related to the day-to-day operations of the organization, such as improving efficiency or increasing productivity. Process goals focus on specific processes within the organization, such as reducing waste or improving quality control. Improvement goals are related to continuous improvement efforts, such as implementing new technologies or processes to enhance operations.

By setting BOPI goals, organizations can establish clear targets and objectives for each area of operation, providing a framework for decision-making and ensuring that resources are allocated effectively. BOPI goals also enable organizations to monitor progress and measure success, helping to identify areas for improvement and to adjust strategies accordingly.

BOPI goals are a key aspect of the Six Sigma methodology. BOPI refers to opportunities for improvement that have a significant impact on overall performance. These goals are identified through the analysis of data and processes to determine where improvements can be made.

In Six Sigma, BOPI goals are used to set targets for improvement initiatives. By focusing on the most significant opportunities for improvement, organizations can achieve significant gains in efficiency, productivity, and quality.

The utilization of BOPI (Business Objectives and Performance Indicators) goals is intricately linked to the DMAIC (Define, Measure, Analyse, Improve, Control) process employed within the realm of Six Sigma. Allow me to elaborate on this matter with utmost precision and clarity.

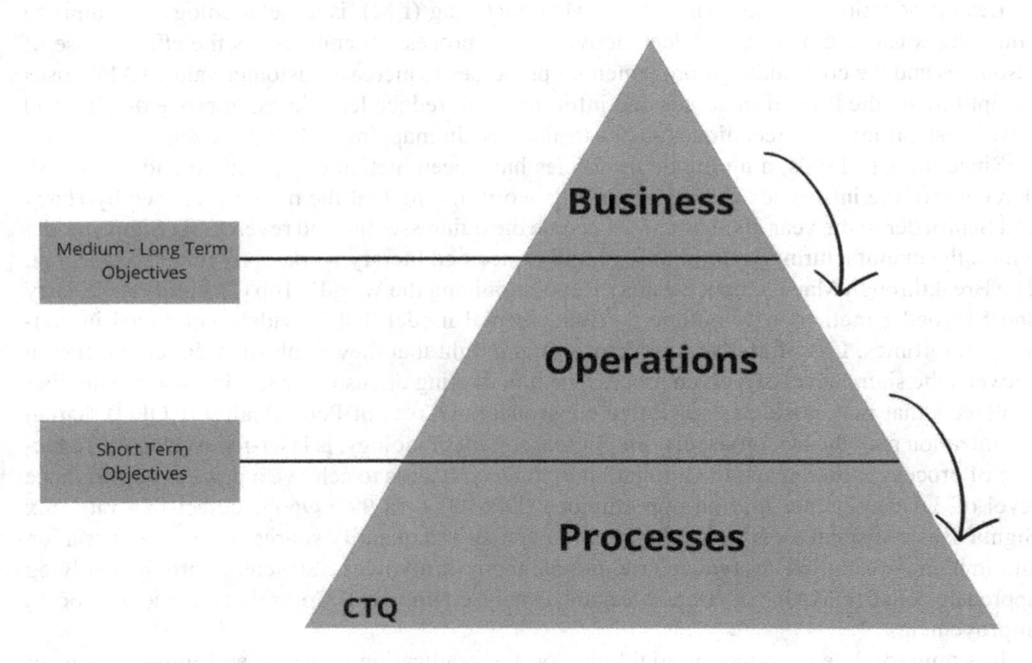

Figure 3.12 BOPI goals.
Source: authors' own elaboration

During the Define phase, the BOPI goals are meticulously identified and prioritized based on their potential impact on performance. This crucial step ensures that the objectives are aligned with the overarching business strategy, thereby setting the stage for success. The selection process is conducted with utmost care, taking into consideration various factors such as market trends, customer demands, and organizational capabilities.

Moving on to the Measure and Analyse phases, a comprehensive collection of data is undertaken to ascertain the current state of the process. This meticulous examination allows for a thorough understanding of the existing performance levels, potential bottlenecks, and areas for improvement. The data is meticulously analysed using statistical tools and techniques, enabling the identification of patterns, trends, and root causes of any inefficiencies or deviations from the desired outcomes.

Once the Measure and Analyse phases have been completed, we proceed to the Improve phase. This is the point where theory or plans are put into practice, as improvement initiatives are meticulously developed and implemented to achieve the BOPI goals. These initiatives are carefully crafted, taking into account the insights gained from the previous phases. The implementation process is executed with precision, ensuring that all relevant stakeholders are involved and that the necessary resources are allocated appropriately.

Finally, we arrive at the Control phase, which is of paramount importance in sustaining the achieved improvements over time. During this phase, processes are vigilantly monitored to ensure that the desired outcomes are consistently met. Any deviations or potential risks are promptly identified and addressed, thereby safeguarding the integrity of the improved processes. This ongoing monitoring and control mechanism ensures that the organization continues to reap the benefits of the implemented improvements, thereby enhancing overall performance and competitiveness.

In conclusion, the utilization of BOPI goals within the DMAIC process is a meticulously designed approach that enables organizations to identify, measure, analyse, improve, and control their performance. By adhering to this structured methodology, organizations can confidently navigate the path towards operational excellence and achieve their desired business outcomes. The utilization of BOPI (Breakthrough Objectives and Priorities for Improvement) goals is an exceedingly crucial and indispensable component of the Six Sigma methodology. These goals support companies to concentrate their efforts on enhancing performance by targeting the most significant opportunities for improvement. In the classical approach to implementing Six Sigma strategic projects, there exist five distinct phases, which are commonly referred to as DMAIC (Define, Measure, Analyse, Improve, Control). This acronym, widely recognized and acknowledged in the literature (Pande et al., 2018; Breyfogle, 2003; Pyzdek, 2009), succinctly captures the essence of each phase.

To further elucidate the connection between the five phases of DMAIC and Deming's (1982) classical approach, it is worth noting that they can be seamlessly linked to the Plan-Do-Check-Act (PDCA) cycle. This linkage, as expounded upon by Harry and Schroeder (2000), underscores the versatility and applicability of the five phases across various levels of an organization, encompassing business activities, operations, and processes. In fact, each phase of DMAIC represents the logical progression and deployment of the preceding phase, aligning harmoniously with the classical distribution of the management system, as posited by Kaplan (1996).

In summary, the utilization of BOPI goals within the framework of Six Sigma is an indispensable practice that empowers organizations to focus their improvement endeavours on the most significant opportunities for performance enhancement. The five phases of DMAIC, which are intricately connected to Deming's classical approach, provide a robust and comprehensive roadmap for implementing strategic projects at different levels of an organization. This alignment with the classical distribution of the management system further solidifies the efficacy and relevance of the DMAIC methodology.

Integration of Lean Six Sigma

Lean Six Sigma is a methodology that is based on the collaborative effort of a team to improve business performance by systematically removing waste and combining lean manufacturing/lean enterprise with Six Sigma to eliminate the seven types of waste (Muda): inventory, motion, waiting, overproduction, overprocessing, defects, and skills. Lean Six Sigma combines lean production (originally developed by Toyota) and Six Sigma (originally developed by Motorola). It is a management approach for improving business performance that has blended the two specializations of Lean and Six Sigma. Both management strategies are widely used globally and have proven success in various industries and services.

Six Sigma is a data-driven approach to solving complex business problems and is only pursued after an organization has implemented a lean commitment. It uses a methodological approach consisting of five phases – define, measure, analyse, improve, and control – to help understand the variables that impact a process in order to optimize it. Lean Six Sigma is a powerful and proven method for improving a company's efficiency and effectiveness. The key principles of Lean Six Sigma are customer focus; understanding how work is done (value stream); managing, improving process flow; removing unnecessary steps and waste; managing through facts and reducing variation; engaging and empowering people in the process;

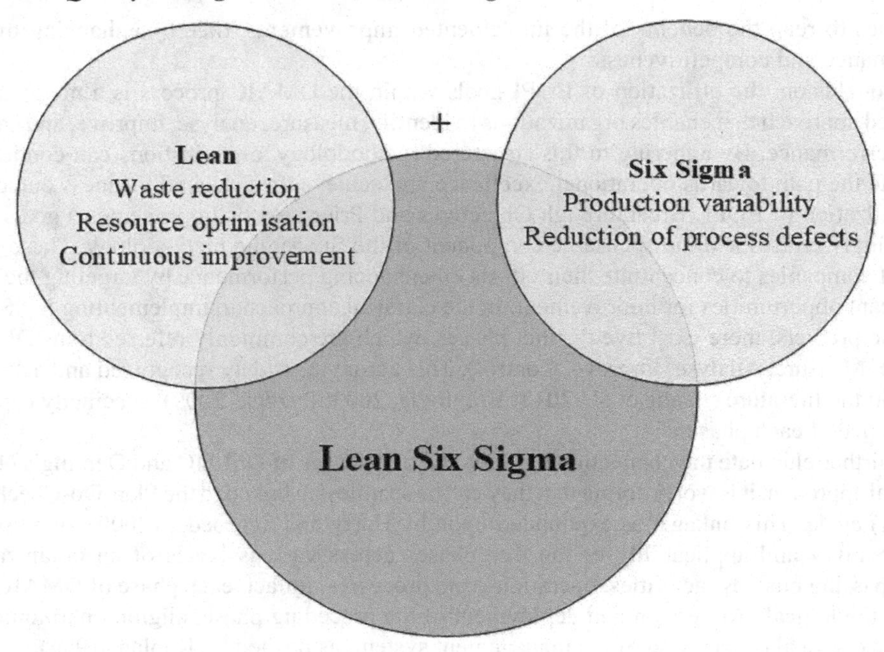

Figure 3.13 Lean Six Sigma.
Source: authors' own elaboration

and systematically pursuing improvement activities. Lean Six Sigma can help achieve rapid improvements, whether in a production or service context. The excellence it promotes permeates the entire operation and leads to customer and employee satisfaction. This fact is evidenced by many companies – from accounting firms to human resource agencies – that achieve excellent results by following this approach.

Key Takeaways

- **Unravelling Lean Thinking**: Lean Thinking is a management philosophy that emphasizes the elimination of waste and the continuous improvement of processes. Understanding its evolution provides insights into its core principles.
- **Defining Lean Thinking 5.0**: Lean Thinking 5.0 represents the latest evolution of Lean methodologies, incorporating modern business practices and technologies to enhance efficiency and effectiveness.
- **Birth of Lean Production—Toyota Motor Company and Toyota Production System**: The origins of Lean Production can be traced back to Toyota Motor Company and the development of the Toyota Production System, which revolutionized manufacturing through waste reduction and improved productivity.
- **Eliminating the 7 Muda**: Lean Thinking identifies and targets seven types of waste (Muda) in processes, emphasizing their elimination for optimal efficiency.
- **Toyota's House of Lean**: Toyota's House of Lean serves as a visual representation of the key elements and principles that constitute a Lean organization, guiding effective implementation.

- **Comparing Lean Production and TQM**: Lean Production and Total Quality Management (TQM) share common goals of continuous improvement, but they differ in focus and approach. Understanding these differences aids in choosing the most suitable methodology for specific contexts.
- **Tools of Lean Production**: Practical tools and techniques, such as value stream mapping and kanban, are of substantial importance in implementing Lean Production principles and optimizing processes.
- **UNI 11063:2017 and UNI 10147:2003**: The UNI 11063:2017 standards provide a framework for applying Lean Thinking in a structured manner, ensuring consistency and quality in its implementation. The UNI 10147:2003 standards contribute to the understanding and application of Lean Thinking methodologies, providing guidelines for organizations seeking to improve their processes.
- **The Five Lean Principles**: The Five Lean Principles – value, value stream, flow, pull, and perfection – serve as foundational concepts for organizations adopting Lean Thinking, guiding them towards continuous improvement.
- **Lean Approach—Push versus Pull**: Understanding the difference between the Push and Pull approaches in Lean Thinking helps organizations optimize their production processes based on customer demand.
- **Embracing Digital Lean 4.0**: The integration of digital technologies in Lean Thinking, known as Digital Lean 4.0, enhances agility and responsiveness in today's technologically driven business landscape.
- **The Six Sigma Model**: The Six Sigma Model, with its focus on reducing defects and variations, complements Lean Thinking, providing a comprehensive approach to quality improvement.
- **Integration of Lean Six Sigma**: The integration of Lean Thinking and Six Sigma combines the strengths of both methodologies, creating a powerful framework for achieving operational excellence and superior quality.

Review Questions

1. What is Lean Thinking, and how has it evolved over time?
2. Why is Lean Thinking essential for operational excellence?
3. How does Lean Thinking 5.0 differ from earlier versions, and what are the key characteristics of this latest evolution?
4. In what ways does Lean Thinking 5.0 incorporate modern business practices and technologies?
5. What role did Toyota Motor Company play in the development of Lean Production?
6. Define the concept of the 7 Muda in Lean Thinking.
7. Why is the elimination of waste crucial for achieving Lean objectives?
8. What components and principles make up Toyota's House of Lean?
9. How does the House of Lean serve as a guide for organizations implementing Lean practices?
10. Find the parallels and distinctions among Lean Production and Total Quality Management (TQM).
11. When might an organization choose Lean Production over TQM, or vice versa?
12. Provide examples of practical tools and techniques used in Lean Production.
13. How do these tools contribute to process optimization and waste reduction?

14. What is the significance of UNI 11063: 2017 standards in the context of Lean Thinking?
15. Explore the implications of UNI 10147: 2003 standards in the realm of Lean Thinking.
16. Outline the Five Lean Principles and explain their importance in Lean Thinking.
17. Define the Push and Pull approaches in Lean Thinking.
18. What is Digital Lean 4.0, and how does it contribute to Lean Thinking in the digital age?
19. Explain the key principles of the Six Sigma Model.
20. How does the Six Sigma Model complement Lean Thinking in quality improvement efforts?
21. Why are Lean Thinking and Six Sigma considered an influential approach?
22. What benefits can organizations gain from implementing Lean Six Sigma together?

4 Quality Management Tools in TQM and Lean Thinking

Learning Objectives

- Understand the principles and application of the PDCA Methodology (Deming Cycle) in the context of Total Quality Management (TQM) and Lean Thinking.
- Explain the DMAIC Methodology and its role in the Six Sigma approach to quality management.
- Evaluate the A3 Methodology and its significance in problem-solving within the TQM and Lean Thinking frameworks.
- Explore various Employee Involvement Techniques and their application in effective problem-solving within an organizational context.
- Analyse the Ishikawa Diagram (Fishbone Diagram) as a quality management tool and understand its practical application in identifying and solving quality-related issues.
- Examine the CEDAC (Cause-and-Effect Diagram with the Addition of Cards) method and its role in visualizing and addressing complex problems in TQM and Lean contexts.
- Understand the concept and implementation of Quality Circles as a means of fostering employee involvement and continuous improvement in quality management.
- Explore the 6M Method and its relevance in comprehensively assessing and managing factors affecting quality in various processes.

PDCA Methodology (Deming Cycle)

In the context of Total Quality Management (TQM), it is worth delving into a separate discussion on the practice known as Kaizen. The term "Kaizen" is derived from the combination of two words: "Kai", which signifies "improvement", and "Zen", which can be translated as "cycle". This particular practice places emphasis on the continuous enhancement of performance across all areas of the company, ranging from production to management. It is a daily process that extends beyond mere productivity improvement. When correctly implemented, Kaizen humanizes the workplace, eliminates excessive workload (referred to as "muri"), encourages individuals to experiment with their work using the scientific method, and identifies and eliminates waste in business processes. The underlying objective is to nurture the human resources within the company, as the successful execution of Kaizen necessitates the participation of all workers in the pursuit of improvement.

DOI: 10.4324/9781032726748-4

The fundamental concept behind Kaizen is closely intertwined with the Deming cycle, also known as the Plan–Do–Check–Act (PDCA) cycle. In order to enhance a particular aspect, an individual conceives an idea to improve it (ACT). This idea is then subjected to a series of rigorous tests and simulations to ascertain its validity (DO). The obtained results are meticulously evaluated to determine whether the idea has successfully achieved its intended objective (CHECK). If this is indeed the case, the existing standard procedures are modified and the new method is adopted (ACT).

The implementation of this systematic process not only yields improvements in overall productivity, but also has the potential to enhance worker satisfaction. Employees, feeling a heightened sense of responsibility and motivation, subsequently exhibit an even greater level of dedication, leading to further enhancements in the quality of their work performance.

For a Lean Manufacturing (LM) system to effectively provide the inherent benefits it promises, it is imperative that all the aforementioned tools and techniques be wholeheartedly embraced and implemented throughout the entire organization. The delicate equilibrium of this type of production necessitates that all processes be executed with utmost smoothness and precision, in order to prevent any setbacks that could potentially jeopardize the integrity of the entire system.

Upon careful analysis of the tools employed by LM, it is with utmost certainty that we affirm the implementation of said tools allows for a multitude of benefits. Firstly, it enables a reduction in lead time, thereby enhancing the fluidity of production processes. This reduction in lead time not only optimizes efficiency but also contributes to a more streamlined workflow.

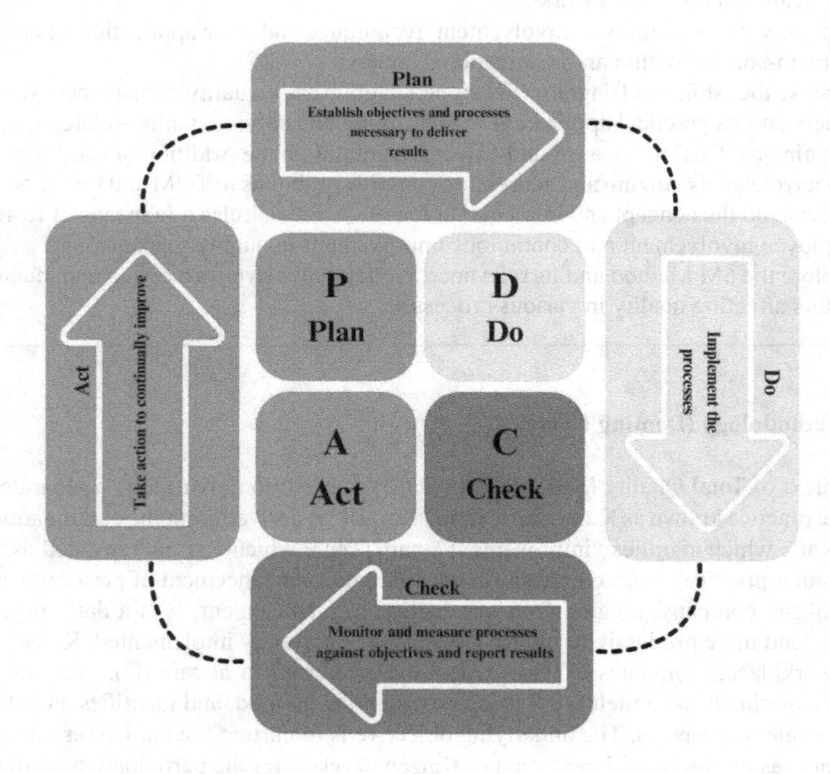

Figure 4.1 The PDCA Cycle for continuous improvement.

Source: authors' own elaboration

Furthermore, the implementation of these tools results in a rise in the overall quality of products and worker productivity. By utilizing LM, companies are able to effectively manage supplies, ensuring that the necessary resources are readily available when needed. This, in turn, facilitates a smoother production process and enhances the quality of the final products.

Moreover, LM aids in the better management of maintenance, resulting in fewer breakdowns and disruptions to the production line. This reduction in maintenance-related issues allows for a more efficient utilization of resources and a decrease in costly downtime.

Additionally, LM provides greater flexibility of machinery, enabling companies to adapt to changing demands and requirements. This flexibility not only enhances productivity but also minimizes downtime, as machinery can be easily reconfigured to accommodate different production needs.

Furthermore, LM encompasses a broader quality management approach, encompassing all variables that affect product quality. By implementing LM, businesses can pinpoint and resolve any potential issues or bottlenecks in the production process, thereby ensuring consistent and high-quality products.

Moreover, LM facilitates a more effective and efficient organization of the workplace. By optimizing workflows and streamlining processes, LM enables companies to achieve a higher level of organization, resulting in improved productivity and overall operational efficiency.

Lastly, the implementation of LM fosters a culture of constant improvement in the quality of all company processes. By utilizing LM, companies are able to find areas for enhancement and implement essential modifications, thereby continuously enhancing the quality of their operations.

It is worth noting that the increasing attention to sustainability has progressively led to the involvement of another management variable in manufacturing studies, namely the natural environment. This recognition of the importance of environmental sustainability further highlights the comprehensive nature of Lean Manufacturing (LM) and its ability to address various aspects of modern manufacturing practices. In particular, the most encouraging methods involve preventive tactics that view resource efficiency as the origin of all technical, economic, and environmental advantages (Porter and van der Linde, 1995; WBCSD, 2010).

In this context, we would like to take a moment to discuss the concept of "Clean(er) Production" (CP), which encompasses all these approaches. CP represents an important opportunity for companies to introduce new solutions that improve the environmental performance of products and processes, ultimately enhancing the competitiveness of businesses. The first efforts in this direction focused primarily on the technological dimension of innovation, seeking ways to minimize waste and reduce pollution.

However, it is important to note that CP goes beyond just technological advancements. It also encompasses changes in organizational practices, such as the adoption of environmental management systems and the integration of sustainability principles into the decision-making process. By implementing CP, companies can not only diminish their environmental footprint but also achieve cost savings through improved resource efficiency.

Moreover, CP offers a range of benefits that extend beyond the environmental realm. For instance, by adopting cleaner production practices, companies can enhance their reputation and brand image, attracting environmentally conscious consumers. Additionally, CP can lead to increased employee satisfaction and motivation, as employees are more likely to be proud of working for a company that prioritizes sustainability.

In conclusion, the increasing attention to sustainability has brought the natural environment into the realm of manufacturing studies. Lean Manufacturing, with its comprehensive approach, has recognized the importance of environmental sustainability. Clean(er) Production represents a valuable opportunity for companies to improve their environmental performance, enhance competitiveness, and reap a multitude of benefits. In this regard, the Deming cycle, also known as the Plan–Do–Check–Act cycle, is not only a method, but it serves as the very foundation of every improvement project. This cycle, with its inherent cyclicality, constantly produces excellence. However, it is important to note that the most challenging phase of this cycle is that of standardization. Standardization plays a crucial role in preventing regression and the loss of hard-earned results. Once the initial objectives have been successfully achieved, new objectives are set to further elevate the quality level and drive continuous improvement.

In the realm of problem-solving within the industrial, manufacturing, and service sectors, various techniques have been adopted by different schools of thought. It is worth mentioning that there is a significant overlap among these methodologies. However, it is essential to recognize that each methodology may be particularly suitable for addressing a specific problem.

One such methodology is the PDCA methodology, as introduced by N. R. Tague (2005). The PDCA methodology, which stands for Plan–Do–Check–Act, offers a general sequence of work that can be effectively employed to resolve any non-compliance issue across various work areas. These four segments, namely Plan–Do–Check–Act, form an iterative process that ensures thoroughness and effectiveness in problem-solving. By adhering to this methodology, organizations can confidently tackle any non-compliance issue that may arise.

The concept in question, known as the Plan–Do–Check–Act (PDCA) cycle, is a highly effective method for problem-solving and continuous improvement. It is worth noting that PDCA is also referred to as the Deming cycle or Deming wheel, paying homage to the esteemed management consultant, Dr William Deming. It is important to acknowledge that Dr Deming built upon the original idea put forth by Walter Shewhart, further refining and enhancing its efficacy. The primary purpose of PDCA is to ensure that customer quality requirements are met, making it an invaluable tool for the development of new products that align with said requirements. Additionally, PDCA fosters a sense of teamwork among the various functions within a company, facilitating seamless collaboration in areas such as product design and development, production, sales, and market research. By implementing PDCA, organizations can effectively navigate the intricacies of their operations, ultimately resulting in improved customer satisfaction and overall success.

The process known as "Plan" is an essential step in problem-solving. It entails meticulously defining and analysing the problem at hand, with the ultimate goal of identifying its root cause. By thoroughly understanding the underlying issues, one can effectively devise a solution that addresses the core problem.

Moving on to the next step, "Do" it is crucial to advance a detailed action plan that summaries the necessary steps to implement the solution. This plan should be comprehensive and leave no room for ambiguity. By meticulously following this action plan, one can ensure that the solution is executed in a systematic and efficient manner.

Once the solution has been implemented, it is imperative to proceed to the "Check" phase. During this stage, it is essential to exercise control and monitor the results of the actions taken. This evaluation process serves multiple purposes. Firstly, it allows for the identification of any gaps that may exist between the desired outcome and the current state. Secondly, it provides valuable insights that can inform the selection of potential further actions. By carefully assessing the outcomes, one can make informed decisions on how to proceed.

Finally, we arrive at the "Act" phase, which focuses on standardizing the various solutions that have been implemented. This standardization is crucial for achieving a sustainable solution. By establishing a consistent approach, one can ensure that the problem is not only resolved in the short term but also prevented from recurring in the future.

In summary, by diligently following the steps of "Plan", "Do", "Check", and "Act", one can effectively address problems and achieve sustainable solutions. It is imperative to approach each step with meticulous attention to detail and a commitment to excellence.

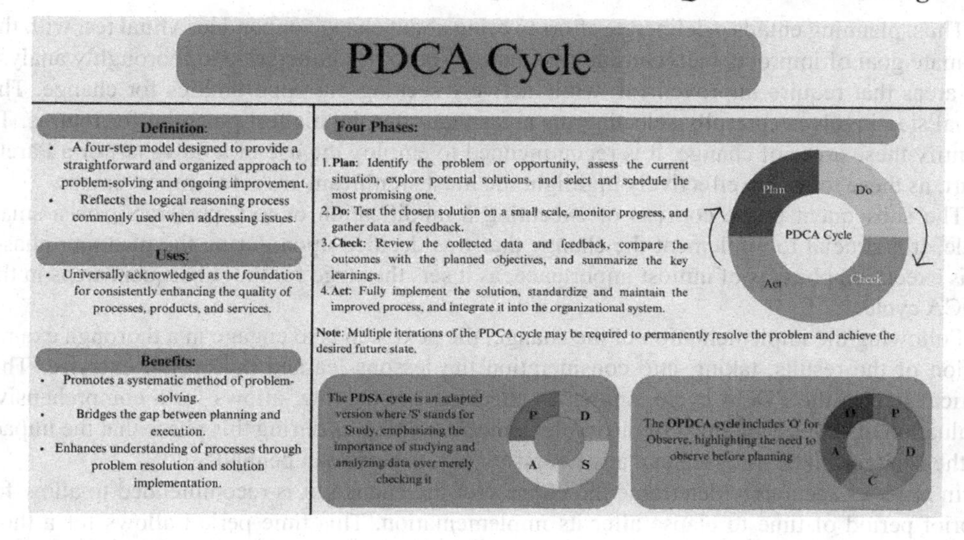

Figure 4.2 PDCA Cycle.
Source: authors' own elaboration

This particular model has been extensively implemented with remarkable success in a multitude of business domains. These domains encompass, but are not restricted to, production management, supply chain management, project management, and human resource management. It is worth noting that each PDCA (Plan–Do–Check–Act) cycle, upon reaching its culmination, leads to the initiation of a slightly more intricate project. This transition, known as the rollover function, plays a pivotal role in the overall continuous improvement process. The significance of this function cannot be overstated, as it serves as an integral component in the pursuit of organizational excellence. To further illustrate this point, a visual representation of the improvement ramp can be observed in the accompanying figure, providing a comprehensive depiction of the gradual progression towards enhanced performance and efficiency.

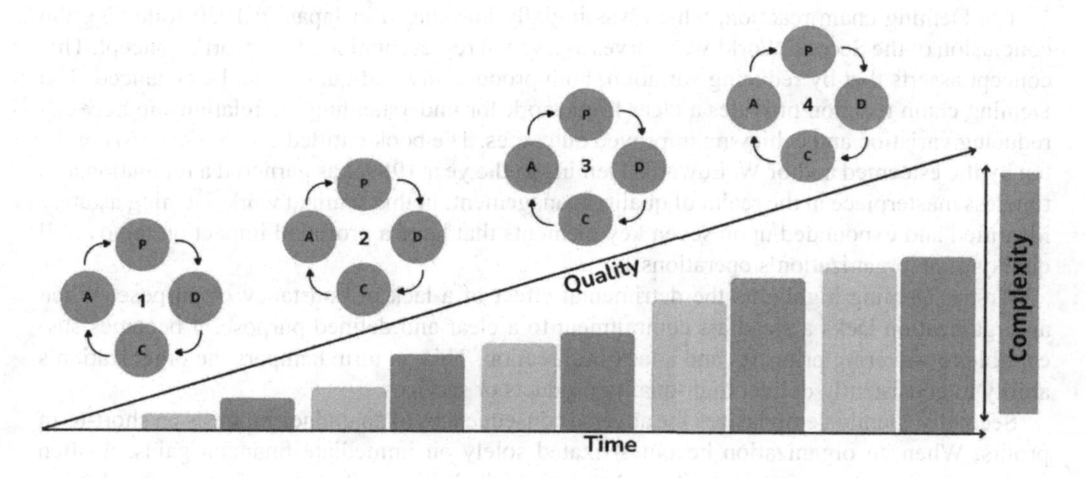

Figure 4.3 Improvement ramp.
Source: authors' own elaboration

Thus, planning entails a deliberate effort to bring about change or conduct a final test with the ultimate goal of improvement. During this phase, it becomes imperative to thoroughly analyse the areas that require improvement, while actively seeking out opportunities for change. The initial step involves carefully selecting the areas that offer the highest potential for returns. To identify these areas of change, it is recommended to employ the use of a flowchart or a Pareto chart, as these tools can effectively highlight the most significant areas for improvement.

The subsequent phase consists of executing a modification or test, preferably on a small scale. It is crucial to implement the change that was decided upon during the planning phase. This execution phase is of utmost importance, as it sets the stage for the subsequent steps in the PDCA cycle.

Following the implementation of the change, the next step is to engage in a thorough examination of the results, taking into consideration the lessons learned during the exercise. This critical step in the PDCA cycle, known as checking or studying, allows for a comprehensive evaluation of the effectiveness of the implemented change. It is during this phase that the impact of the change can be assessed and any necessary adjustments can be made.

In order to accurately determine the efficacy of the change, it is recommended to allow for a brief period of time to elapse after its implementation. This time period allows for a thorough assessment of how well the change is functioning and whether it is achieving the desired outcomes. By carefully evaluating the results, one can confidently ascertain the success of the change and make any necessary refinements.

Moreover, to effectively monitor the level of improvement, it is imperative to determine the appropriate metrics. One method that proves to be quite useful for this purpose is the utilization of run charts. These charts provide valuable insights and aid in the assessment of progress.

Once the change has been planned and implemented, it is crucial to closely monitor its effects. This monitoring phase allows for a comprehensive evaluation of the change's impact. At this point, a decision must be made regarding the continuation of the change. If the change has resulted in a noticeable improvement or has yielded desirable outcomes, it may be deemed worthwhile to expand the implementation of this particular change. This expansion could involve applying the change to a diverse area or slightly growing its complexity. This decision-making process brings us back to the Planning phase, effectively initiating the improvement ramp once again.

The Deming chain reaction, which was initially introduced in Japan in 1950 following the conclusion of the Second World War, serves as a visual representation of Shewart's concept. This concept asserts that by reducing variation, both productivity and quality can be enhanced. The Deming chain reaction provides a clear framework for understanding the relationship between reducing variation and achieving improved outcomes. The book entitled *Out of the Crisis*, written by the esteemed author W. Edwards Deming in the year 1989, has garnered a reputation as a timeless masterpiece in the realm of quality management. In this seminal work, Deming astutely identified and expounded upon seven key elements that have a profound impact on the overall quality of an organization's operations.

Firstly, Deming highlights the detrimental effect of a lack of constancy of purpose. When an organization lacks a steadfast commitment to a clear and defined purpose, it becomes susceptible to wavering priorities and a lack of direction. This, in turn, hampers the organization's ability to consistently deliver high-quality products or services.

Secondly, Deming emphasizes the adverse consequences of an undue emphasis on short-term profits. When an organization becomes fixated solely on immediate financial gains, it often neglects the long-term sustainability and growth of the business. This myopic focus can lead to compromised quality standards and a decline in overall customer satisfaction.

Thirdly, Deming cautions against an excessive dependence on performance evaluations as a means of assessing and improving quality. While performance evaluations can provide valuable insights, an overreliance on them can create a culture of fear and competition, stifling creativity and collaboration. This, in turn, can hinder the organization's ability to achieve and maintain high levels of quality.

Fourthly, Deming highlights the detrimental impact of a high degree of mobility within management. When there is a constant turnover of key decision-makers, establishing and upholding consistent quality standards becomes challenging. The lack of continuity and familiarity with the organization's processes and goals can impede the implementation of effective quality management practices.

Fifthly, Deming draws attention to the excessive emphasis placed on visible figures, such as financial metrics or production quotas. While these figures can provide a snapshot of an organization's performance, they often fail to capture the underlying factors that contribute to quality. Relying solely on visible figures can lead to a narrow focus on quantity rather than quality, ultimately compromising the organization's ability to deliver superior products or services.

Sixthly, Deming addresses the issue of excessive medical costs for employee healthcare. When an organization incurs exorbitant expenses related to employee health, it can strain financial resources that could otherwise be allocated towards quality improvement initiatives. By effectively managing healthcare costs, organizations can redirect these resources towards enhancing quality and fostering a healthier and more productive workforce.

Lastly, Deming sheds light on the detrimental impact of excessive warranty costs and legal expenses. When an organization faces a high volume of warranty claims or legal disputes, it not only incurs significant financial burdens but also tarnishes its reputation. These costs and legal battles divert resources away from quality improvement efforts, hindering the organization's ability to consistently deliver products or services that meet or exceed customer expectations.

In conclusion, Deming's book *Out of the Crisis* serves as a comprehensive guide to understanding and addressing the various elements that impact the quality of an organization's operations. By meticulously examining and providing detailed insights into these seven key elements, Deming equips organizations with the knowledge and understanding necessary to cross the complex scenery of quality management. In relation to this matter, it is worth noting Deming's Theory of Deep Knowledge, which posits that a production system is composed of numerous interconnected subsystems. It is the duty of management to establish the purpose of said system and to optimize it accordingly. Deming ardently advocated for the improvement of the system rather than the criticism of the workers. He firmly believed that the workers were already exerting their utmost effort within the parameters set by management. However, the absence of clear guidance has resulted in subpar outcomes. Consequently, Deming deduced that it falls upon management to provide the necessary direction to the workers. Nevertheless, Deming asserted that this cannot be achieved through the utilization of goals imposed by management, which he referred to as "management by fear", nor through the implementation of annual performance reviews, which he vehemently condemned. The plan he proposed is meticulously elucidated in the 14 points, which he firmly believed must be implemented in their entirety in order to yield the desired effectiveness. According to Deming, the omission of even a single point would impede the efficacy of the remaining 13.

DMAIC Methodology

The DMAIC cycle is the quintessential tool employed to guide Six Sigma projects. The concept of Six Sigma is an exceedingly rigorous, targeted, and remarkably effective implementation of quality principles and techniques. A company's performance is meticulously measured by the sigma level of its business processes. Traditionally, companies have deemed level 3 or 4 Sigma performance as the norm, despite the rather disconcerting fact that these processes have been responsible for generating a staggering number of problems. To be precise, between 6,200 and 67,000 problems per million opportunities have been observed. However, in response to the ever-increasing expectations of customers and the escalating complexity of modern products and processes, the Six Sigma standard has been established. This standard dictates that there should be no more than 3.4 problems per million opportunities.

Six Sigma is a highly effective methodology that draws upon decades of proven methods. It is important to note that Six Sigma distinguishes itself from Total Quality Management (TQM) by discarding the unnecessary complexity that often accompanies the latter. While TQM boasts an extensive repertoire of over 400 tools and techniques, Six Sigma selectively adopts only those that have been thoroughly tested and validated. To safeguard the successful implementation of Six Sigma, organizations invest in training their internal technical leaders, commonly referred to as "Six Sigma black belts".

The DMAIC methodology, an integral part of Six Sigma, is a data-driven improvement cycle specifically designed to enhance and stabilize business processes. In contrast to the PDCA methodology, DMAIC follows a structured sequence of five distinct steps: Define, Measure, Analyse, Improve, and Control. This meticulous approach to problem-solving aligns closely with the strategic framework advocated by the esteemed quality management expert, Juran. Furthermore, projects within the DMAIC methodology are thoughtfully classified based on their potential to generate substantial cost savings. By embracing Six Sigma and diligently adhering to the DMAIC methodology, organizations can confidently navigate the path towards process improvement and operational excellence. The DMAIC methodology is summarized in Figure 4.4.

Phase 1 – Definition

In this initial phase, we embark upon the crucial task of defining the quality priorities of our esteemed customers. It is of utmost importance to identify the product/service attributes that hold the highest significance in their evaluation of product quality. These paramount attributes, known as "critical-to-quality" (CTQ) characteristics, serve as the cornerstone of our understanding.

Table 4.1 PDCA vs DMAIC.

PDCA	*DMAIC*
PDCA is a repetitive four-phase model (Plan–Do–Check–Act) used to achieve continuous improvement in business process management.	DMAIC is a data-driven improvement cycle used to enhance and stabilize business processes, consisting of five phases: Define, Measure, Analyse, Improve, Control.
The PDCA was created in the 1950s.	DMAIC was created in the 1980s as an integral part of Six Sigma.
PDCA can be a standalone concept but, it is primarily used in conjunction with another famous Japanese technique called Kaizen.	*DMAIC is primarily used in conjunction with the Six Sigma concept.*

Source: author's own elaboration

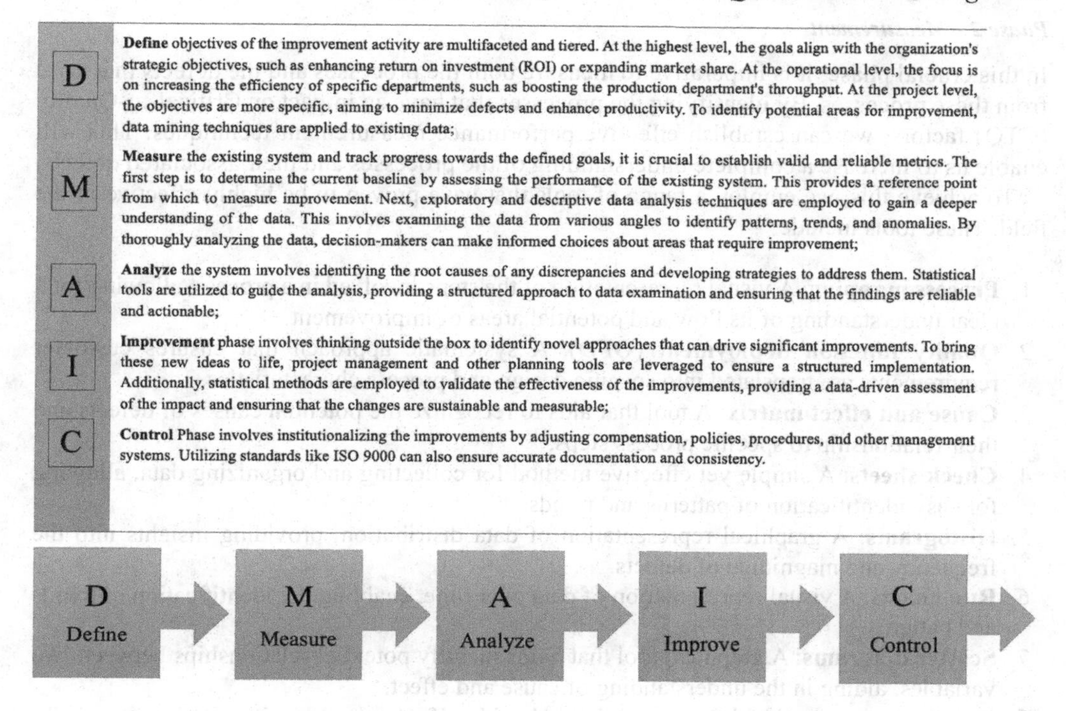

Figure 4.4 Diagram of the five stages of the DMAIC methodology.

Source: authors' own elaboration

To ensure that our perception of quality attributes remains up to date, we diligently conduct surveys among our valued customers. This allows us to stay attuned to their evolving needs and preferences. In this phase, we employ the formidable tool of Quality Function Deployment (QFD) to facilitate the implementation of our objectives.

To aid us in this endeavour, we utilize a range of indispensable tools. Firstly, we engage in the process of brainstorming, which encourages the generation of innovative ideas and perspectives. Additionally, we employ Pareto analysis, a method that enables us to identify and prioritize the most significant factors contributing to quality.

Furthermore, we rely on process mapping to comprehensively understand the intricacies of our operations. By employing project management fundamentals, we ensure a systematic approach towards achieving our goals. The IPO Diagram, SIPOC Diagram, and Flow Diagram serve as invaluable aids in visualizing and comprehending the various aspects of our quality priorities.

To further refine our understanding of the CTQ characteristics, we employ the CTQ Tree, which provides a structured framework for analysis. Lastly, the Project Charter serves as a guiding document, outlining the objectives, scope, and deliverables of our quality-focused endeavours.

By adhering to these meticulous steps and utilizing the aforementioned tools, we establish a solid foundation for the subsequent phases of our quality journey.

Phase 2 – Measurement

In this crucial phase, it is imperative to measure both the processes and the defects that arise from these processes. By identifying the processes that have an impact on Critical-to-Quality (CTQ) factors, we can establish effective performance measurement techniques. This will enable us to increase a complete understanding of the processes and their associated defects.

To achieve this, we employ a range of tools that have proven to be highly effective in the field. These tools include:

1. **Process mapping**: A visual representation of the steps involved in a process, allowing for a clear understanding of its flow and potential areas of improvement.
2. **Quality function deployment (QFD)**: A systematic approach that ensures customer requirements are translated into specific design and process characteristics.
3. **Cause and effect matrix**: A tool that aids to recognize the potential causes of defects and their relationship to specific process steps.
4. **Check sheets**: A simple yet effective method for collecting and organizing data, allowing for easy identification of patterns and trends.
5. **Histograms**: A graphical representation of data distribution, providing insights into the frequency and magnitude of defects.
6. **Run charts**: A visual representation of data over time, enabling the identification of trends and patterns.
7. **Scatter diagrams**: A graphical tool that helps identify potential relationships between two variables, aiding in the understanding of cause and effect.
8. **Pareto charts**: A prioritization tool that helps identify the most significant factors contributing to defects, allowing for focused improvement efforts.
9. **Control charts**: Statistical tools that monitor process performance over time, enabling the identification of variations and the implementation of corrective actions.
10. **Flow process charts**: Detailed visual representations of a process, highlighting the sequence of steps and potential areas for improvement.
11. **Creativity techniques**: Various methods that stimulate innovative thinking and problem-solving, promoting a culture of continuous improvement.
12. **Seven quality control tools**: A comprehensive set of tools that includes check sheets, histograms, Pareto charts, cause and effect diagrams, control charts, scatter diagrams, and flowcharts.
13. **Calculation of process sigma and process capability studies**: Statistical methods that assess the capability of a process to meet customer requirements and identify areas for improvement.
14. **Measurement studies**: Rigorous analysis of measurement systems to ensure accuracy and reliability of data.
15. **Analysis of variance (ANOVA)**: A statistical technique used to analyse the differences between groups and determine the factors that contribute to variations.

By utilizing these tools and techniques, we can gather valuable insights into our processes and defects, enabling us to make informed decisions and drive continuous improvement.

Phase 3 – Analysis

In this crucial phase, we shall delve into the intricate process and meticulously identify the most probable causes of defects. Our aim is to pinpoint the key variables that are most likely responsible for the variation in the process. To achieve this, we shall employ a range of powerful tools that have been proven to yield valuable insights. Allow me to elaborate on these tools:

1. **Gap analysis and improvement objectives**: By conducting a comprehensive gap analysis, we can identify the areas where the process falls short of expectations. This will enable us to set clear improvement objectives that will guide our efforts.
2. **Process map analysis**: Through a meticulous examination of the process map, we can gain a deep understanding of the various steps involved and identify any potential bottlenecks or inefficiencies.
3. **Data stratification**: By stratifying the data, we can categorize it into meaningful groups, allowing us to discern patterns and trends that may be contributing to defects.
4. **Advanced analytical tools**: Leveraging sophisticated analytical tools, we can uncover hidden insights and correlations within the data that may not be immediately apparent.
5. **Regression analysis**: By employing regression analysis, we can determine the relationship between the key variables and the occurrence of defects, providing us with valuable insights into their impact.
6. **RU/CS analysis**: This analysis technique, which stands for Root Cause and Corrective Action Analysis, enables us to find the underlying root causes of defects and develop effective corrective actions.
7. **SWOT analysis**: By conducting a SWOT analysis, we can assess the strengths, weaknesses, opportunities, and threats associated with the process, helping us to identify areas for improvement.
8. **PESTLE analysis**: This analysis method, which stands for Political, Economic, Sociocultural, Technological, Legal, and Environmental analysis, allows us to assess the external factors that may be influencing the process and contributing to defects.
9. **The five whys**: By repeatedly asking "why" at least five times, we can uncover the underlying causes of defects, enabling us to address them at their root.
10. **Interrelationship diagram**: This diagramming technique helps us visualize the complex relationships between various factors and variables, aiding in the identification of potential causes of defects.
11. **Overall equipment effectiveness**: By assessing the overall equipment effectiveness, we can determine the extent to which the equipment is contributing to defects and identify areas for improvement.
12. **TRIZ**: Innovative Problem Solving: TRIZ, a powerful problem-solving methodology, provides us with a systematic approach to generate innovative solutions to complex problems, including those related to defects.
13. **ANOVA**: Analysis of Variance (ANOVA) allows us to compare the means of multiple groups and determine if there are significant differences that may be contributing to defects.
14. **Hypothesis testing**: By formulating and testing hypotheses, we can determine if there is a statistically significant relationship between certain variables and the occurrence of defects.

In conclusion, by employing these comprehensive and powerful tools, we can confidently analyse the process and identify the most probable causes of defects. This will enable us to take targeted actions and make informed decisions to improve the overall quality and efficiency of the process.

Phase 4 – Enhance

In this crucial phase, we shall focus on enhancing the performance of our processes and eliminating any potential causes of defects. To achieve this, we must establish key variable specification limits and validate the system that measures variable deviations. By doing so, we can ensure that our variables remain within the specified limits.

To facilitate these process improvements, we have a range of powerful tools at our disposal. Allow me to elaborate on each of these tools, as they will undoubtedly prove invaluable in our pursuit of excellence:

1. **Affinity diagram**: This tool enables us to organize and categorize a large amount of information, helping us recognize patterns and relationships that may not be immediately apparent.
2. **Nominal group technique**: This technique fosters collaboration and encourages the active participation of team members. By engaging in structured discussions and carefully considering each individual's input, we can generate innovative ideas and make informed decisions.
3. **SMED (single minute exchange of die)**: This technique focuses on reducing the time required for equipment changeovers. By streamlining these processes, we can minimize downtime and increase overall productivity.
4. **Five S**: This methodology emphasizes the importance of workplace organization and cleanliness. Thus, we can create an environment that promotes efficiency and reduces the risk of errors.
5. **Mistake proofing**: This technique aims to design processes and systems in a way that prevents errors from occurring in the first place. By implementing foolproof mechanisms, we can lessen the likelihood of defects and better the general quality.
6. **Value stream mapping**: This tool allows us to visualize the flow of materials and information throughout our processes. Thus, we can make targeted improvements to enhance overall value delivery.
7. **Brainstorming**: This creative technique encourages the generation of a wide range of ideas and solutions. By fostering an open and non-judgmental environment, we can tap into the collective wisdom of our team and uncover innovative approaches.
8. **Mind mapping**: This visual tool helps us organize our thoughts and ideas in a structured manner. By creating interconnected diagrams, we can explore relationships and uncover new insights.
9. **Force field diagram**: This tool enables us to analyse the driving and restraining forces that impact our processes. By understanding these forces, we can develop strategies to strengthen the positive influences and mitigate the negative ones.
10. **DOE techniques (design of experiments)**: This statistical approach allows us to systematically investigate the impact of various factors on our processes. By conducting controlled experiments, we can identify the optimal settings and make data-driven decisions.
11. **Hypothesis testing**: This rigorous statistical method allows us to test the validity of our assumptions and draw meaningful conclusions. By subjecting our hypotheses to rigorous scrutiny, we can ensure that our decisions are based on sound evidence.
12. **Confirmation or validation studies**: These studies provide a means to verify the effectiveness of our process improvements. By conducting thorough assessments and collecting relevant data, we can validate the positive impact of our efforts.

In conclusion, by utilizing these powerful tools and techniques, we can enhance our processes, maintain variables within specification limits, and ultimately achieve superior performance.

Let us embark on this journey of continuous improvement with confidence and determination. Together, we shall overcome any challenges and emerge triumphant in our pursuit of excellence.

Phase 5 – Control

In order to maintain the improvements achieved, it is imperative to exercise control over the modified process. This control should be implemented through regular monitoring intervals, ensuring that key variables do not exhibit any unacceptable variation beyond the specified limits.

To effectively exercise control, it is recommended to utilize a range of key control tools. These tools include:

1. **Gantt chart**: A visual representation of the project schedule, displaying the start and end dates of each task, as well as their dependencies. This tool allows for a comprehensive overview of the project timeline, aiding in the identification of potential bottlenecks or delays.
2. **Activity network diagram**: A graphical representation of the project's activities and their interdependencies. This tool provides a clear visualization of the critical path, enabling project managers to prioritize tasks and allocate resources accordingly.
3. **Radar map**: A visual tool that allows for the assessment of multiple performance indicators simultaneously. By plotting various metrics on a radar map, project managers can quickly identify areas of strength and weakness, facilitating targeted improvement efforts.
4. **PDCA cycle**: An iterative four-step management method consisting of Plan–Do–Check–Act. This cycle promotes continuous improvement by encouraging a systematic approach to problem-solving and decision-making.
5. **Tracker milestone diagram**: A graphical representation of project milestones, highlighting key deliverables and their respective deadlines. This tool aids in tracking progress and ensuring that project milestones are achieved within the specified timeframes.
6. **Earned value management**: A technique that integrates project scope, schedule, and cost to assess project performance. By comparing the planned value, earned value, and actual cost, project managers can evaluate the project's progress and make informed decisions to keep it on track.

By employing these key control tools, project managers can confidently exercise control over the modified process, ensuring that improvements are sustained over time.

A3 Methodology

The A3 methodology is an extensively utilized tool in the realm of problem-solving within the Lean philosophy. Originating from Toyota's esteemed Toyota Production System (TPS), it serves as an integral component of their continuous improvement efforts. The name "A3" is derived from the dimensions of the paper format it employs, measuring at 420 × 297mm. This methodology proves invaluable in resolving problems, facilitating thorough cause analysis, and aiding in the development of effective solutions. It should be regarded as a supportive framework for the creation of a PDCA (Plan–Do–Check–Act) plan, aligning with the pursuit of continuous improvement. To ensure comprehensive coverage, this report should encompass a series of meticulously structured sections, including a clear definition of the problem at hand and a detailed description of the current situation. Additionally, a thorough analysis of the causes should be conducted, followed by the identification of appropriate countermeasures. Furthermore, an implementation plan should be meticulously outlined, and finally, the monitoring and verification of the results should be diligently carried out. By adhering to these guidelines, one can effectively utilize the A3 methodology to address and resolve complex problems within their organization.

This tool leads users to follow in a structured way the various steps provided by the scientific method and effective problem-solving. The goal is to guide the project leader in solving the problem using a synthetic and shared communication tool.

From the perspective of a Lean process, A3 is a real tool capable of facilitating the problem-solving process and project management, just as expected in the PDCA cycle. The A3 model actually incorporates the PDCA cycle, as it is expected to be reviewed and updated several times until the problem is solved.

Furthermore, A3 represents an effective means of communication between people. The rigorous and well-defined structure helps the resources working in the company to have a common language and logic.

Finally, given the reduced size of the format (A3), it is an excellent synthesis tool to be used during presentations or work tables. Summarizing a project shows that one has really understood the situation and is extremely effective. It can be compared to the Project Charter used in Six Sigma, which follows the DMAIC method.

The primary objectives of A3 are as follows. Firstly, it is imperative to conduct effective problem-solving. This involves thoroughly analysing the issue at hand, considering all relevant factors, and devising a well-thought-out plan of action to address it. By approaching problem-solving in a systematic and meticulous manner, the team can ensure that no stone is left unturned in their pursuit of a solution.

Secondly, it is crucial to recognize and solve the root causes of the problem. Merely addressing the symptoms or surface-level issues will not lead to a sustainable resolution. By delving deep into the underlying causes, the team can effectively eliminate the source of the problem, preventing its recurrence in the future.

Furthermore, A3 provides a common language for the team to communicate and collaborate effectively. This shared understanding ensures that everyone is on the same page, facilitating seamless communication and minimizing misunderstandings or misinterpretations.

Additionally, A3 serves as a valuable working tool for the team. It provides a structured framework that guides the team through the problem-solving process, ensuring that no essential steps are overlooked. This tool acts as a compass, directing the team towards the most efficient and effective path to resolution.

Moreover, A3 encourages the team to synthesize and present their findings in a clear and concise manner. By distilling complex information into a digestible format, the team can effectively communicate their insights and recommendations to stakeholders, fostering a shared understanding and facilitating informed decision-making.

Furthermore, A3 serves as a means to communicate the project status. By documenting the progress made, challenges encountered, and milestones achieved, the team can keep all relevant parties informed and engaged throughout the problem-solving journey.

Additionally, A3 acts as a guide to solve the problem at hand. It provides a step-by-step roadmap, outlining the necessary actions and considerations at each stage of the process. By adhering to this guidance, the team can cross the complexities of problem-solving clearly.

Lastly, it is essential to emphasize that A3 necessitates following all PDCA (Plan–Do–Check–Act) steps. This iterative approach ensures that the team continuously evaluates and refines their problem-solving efforts, driving continuous improvement and increasing the likelihood of achieving a successful outcome.

In conclusion, by adhering to the objectives of A3, the team can approach problem-solving with a structured and systematic mindset, effectively addressing root causes, communicating effectively, and ultimately achieving resolution.

Employee Involvement Techniques and Problem-Solving

Ishikawa Diagram or Fishbone Diagram

The Ishikawa diagram (named after Kaoru Ishikawa who invented it in 1969) is a tool used to graphically illustrate the major causes and sub-causes of certain phenomena that generate a certain effect or problem. The effective contribution of the Ishikawa Diagram is to be almost the only operational tool for analysing the causal relationships of the phenomena, which makes it also known as the Cause-Effect Diagram.

The diagram was then adopted by Dr W. Edwards Deming as a useful tool for improving Quality. Both Ishikawa and Deming used the diagram as one of the very first tools for managing Quality Systems.

The diagram presented here is based on a fundamental principle, one that asserts that the identification of symptoms is the initial and crucial step towards resolving a problem. It can, therefore, be aptly described as a logical and meticulously structured representation of the existing relationships between an effect and its underlying causes, commonly referred to as the "whys".

This diagram, known as the Ishikawa diagram, owes its creation to the esteemed Kaoru Ishikawa, the visionary behind the concept of Total Quality. Ishikawa firmly believed that when confronted with a particular situation, it is imperative to inquire "why" on no less than four occasions. By doing so, one can effectively discern the genuine causes of the problem at hand.

It is worth noting that the Ishikawa diagram serves as a powerful tool for problem-solving, as it enables individuals to visually analyse the intricate connections between various factors contributing to a particular effect. By employing this diagram, one can meticulously examine the potential causes and identify the root of the problem with utmost precision.

In light of this, it is highly recommended that individuals facing challenges or seeking to address issues utilize the Ishikawa diagram as a means of unravelling the underlying causes. By following Ishikawa's guidance and posing the question "why" four times, one can effectively navigate through the complexities of a problem and arrive at a comprehensive understanding of its true origins.

In conclusion, the Ishikawa diagram, with its logical structure and emphasis on cause-and-effect relationships, stands as a testament to the ingenuity of Kaoru Ishikawa. By employing this powerful tool, individuals can confidently approach problem-solving endeavours, armed with the knowledge and insight necessary to recognize and address the root causes of any given issue. This tool, which is commonly referred to as the cause-and-effect diagram, serves as a valuable resource for identifying the potential causes that give rise to a particular effect. It operates by presenting these causes at varying levels of detail, with each level being represented by interconnected branches that progressively increase in detail as they move away from the central point. It is worth noting that each outer branch is the cause of an inner branch, creating a hierarchical structure. The diagram itself derives its name from the resemblance of its branches to the individual bones of a fish, with the head of the fish symbolizing the core problem. By utilizing this tool, individuals are enabled to address the root of the problem with precision and efficacy. The primary bone, originating from the cranium, extends outward, giving rise to a multitude of secondary bones, commonly referred to as branches. These branches, in turn, contribute to the formation of what is known as the "effect", which occupies a position of utmost importance at the head of the diagram. It is crucial to recognize that the fishbone diagram serves as a comprehensive representation of the potential causes and sub-causes of the aforementioned effect. This diagram finds its most frequent utilization during the measurement and analysis phase of a project. Its versatility is evident in its application within various teams, such as those dedicated to Six Sigma, Total Quality Management (TQM), and continuous improvement. Specifically, it serves as a valuable tool during brainstorming exercises, enabling teams to identify the root causes of a problem and subsequently devise effective solutions. By focusing on causes rather than symptoms, these teams are able to address the underlying issues with precision and efficacy. It is imperative, therefore, that one recognizes the significance of employing the fishbone diagram in order to achieve optimal results in problem-solving endeavours.

The methodology employed to construct this diagram is exceedingly straightforward. In order to proceed, one must adhere to the following steps:

Firstly, it is imperative to make a judicious selection of the problem that is to be subjected to analysis. This decision should be made with utmost care and consideration, as it will serve as the foundation upon which the subsequent steps are built.

Secondly, one must meticulously determine the specific characteristics or effects of the aforementioned problem that are to be examined. To ensure that the analysis is focused on the most salient points, it is highly recommended to employ a Pareto chart. This invaluable tool will aid in prioritizing the effects, thereby facilitating a more efficient and effective analysis.

Thirdly, the chosen effects should be transcribed onto the right-hand side of a sheet of paper, or alternatively, onto a whiteboard or poster. It is of paramount importance to enclose these effects within a rectangular border, thereby creating the head of the fish skeleton. This visual representation will serve as the central point of reference throughout the analysis.

Fourthly, extending from the aforementioned "head", a long line should be drawn. This line will serve as the backbone of the fish skeleton, connecting the head to the subsequent steps of the analysis.

Fifthly, one must diligently identify all conceivable causes that may have contributed to the occurrence of the effect under scrutiny. To accomplish this, various brainstorming techniques can be employed, thereby ensuring a comprehensive exploration of all potential causes.

Sixthly, it is crucial to delve into the intricacies of the analysis in order to ascertain the root cause of the problem. This meticulous examination will enable a deeper understanding of the underlying factors that have led to the manifestation of the effect.

Lastly, it is important to note that the list of causes will invariably encompass primary, secondary, and even tertiary causes. To facilitate a more coherent and organized analysis, it is highly advisable to group these causes in a logical and sequential manner.

By adhering to these meticulously outlined steps, one can confidently navigate the process of constructing a comprehensive and insightful diagram, thereby facilitating a more thorough understanding of the problem at hand.

In order to effectively analyse and categorize causes, it is imperative to employ a systematic approach. Begin by drawing a vertical line from the central line, which serves as the backbone of our diagram. These vertical lines, or "bones", should be alternated above and below the central line. Each bone should be assigned a name, representing a cause or set of causes, and this name should be written at the end of the bone. To ensure clarity and organization, enclose each name within a rectangle.

Once the diagram is complete, it is crucial to thoroughly examine it. Pay close attention to the effects associated with each cause. If certain causes appear to have more significant effects than others, it is advisable to highlight them. This will serve to emphasize their importance and make them more readily apparent to the observer.

By following these steps and employing a meticulous approach, one can effectively analyse and categorize causes, thereby gaining valuable insights into the relationships between various factors.

The process of creating a fishbone diagram is predicated upon a series of logical deductions that ultimately culminate in the identification of the problem and its underlying causes. This methodical approach can be broken down into several distinct steps, each of which plays a critical part in the overall process.

First and foremost, it is imperative to describe the problem. This involves articulating the issue in a concise and precise manner, leaving no room for ambiguity or confusion. By establishing a clear understanding of the problem, one can effectively move forward in the problem-solving process.

Next, it is essential to identify the expectations associated with the problem, as well as the objectives and targets that need to be achieved. This step provides a framework for the subsequent analysis.

In order to gain a comprehensive understanding of the problem, it is necessary to compile a comprehensive list of both known and unknown factors that may influence the issue at hand. This exhaustive list ensures that no potential cause is overlooked and allows for a thorough examination of all possible contributing factors.

Furthermore, it is vital to include the entire team in the problem-solving process. By encouraging active participation from all team members, a diverse range of hypotheses, ideas, suggestions, and questions can be generated. This collaborative approach develops a sense of ownership and collective responsibility, leading to more effective problem-solving outcomes.

Once the data, information, methods, and tools have been gathered, it is imperative to validate their accuracy and reliability. This step ensures that the analysis is based on sound and trustworthy information, thereby enhancing the overall credibility of the problem-solving process.

Finally, the process of creating a fishbone diagram often leads to the acquisition of new knowledge, the formulation of new decisions, and the generation of new ideas. These newfound insights can be instrumental in resolving the problem at hand and should be embraced as valuable contributions to the overall problem-solving endeavour.

In conclusion, the creation of a fishbone diagram is a meticulous and systematic process that needs consideration to details and a comprehensive understanding of the problem. By following the aforementioned steps, one can effectively navigate through the problem-solving process and arrive at well-informed decisions and solutions.

There exist multiple methodologies for interpreting a fishbone diagram. The most expeditious approach is to select the top five causes that have been identified on the diagram and possess a high score. The selection of these causes should be carried out through a democratic process, wherein the working group participants cast their votes or, alternatively, through a method that allows for the assignment of a numerical score. Once the causes have been identified, they should be encircled on the diagram, and the corresponding achieved score should be meticulously inscribed next to each one.

Upon reaching this juncture, the working group is now prepared to embark upon an investigation of the five causes that have been identified as the primary ones.

The fishbone diagram can take different forms, including:

- **Analysis of the dispersion of a phenomenon**, which is used to search for the causes of the quality dispersion of products.
- **Classification diagrams for a production process**, where the main line follows the production process and all factors that can influence quality converge on it in the order they occur. The investigation is facilitated by the fact that each phase is viewed separately.
- **Enumeration of causes diagrams**, where the causes are first listed simply, and then an attempt is made to divide them into homogeneous groups, among which a hierarchical order is established.

A basic diagram template of the fishbone diagram is shown in Figure 4.5.

CEDAC

CEDAC (Cause Effect Diagram with Additional Cards) is a tool for involving operational staff aimed at stimulating and directing the formulation of ideas to solve a particular problem in a correct way.

CEDAC is a specific application of the cause-and-effect diagram. The diagram is physically represented by a board displayed in the office where the problem originates. This board continuously monitors the problem and is accessible to all involved collaborators, who, through self-adhesive cards, signal hypotheses of very specific and targeted solutions and interventions by placing them in correspondence with the specific cause on which they intend to impact. These cards, of different colours, highlight obstacles, real causes, and consequent improvement ideas to overcome the problem. From time to time, experimentation with the different suggestions collected and eventual standardization is carried out.

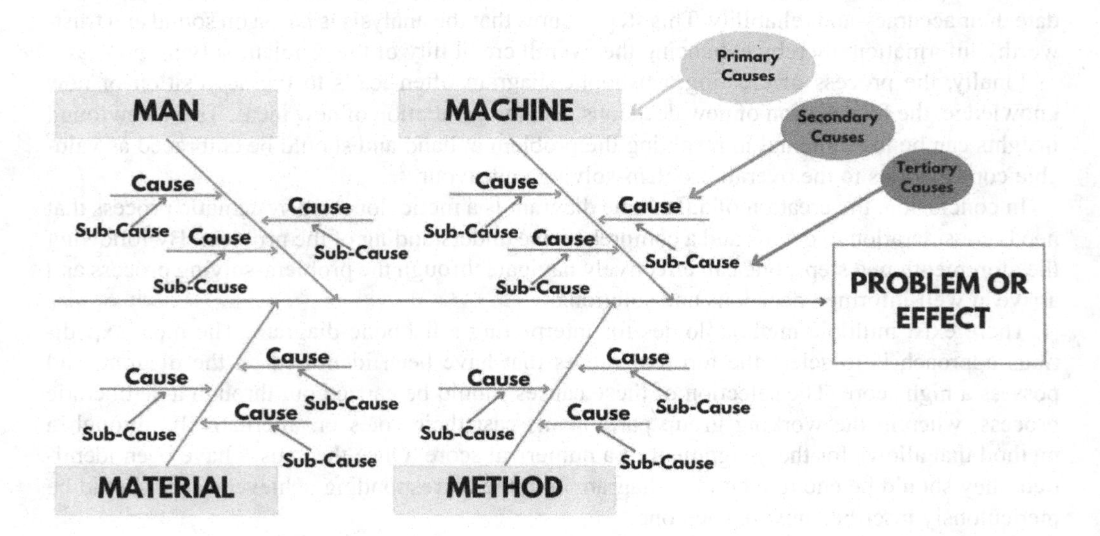

Figure 4.5 Fishbone diagram.

Source: authors' own elaboration

Operationally, the following procedure is followed: the effect (the "waste") is written in a rectangle on the right, and a line is drawn to the left. On this line, all interested staff will apply cards with the causes and any differently coloured cards with proposed solutions. After a certain time, the cards are collected and analysed, and the problem solution is thus set as a result of the work of the entire group, not the ideas of the most authoritative.

Unlike improvement groups, this tool allows identifying solutions to known problems without resorting to frequent meetings.

This investigation can be conducted by employing various other tools that are available within the realm of Quality.

It is important to note that there are numerous variations in the application of the cause-and-effect diagram. However, the two most prevalent types are the 6M diagram and the cause-and-effect diagram assisted by cards (CEDAC). These methodologies offer distinct advantages and can be employed depending on the specific requirements of the situation at hand.

The CEDAC system was ingeniously devised by Ryuji Fukuda during the 1970s. This remarkable system, inspired by Ishikawa's renowned fishbone diagram, was specifically designed to tackle and eradicate any form of "waste" that may plague a company. By waste, we refer to a wide range of undesirable elements such as product or process defects, excessive inventory, wasted time, and inefficiency, among others. The core principle of the CEDAC system is to actively engage and involve all members of the company's staff in this noble endeavour.

To fully comprehend the intricacies of the CEDAC system, it is essential to delve into its fundamental components. Firstly, we have the window analysis, a meticulous examination of the company's operations to identify areas where waste may be lurking. This comprehensive analysis serves as the foundation for the subsequent steps in the CEDAC system. Secondly, we have the window development, a crucial phase where the identified areas of waste are meticulously addressed and refined. This step is of paramount importance as it lays the groundwork for the ultimate elimination of waste. Lastly, we have the CEDAC chart, a visual representation that serves as a roadmap for the entire process. This chart provides a clear and concise overview of the identified areas of waste, the corresponding actions taken, and the progress made.

It is imperative to emphasize the significance of the CEDAC system in the pursuit of operational excellence. By involving all staff members, the system fosters a sense of collective responsibility and ownership, ensuring that no stone is left unturned in the quest to eliminate waste. The CEDAC system empowers individuals at all levels of the organization, thereby generating a culture of continuous improvement.

In conclusion, the CEDAC system is a formidable tool that equips companies with the means to address and eliminate various forms of waste. Its meticulous approach, encompassing window analysis, window development, and the CEDAC chart, ensures a comprehensive and systematic approach to waste reduction. By embracing this system, companies can unlock their full potential and pave the way for enhanced efficiency, productivity, and overall success.

Window Analysis

Window analysis is a highly effective benchmarking technique that is employed to address waste, dysfunction, or critical situations within a department. This method involves comparing the department in question with either a similar department within the same organization or with a department in another company. The analysis is conducted using a meticulously designed matrix that serves to identify the precise nature of the problem at hand. Once the problem has been accurately identified, the next step is to determine whether the method for resolving the issue has already been defined but not yet implemented, or if it remains unknown. In the former scenario, the recommended course of action is to apply window development, while in the latter case, the utilization of a CEDAC chart is advised. By following these prescribed steps, organizations can effectively address and rectify any challenges they may encounter, thereby ensuring optimal performance and success.

Window Development

Window development is an intricate and comprehensive framework encompassing a multitude of control techniques, error analysis, day-to-day management, and visual management. This approach is characterized by its dynamic graphical visualization of data and trends, providing a powerful tool for effective decision-making and oversight.

Control techniques within window development are meticulously designed to ensure optimal performance and efficiency. By implementing these techniques, organizations can effectively monitor and regulate various processes, guaranteeing that they operate within predefined parameters. This level of control allows for the identification and rectification of any deviations or anomalies, thereby minimizing the risk of errors and maximizing overall productivity.

Error analysis is an integral component of window development, as it enables organizations to identify and address any issues or discrepancies that may arise. Through a systematic and thorough examination of data, organizations can pinpoint the root causes of errors and implement appropriate corrective measures. This proactive approach not only minimizes the occurrence of errors but also enhances the overall quality and reliability of the system.

Day-to-day management is a crucial aspect of window development, as it ensures the smooth and efficient operation of the system on a daily basis. This involves the continuous monitoring and evaluation of various parameters, such as performance metrics, resource allocation, and task prioritization. By diligently managing these aspects, organizations can optimize their operations, streamline processes, and achieve their desired outcomes.

Visual management, a key feature of window development, provides organizations with a dynamic and intuitive means of comprehending complex data and trends. This visual representation of data empowers organizations to effectively communicate information, monitor progress, and drive continuous improvement. In conclusion, window development is a comprehensive framework that encompasses a wide range of control techniques, error analysis, day-to-day management, and visual management. By implementing this approach, organizations can enhance their decision-making capabilities, improve operational efficiency, and achieve their desired outcomes. It is imperative for organizations to embrace window development as a powerful tool for effective management and oversight.

CEDAC Chart

The CEDAC chart is a valuable tool for problem-solving in a work team. When faced with an unknown problem, it is crucial to prepare a chart that visually represents the cause-and-effect relationship. This can be achieved by creating a fishbone diagram, where the effect, or "waste", is written in a rectangle on the right side of the chart. A line is then drawn to the left, serving as a pathway for interested personnel to apply cards containing causes and proposed solutions.

To ensure a comprehensive analysis, it is important to involve all members of the work team. Each person should be encouraged to contribute their insights by applying cards with causes and solutions to the designated line. These cards can be of different colours to distinguish between causes and proposed solutions. By allowing each team member to participate, the CEDAC chart fosters a collaborative environment where the problem is addressed as a collective effort, rather than relying solely on the ideas of the most authoritative individuals.

After a designated period of time, the cards are collected and meticulously analysed. This step is crucial in order to identify the root cause of the problem and determine the most effective solutions. By involving the entire group in this process, the CEDAC chart ensures that all perspectives are considered, leading to a more comprehensive and well-rounded problem-solving approach.

In summary, the CEDAC chart is an invaluable tool for problem-solving in a work team. By visually representing the cause-and-effect relationship through a fishbone diagram and involving all team members in the analysis process, it promotes collaboration and ensures a thorough examination of the problem at hand.

It is imperative that we address the matter at hand with utmost seriousness and precision. The suggestion, in its essence, serves as a means to either propose alterations to our current operating methods or to report causes that may only be discernible by those directly involved in the field. To ensure clarity and efficiency, it is of utmost importance that we distinguish the colour coding of suggestion cards into two distinct groups: the suggestions themselves and the causes they stem from.

During our periodic meetings, it is incumbent upon the work team to meticulously analyse the reports we receive. We must delve deep into the intricacies of each report, defining priorities and methods of implementation. It is crucial that we provide direct feedback to those who have a vested interest in the matter at hand. Furthermore, we must quantify the improvements achieved, utilizing graphical representations as they have proven to be the most effective means of conveying information.

In order to facilitate this process, I propose the utilization of a CEDAC diagram. This diagram, prominently displayed in our meeting room, shall feature an empty and clearly visible fishbone structure. Each member of the team shall be required to contribute by posting potential causes and solutions on notes or Post-it, taking into consideration each of the relevant categories.

By adhering to these guidelines, we shall ensure that our discussions are conducted with the utmost professionalism and efficiency. It is my firm belief that by implementing these measures, we shall be able to address any challenges that may arise and achieve the desired outcomes.

A CEDAC diagram that consists of two main formats: the type of dispersion analysis and the type of process classification. The type of dispersion analysis is typically employed after the completion of the 6M or CEDAC diagrams. Once the main causes have been identified, they are treated as separate branches, and their sub-causes are carefully identified. This method allows for a thorough examination of the causes and their underlying factors, ensuring a comprehensive understanding of the problem at hand.

On the other hand, the type of process classification focuses on the main stages of the process rather than the main categories of the cause. This approach is particularly useful when the problem encountered cannot be isolated to a single department. By analysing the process stages, one can gain insights into the overall flow and identify potential areas of improvement or inefficiencies.

It is important to note that both formats of CEDAC serve distinct purposes and can be utilized depending on the nature of the problem being addressed. By employing these tools, organizations can effectively analyse and classify causes, leading to more targeted and efficient problem-solving strategies.

In conclusion, when faced with complex issues, it is recommended to utilize the type of dispersion analysis or the type of process classification within the framework of a CEDAC. These methods provide a systematic approach to understanding and addressing the underlying causes, ultimately leading to improved processes and outcomes.

Quality Circles

The concept of Quality Circles or Quality Control Circles was first introduced in Japan by Dr Ishikawa in April 1962. This groundbreaking idea was presented in the inaugural issue of JUSE's Gemba QC magazine, establishing Dr Ishikawa as a pioneer in the field. Since then, this concept has spread to an impressive number of 130 countries over the course of 36 years. However, it is important to note that the concept of Quality Circles has truly taken root and flourished in ASEAN countries such as Japan, South Korea, China, Taiwan, and others.

A Quality Control Circle (QCC) or Quality Circle (QC) is a small team of individuals who voluntarily come together from the same work area. These dedicated individuals regularly convene to identify, investigate, analyse, and ultimately solve the various work-related problems they encounter. The QCCs operate under a democratic process, fostering a participative management culture within the organization. In line with the QCC philosophy, members actively share their ideas and expertise with management, creating an environment where collective intelligence is harnessed to tackle work-related challenges. By pooling their minds and resources, the members of the circle are able to effectively address and resolve these issues.

It is crucial to recognize the significance of Quality Circles in promoting a culture of continuous improvement within organizations. By encouraging collaboration and open communication, Quality Circles empower employees to actively contribute to the betterment of their work environment. This approach not only enhances problem-solving capabilities but also fosters a sense of ownership and engagement among team members. As a result, organizations that embrace Quality Circles often experience increased productivity, improved quality, and heightened employee satisfaction.

In conclusion, the concept of Quality Circles, pioneered by Dr Ishikawa, has made a profound impact on the global stage. While it has been introduced in numerous countries, it is in ASEAN countries like Japan, South Korea, China, Taiwan, and others where Quality Circles have truly taken hold. By adopting a democratic process and promoting a participative management culture, Quality Circles enable teams to effectively address work-related challenges and drive continuous improvement. It is imperative for organizations to recognize the value of Quality Circles and embrace this approach to unlock the full potential of their workforce.

The circle, in its capacity as a problem-solving entity, diligently presents solutions and, upon receiving the necessary approval, proceeds to implement them. Furthermore, the circle takes on the responsibility of conducting thorough reviews and diligently following up on the implementation process.

The concept of Quality Circles, which we shall now delve into with utmost precision, can be distilled into three key attributes. Firstly, Quality Control is a manifestation of management participation, thereby emphasizing the importance of managerial involvement in the process. Secondly, Quality Control serves as a technique for the development of human resources, highlighting the significance of nurturing and enhancing the capabilities of our workforce. Lastly, Quality Control is a problem-solving technique, underscoring its pivotal role in addressing and resolving issues that may arise.

The foundation of this concept rests upon the notion that suggestions pertaining to the workplace should originate from those who are directly involved in the work itself, possessing an unparalleled depth of knowledge and understanding. It is predicated on the belief that individuals who are intimately acquainted with the problem at hand possess a superior grasp of its intricacies, enabling them to discern viable solutions from those that are not feasible. Moreover, the concept posits that collective collaboration among a group of individuals working in unison will yield superior outcomes compared to the efforts of an individual working in isolation.

In light of these principles, it is incumbent upon us to embrace the concept of Quality Circles and actively engage in its implementation. By doing so, we shall harness the collective wisdom and expertise of our workforce, thereby propelling our organization towards greater heights of success and achievement.

- Quality Circles exemplify a meticulously structured approach to participative management. They are based on the philosophy that granting individuals autonomy and control over decisions impacting them enhances their pride and interest in their work. Consequently, this fosters a deep sense of belongingness towards the organization among employees.
- The concept of Quality Circles is firmly grounded in the principles of intrinsic motivation. By providing employees with greater satisfaction and involvement in the decision-making process, Quality Circles offer them the opportunity to fulfil their higher-order needs. These needs encompass a high-performance orientation and the preservation of individual dignity, among others.
- It is an undeniable truth that every employee possesses an innate desire to actively contribute to better the place of work. By embracing the concept of Quality Circles, organizations can tap into this inherent motivation and harness the collective power of their workforce to drive positive change.
- In conclusion, the implementation of Quality Circles is not only a prudent choice but also a strategic imperative for companies seeking to optimize their performance. By empowering employees with autonomy, involving them in decision-making processes, and satisfying their higher-order needs, organizations can cultivate a culture of excellence and foster a deep sense of commitment among their workforce. It is imperative that organizations recognize the immense value that Quality Circles bring and take decisive action to integrate them into their management practices.
- The recognition of the importance of human resource development is a crucial aspect that cannot be overlooked. It entails the meticulous process of enhancing the skills, abilities, confidence, and creativity of individuals through various means such as education, training, work experience, and active participation to reach improved performances and increased productivity.
- Furthermore, it is imperative to foster an environment that encourages individuals to willingly dedicate their time and efforts towards enhancing organizational performance. By allowing employees to voluntarily contribute to the betterment of the organization, a sense of ownership and commitment is instilled, leading to a more engaged and motivated workforce.

In addition, recognizing the significance of each member's role in achieving organizational objectives is paramount. Quality Circles, a concept that was conceived and implemented with great success, serve as an effective means to facilitate the sharing of responsibilities, knowledge, and the exploration of teamwork. By encouraging collaboration and cooperation among all members of the organization, quality, productivity, and perfection can be achieved.

Finally, it is important to recognize that engaged and respected employees are highly productive. When employees feel valued and respected, they tend to deliver top-quality work. By fostering an environment of belonging and appreciation, organizations can unlock their workforce's full potential, resulting in outstanding results.

In conclusion, the importance of human resource development, voluntary dedication, recognizing individual roles, and fostering employee involvement cannot be overstated. By implementing these principles, organizations can create a thriving and high-performing workforce that is capable of achieving organizational objectives with utmost excellence.

The characteristics of Quality Circles as a management tool for improving quality are as follows:

1. Quality Circles consist of small groups of employees or workers, typically ranging in size from 2 to 3 persons. This limited number of participants ensures that each member can actively contribute and engage in the circle's activities.
2. Membership in a Quality Circle is entirely voluntary. This means that individuals who choose to join a circle do so out of their own free will, demonstrating their commitment and dedication to the improvement of quality within the organization.
3. Each Quality Circle is led by an area supervisor who possesses the necessary skills and knowledge to guide the group effectively. These supervisors are typically selected based on their expertise and experience in the field, ensuring that they can provide valuable insights and guidance to the circle members. Moreover, these circles are centrally coordinated within the organization by a trained facilitator, who plays a crucial role in ensuring the smooth functioning and coordination of the circles.
4. Regular meetings are an integral part of Quality Circles. Circle members convene according to a pre-established schedule, allowing for consistent and structured discussions. This regularity ensures that issues are promptly addressed and progress is made towards the improvement of quality and productivity.
5. Circle members undergo specialized training in problem-solving and analysis techniques. This training equips them with the necessary skills and knowledge to effectively fulfil their role within the circle. By honing their problem-solving abilities, members can identify and address work-related issues with precision and efficiency.
6. The fundamental purpose of Quality Circles is to identify and solve work-related problems in order to enhance quality and productivity. By actively engaging in problem-solving activities, circle members contribute to the overall improvement of the organization. Their efforts are focused on identifying areas for enhancement, implementing effective solutions, and continuously monitoring and evaluating the outcomes. In summary, Quality Circles serve as a valuable management tool for improving quality within an organization. Through their small group structure, voluntary membership, competent leadership, regular meetings, specialized training, and problem-solving focus, these circles empower employees to actively contribute to the enhancement of quality and productivity. By implementing Quality Circles, companies can foster a culture of continuous improvement and achieve long-term success.
7. The Quality Circle (QC) program, which is implemented within the organization, provides an exceptional platform for members to unleash their hidden talents, creative abilities, and skills. By engaging in this program, members are able to tackle challenging tasks, thereby fostering their personal development and growth.
8. Furthermore, the QC program also plays a pivotal role in promoting mutual development among our esteemed members. Through cooperative participation, members are able to collaborate effectively, exchange valuable insights, and collectively enhance their skills and knowledge. This collaborative approach not only strengthens the bonds within the organization but also fosters a sense of unity among the members.

9. The work conducted within the QC program is characterized by a multitude of attributes that greatly contribute to the enrichment of members' professional lives. These attributes include high skill variety, task identity, task significance, autonomy, goal setting, and feedback. By engaging in work that encompasses such attributes, members are able to experience a heightened sense of job enrichment. This, in turn, leads to increased job satisfaction and a greater sense of fulfilment in their roles within our esteemed organization.
10. Moreover, the QC program also plays a pivotal role in enhancing the job satisfaction of our esteemed members. By engaging in work that involves identifying and solving complex problems, members are able to experience a profound sense of accomplishment. This sense of achievement stems from their ability to overcome intricate challenges and contribute to the overall success of our organization. Through the QC program, our esteemed members are empowered to make a tangible impact and derive immense satisfaction from their contributions. The QC program within our esteemed organization provides an exceptional platform for our members to unleash their hidden talents, foster personal development, and contribute to the overall success of our organization. By promoting mutual development, job enrichment, and job satisfaction, this program plays a vital role in nurturing the growth and success of our esteemed members. We strongly encourage all members to actively participate in the QC program and seize the remarkable opportunities it presents. Together, we can achieve greatness and propel our organization to new heights.
11. The organization offers its esteemed members a valuable opportunity to receive public recognition from the management. This recognition is bestowed upon them in the form of a corporate presentation that showcases their exemplary work. The presentation serves as a platform for the members to showcase their achievements and receive acknowledgement from the higher-ups. This recognition not only highlights their contributions but also reinforces their professional standing within the organization.
12. Members of the quality circle are bestowed with well-deserved recognition in the form of cherished mementos, esteemed certificates, and valuable privileges. These tokens of appreciation serve as a testament to their exceptional contributions. Furthermore, in certain instances, these dedicated individuals are also entitled to partake in the fruits of their labour, as they are granted a share of the productivity gains that may arise from their diligent efforts.
13. It is important to acknowledge that the participation in a quality circle not only fosters a sense of accomplishment but also significantly bolsters one's self-esteem and self-confidence. This is primarily achieved through the invaluable acceptance of their recommendations by the esteemed management. By embracing and implementing the suggestions put forth by these esteemed individuals, management not only validates their expertise but also empowers them to further excel in their endeavours.
14. The involvement of circle members in organizational decisions and policies is a pivotal aspect of the quality circle framework. These dedicated individuals actively engage in the identification and resolution of work-related predicaments, thereby playing a crucial role in shaping the course of the organization. Moreover, their recommendations are not merely acknowledged but are also embraced and implemented by the management, further solidifying their significance and impact.

To ensure the effectiveness and efficiency of a quality circle, it is imperative to establish an appropriate organizational structure. While the specific configuration may vary across industries, having a fundamental framework to serve as a model can prove to be immensely beneficial. This framework should encompass the necessary elements that facilitate seamless communication, effective problem-solving, and efficient decision-making within the quality circle. By adhering to this well-defined structure, the quality circle can thrive and fulfil its purpose with utmost proficiency.

In a typical organizational setting, the structure of a quality circle is composed of several key elements that work together harmoniously to achieve the desired outcomes. Allow me to elucidate on these elements in a comprehensive manner.

First and foremost, we have the esteemed steering committee, which occupies the highest echelon of the quality circle structure. This committee, under the able leadership of the General Manager, Works Director, or Manager, assumes the responsibility of overseeing and guiding the entire program. The membership of this committee is carefully curated to include top personnel representatives and human resource development managers, who bring their invaluable expertise to the table. Additionally, it may also include a union representative, ensuring a well-rounded representation of all stakeholders.

The steering committee, being the authoritative body, is entrusted with the task of establishing policies that govern the quality circle program. These policies serve as the guiding principles that steer the program towards success. Furthermore, the committee meticulously plans and directs the program as a whole, leaving no stone unturned in ensuring its smooth functioning. This includes the crucial aspect of organizing personnel training, which is vital for the development and enhancement of the skills necessary for effective quality circle participation. To facilitate effective communication and decision-making, the steering committee convenes on a monthly basis. During these meetings, the committee members deliberate on pertinent matters, exchange valuable insights, and collectively make informed decisions that shape the trajectory of the quality circle program.

In summary, the structure of a quality circle in a typical organization is meticulously designed to foster collaboration, ensure effective leadership, and promote the development of personnel. By adhering to the guidance provided by the steering committee, organizations can pave the way for a successful quality circle program that yields tangible results.

Coordinators, whether they are personnel or administrative officials, play a critical part in the smooth functioning of an organization. Their primary responsibility is to coordinate and supervise the work of facilitators, ensuring that the program runs efficiently. In smaller organizations, typically those with fewer than 500 employees, the role of the coordinator is often combined with that of the administrator and facilitator. However, in larger organizations, particularly those with more than 5,000 employees, corporate and unit-level coordination becomes essential, necessitating the presence of multiple facilitators.

The role of coordinators extends beyond mere supervision. They are instrumental in resolving any difficulties that may arise and facilitating interdepartmental arrangements and communication. Their expertise in managing complex situations and ensuring effective collaboration between different departments is invaluable. It is worth noting that the administrator, who holds a position at the middle management level, is an integral part of the coordination process.

Facilitators, on the other hand, are senior supervisory officials or team leaders. A facilitator is typically capable of handling up to ten circles, showcasing their ability to manage multiple teams simultaneously. It is noteworthy that facilitators are usually selected from three key departments: quality control, production, or training.

In conclusion, the coordination and facilitation roles within an organization are of utmost importance. Coordinators, with their administrative expertise, ensure the smooth operation of the program, while facilitators, with their leadership skills, effectively manage quality circles. By understanding the significance of these roles and their respective responsibilities, organizations can foster an environment of efficiency and collaboration.

Circle leaders, the esteemed individuals who hold the position of circle leader, are an integral part of the quality circle structure. These leaders, often found at the lowest level of supervisors, possess the responsibility of organizing and conducting circle activities. In addition to their own group members, circle leaders engage in interactions with each other and the facilitator. It is worth noting that many organizations initially train supervisors to assume the role of circle leaders. However, as the organization progresses towards developing a quality circle culture, even non-supervisory staff members can effectively take on the mantle of a circle leader.

Let us turn our attention to the circle members, the backbone of the quality circle structure. These members, who can be line employees and/or staff, are absolutely indispensable to the program's existence. They form the largest part of the quality circle structure and serve as the lifeblood of quality circles. It is of utmost importance that circle members make every effort to attend all meetings, as their presence is crucial. Furthermore, they should actively contribute suggestions and ideas, wholeheartedly participate in the group process, and take part in training programs to enhance their skills and knowledge. By adhering to these guidelines, circle members can truly fulfil their role within the quality circle structure.

6M Methodology

This particular methodology is widely employed in the realm of cause-and-effect analysis and has proven to be remarkably effective. It encompasses a comprehensive framework that serves as a catalyst for initiating a brainstorming methodology, with the ultimate goal of identifying the underlying causes that have contributed to a given effect or problem. These causes are further categorized into six distinct "M"s, which are Manpower, Machines, Materials, Methods, Mother Nature, and Measurements.

The six Ms in greater detail are:

1. Manpower: It is imperative to thoroughly assess the technical expertise and experience possessed by the personnel involved. Are they adequately equipped to handle the task at hand? Ensuring that the individuals possess the necessary skills and knowledge is of utmost importance.
2. Machines: It is crucial to meticulously examine the stability and functionality of the equipment being utilized. This includes scrutinizing the precision of devices such as GPS systems, as well as assessing the effectiveness of cooling and lubrication functions. Should the machines exhibit signs of rust or erosion, it is highly likely that production efficiency will be compromised. Consequently, it is imperative to devise solutions for regular maintenance and repair of the equipment to mitigate any potential issues.
3. Materials: It is essential to thoroughly evaluate the components of the materials being employed, paying close attention to their physical and chemical properties. Furthermore, it is imperative to verify that the different parts seamlessly align with one another. This meticulous examination ensures that the materials utilized are of the highest quality and are capable of fulfilling their intended purpose. By adhering to this meticulous and comprehensive methodology, one can effectively identify the causes that have led to a given effect or problem. By thoroughly assessing the six Ms, one can address any potential issues and ensure optimal performance and efficiency.
4. Methodology, methods, or techniques influence the outcome of any action. It is imperative to carefully consider various factors such as workflow, technical parameter selection, technical guidance, precision, and execution of work. These elements collectively contribute to the success or failure of the endeavour at hand.
5. Furthermore, it is essential to take into account the impact of the production environment, often referred to as Mother Nature. Factors such as temperature, humidity, noise, vibration,

lighting, and pollution can significantly affect the quality of products or services. By carefully assessing and managing these environmental variables, one can ensure optimal conditions for production.

6. In the realm of measurements, accuracy is of utmost importance. To obtain precise and reliable results, it is necessary to consider several aspects. These include estimation techniques, measurement methods, calibration procedures, meter fatigue, and result readability. By meticulously addressing these factors, one can enhance the accuracy and validity of the measurements taken.

In summary, by diligently considering the methodology, taking into account the impact of the production environment, and ensuring accurate measurements, one can greatly improve the overall outcome of any action or project. It is crucial to pay attention to these details and implement appropriate measures to achieve the desired results.

In a manufacturing facility, there are several factors that contribute to the occurrence of problems and the subsequent decrease in production. These factors include the presence of individuals, the operation of machines, the availability of materials, the structure of monetary systems, the condition of the machines themselves, and the skills possessed by the operators who carry out the process. These elements are the primary causes of the aforementioned problems.

The consequences of these problems are twofold: low worker productivity and low revenue. These effects are directly linked to the causes mentioned above. To gain a comprehensive understanding of the interrelationships between these causes and their potential consequences, the 6M cause-and-effect diagram is an invaluable tool. This diagram provides a visual representation of the various causes that impact a specific event, making it an essential resource for achieving improved results in any given process. It is worth noting that the 6M cause-and-effect diagram is not limited to a specific field, but rather can be applied across a wide range of industries, including both production and services. Its versatility is due to its ability to identify all possible causes and factors that influence a given process. As a result, it is widely utilized in product design and other areas where a comprehensive understanding of the underlying causes is crucial.

In order to effectively utilize the 6M cause-and-effect diagram, it is imperative to identify and analyse the various causes that contribute to a particular event. By doing so, one can develop appropriate outcomes and subsequently implement corrective measures. This tool serves as a fundamental component of problem-solving, enabling individuals to address issues and improve overall performance.

Key Takeaways

- **PDCA Method (Deming Cycle):** The PDCA method is a fundamental tool for continuous improvement. It involves four key phases: Plan–Do–Check–Act. Its application is crucial in achieving and maintaining high-quality standards.
- **DMAIC Methodology:** DMAIC (Define, Measure, Analyse, Improve, Control) is integral to the Six Sigma approach. It provides a structured framework for problem-solving and process improvement.

Each phase plays a specific role in ensuring the effectiveness of quality management initiatives.

- **A3 Methodology:** The A3 methodology is a concise and visual approach to problem-solving.

It emphasizes the importance of clear communication and documentation in addressing issues.

Its use aids in streamlining decision-making processes.

- **Employee Involvement Techniques and Problem-Solving:** Various techniques exist to involve employees in the improvement process. Engaging employees in problem-solving fosters a culture of continuous improvement. Collaborative efforts lead to innovative solutions and increased ownership of quality outcomes.
- **Ishikawa Diagram (Fishbone Diagram):** The Ishikawa Diagram is a powerful tool for identifying and visualizing root causes of problems. It helps teams understand the relationships between various factors affecting quality. Effective utilization of this tool enhances problem-solving capabilities.
- **CEDAC (Cause-and-Effect Diagram with the Addition of Cards):** CEDAC is a method for visually managing complex processes. It involves using cards to identify and address causes and effects systematically. This approach enhances clarity and efficiency in problem resolution.
- **Quality Circles:** Quality Circles involve small groups of employees working collaboratively on quality improvement projects. These circles promote employee involvement, teamwork, and a shared commitment to quality excellence. The approach encourages continuous learning and adaptation.
- **6M Method:** The 6M Method systematically considers six key factors: Manpower, Machine, Material, Method, Measurement, and Mother Nature (Environment). Comprehensive assessment of these factors contributes to a holistic understanding of quality-related issues. Implementing this method aids in creating well-rounded solutions.

Review Questions

1. Explain the four phases of the PDCA method.
2. How does the PDCA cycle contribute to continuous improvement in quality management?
3. Outline the five phases of the DMAIC methodology.
4. In what ways does DMAIC align with the principles of Six Sigma?
5. What is the A3 methodology, and how is it used in problem-solving?
6. Discuss the advantages of using A3 for communication in quality management.
7. List and describe two employee involvement techniques discussed in the chapter.
8. How does employee involvement contribute to effective problem-solving in quality management?
9. Explain the purpose of the Ishikawa Diagram in quality management.
10. Provide an example of how an Ishikawa Diagram can be used to identify root causes of a quality issue.
11. What is CEDAC, and how does it differ from other cause-and-effect diagrams?
12. Define Quality Circles and their role in quality improvement.
13. Explain the significance of the 6M Method in quality management.

5 Statistical Tools in TQM and Lean Thinking

Learning Objectives

- Understand the fundamental role of statistics in Total Quality Management (TQM) and Lean Thinking methodologies.
- Explain the importance of data collection sheets in capturing relevant information for process improvement.
- Analyse data effectively using histograms to identify patterns and variations within a process.
- Utilize fishbone diagrams as a visual tool to identify potential causes of defects or inefficiencies in a process.
- Apply stratification techniques to segment data and uncover underlying trends or patterns that may not be immediately apparent.
- Demonstrate the ability to integrate statistical tools into TQM and Lean Thinking frameworks to drive continuous improvement initiatives.

The Role of Statistics in TQM and Lean Thinking

Total Quality Management, or TQM, is a method of management that involves all members of the company in order to continuously improve and satisfy customers. Conversely, lean thinking is an organized approach to getting rid of waste and increasing process efficiency. Statistics are a key component of TQM and lean thinking, as they help gauge performance, pinpoint areas for development, and inform data-driven choices. Statistics give organisations important insights into consumer preferences, process variability, and overall quality levels. These insights help them use lean thinking and TQM concepts in an efficient manner for long-term success (Pearce et al., 2023). Some statistical methods and tools that are frequently employed in TQM include fishbone diagrams, which graphically depict the possible causes of an issue or result, and Pareto charts, which assist in determining the most important issues or sources of variance in a process. Regression analysis and other statistical methods can be used to ascertain the link between process variables and outcomes, while value stream mapping and other tools are employed in lean thinking to evaluate process flows and pinpoint wasteful regions. Organisations need to use these statistical tools and approaches because they offer insightful analysis and data-driven decision-making. With the help of these technologies, businesses may cut waste, find and solve the core causes of issues, and continuously enhance their operations to achieve

DOI: 10.4324/9781032726748-5

long-term success. Furthermore, the application of lean thinking and statistical techniques in TQM fosters an environment in which learning and development never stop within the company (Hoerl and Snee, 2010).

Statistics, although it is not easy to define it, is properly the application of scientific methods to the programming of data collection, their classification, processing, analysis, synthesis, and presentation, as well as the inference of reliable conclusions from them. In practice, it aims at identifying the most suitable logical-mathematical methods for discovering laws and relationships between phenomena for descriptive, explanatory, predictive purposes, etc. Its goal is the quantitative and qualitative study of collective phenomena (demographic, economic, social, health, etc.), that is, of repeatable individual situations with common characteristics, identifying the regularities that underlie such collective phenomenon.

Within the statistical methodology, there are two fundamental branches: descriptive statistics, which aims to describe, through mathematical means, a real phenomenon by conducting a study on the population (or universe) or on a representative part of it, called a sample, in which the phenomenon under examination manifests itself. It studies the criteria for detecting, classifying, and synthesizing information relating to the reference aggregates. It collects such information in distributions, simple or multiple, represents them graphically, and achieves a synthesis through indexes of the obtained results.

Inferential statistics, which aims at the probabilistic induction of the unknown structure of a population. It deals with solving the so-called inverse problem, that is, based on observations made on a sample of units selected with specific procedures from the population, it arrives at valid conclusions, within given levels of probability of error, for the entire population itself. The theory of probability and statistical sampling, which uses the theory of samples, are the basis of inferential statistics.

Statistical Process Control as a Key Tool in TQM

Businesses may monitor and manage their processes in real-time with statistical process control (SPC), ensuring that they are functioning within reasonable bounds and delivering reliable, high-quality goods and services. SPC assists in locating any discrepancies, allowing companies to act quickly to address the issue and avoid further errors or defects. Furthermore, statistics are essential for measuring and assessing data so that decisions may be made with knowledge and to pinpoint areas that need development. Businesses can evaluate the success of their quality management initiatives and locate any possible bottlenecks or inefficiencies in their operations by employing statistical techniques like regression analysis and hypothesis testing (Newbold et al., 2022). This enables businesses to consistently enhance and optimize their operations, guaranteeing sustained prosperity and contentment from their clientele. Moreover, companies can ensure consistency and dependability in their goods and services by standardising processes and procedures through the implementation of a strong quality management system. In addition, data analysis enables businesses to recognize patterns and trends in consumer behaviour, which enables them to better customize their goods and services to suit the requirements and tastes of their target market. Businesses may increase customer satisfaction and obtain a competitive advantage in the market by knowing their preferences. Furthermore, data analysis can help businesses see possible dangers or problems before they become serious ones, enabling proactive steps to be made to reduce these risks and guarantee smooth operations. This not only increases client confidence but also lowers the possibility that mistakes or flaws will arise in the first place. Businesses may proactively identify areas that need attention and take the required steps to prevent any quality concerns from emerging by regularly monitoring and assessing key performance indicators. In the end, this proactive strategy aids companies in maintaining a competitive advantage in the marketplace and solidifying their image for providing high-quality products or services.

Statistical Tools: Value Stream Mapping, Statistical Process Improvement Techniques in Lean Thinking, and Six Sigma

Statistical Process Control (SPC) and value stream mapping (VSM) have the potential to enhance the extent to which data analysis improves business processes. Businesses can visually map out every step of their process flow with VSM, pinpointing inefficiencies and waste that may be improved. SPC, on the other hand, assists businesses in keeping an eye on and managing process variability to guarantee constant quality and fewer faults. Through the integration of statistical tools and data analysis, organisations can effectively pinpoint opportunities for enhancement and execute efficacious tactics aimed at streamlining their operations and augmenting productivity. An organisation can also foster continual innovation and efficiency by implementing the Japanese concept of kaizen, which is continuous improvement. Companies may develop a culture of continuous improvement that results in long-term success by encouraging staff members to look for ways to improve procedures and get rid of waste. Employers can also empower their staff members and give them a feeling of ownership by using Kaizen concepts, which will boost output and improve overall business success. It can increase the efficiency of data analysis by pinpointing areas in need of development and cutting down on waste. These technologies offer a methodical way to find bottlenecks, cut down on errors, and boost overall productivity. Furthermore, businesses may analyse and assess process variances with the aid of methodologies such as Six Sigma, which enable them to pinpoint the underlying causes of faults and put remedial measures in place. Six Sigma uses a structured problem-solving methodology called DMAIC (Define, Measure, Analyse, Improve, and Control) to help businesses along the path of process improvement. It offers a methodical way to set objectives for the project, assess present performance, analyse data, find areas for development, and put control mechanisms in place to maintain the gains made. This all-encompassing strategy guarantees that businesses have a clear roadmap to follow in order to achieve their desired goals, in addition to identifying opportunities for development. It can be applied to pinpoint inefficient and wasteful areas in an organisation's operations. This data-driven approach makes targeted improvements and cost savings possible, which eventually boosts profitability. Businesses can achieve greater levels of operational efficiency and quality control, which will ultimately result in better customer satisfaction and a competitive edge, by implementing these statistical techniques into their data analysis methods. Additionally, data analysis enables businesses to track key performance indicators (KPIs) and keep tabs on their advancement towards objectives. This gives businesses the ability to make data-driven decisions and modify their strategy accordingly, guaranteeing long-term success and ongoing growth. Furthermore, data analysis can offer insightful information about the behaviour and preferences of customers, enabling businesses to better customize their marketing campaigns and product offers to the demands of their target market. Furthermore, businesses may obtain important insights into future trends and customer behaviour by utilising advanced analytics techniques like predictive modelling and machine learning. This helps them remain ahead of the competition and make proactive business decisions. Through the integration of statistical tools and data analysis, organisations can achieve cost savings and greater productivity by optimising their resources and streamlining operations.

Data Collection Sheet

It constitutes the starting point for the study of a specific phenomenon:

- establishing the characteristics to be observed;
- choosing the sample size to be observed;
- recording the result on specially made sheets.

The information collected can belong to different categories:

- continuous measurement data (e.g., lengths, weights, or diameters);
- countable and therefore discrete data (e.g., quantity, number of defective pieces);
- data on relative judgment or priority;
- data resulting from sequential ordering;
- data resulting from score scales.

For a correct data collection, it is necessary to:

- define the objective to be pursued accurately;
- use the data recording method that is least subject to errors and that allows for the most complete and aggregable collection or other types of processing.

Any decisions on intervention on the analysed phenomenon should be based exclusively on what has been collected. This constitutes the starting point for the study of a specific phenomenon:

- establishing the characteristics to observe;
- choosing the sample size to observe;
- recording the results on specifically designed sheets.

Data Collection Plan / Matrix

Process

Process Owner: ______________
Contact Info: ______________
Location: ______________
Area: ______________

Prepared by: ______________
Approved by: ______________
Approved by: ______________
Approved by: ______________

Page __ of __
Revision: ____

Process Step	QTC		Metric	Data Type	Operational Definition	Specification		How measured and collected	Sample Size	Frequencies	Deliverables	Location of Data	Format of reporting	Standard operating procedure
	KPIV	KPOV				USL	LSL							

Figure 5.1 Data collection sheet.

Source: authors' own elaboration

Therefore, the data collection sheet is a structured form designed specifically to enter the collected data and facilitate analysis. Its great usefulness is due to the fact that it is very generic and, therefore, suitable for use for many purposes. It is the first of the tools in chronological order and is the starting point for any quantitative analysis. When designing a data collection sheet, it is important to clearly define what type of information needs to be collected. It is therefore necessary to consider what the data will be used for, to ensure that all necessary information is collected. In some cases, a testing period may be useful to verify the effectiveness of the sheet.

Header

The header is essential to avoid rendering the data collection sheet unusable in case, for example, changes are made to the production process. It should answer the following questions:

1. Who filled out the data collection sheet.
2. When the collection was made.
3. What was collected (legend of the symbols used).
4. Where the collection was made.
5. Why the data was collected.

It should contain the date, the operator who made the collection and any other notes that can make the data source more understandable. It is necessary to clearly annotate the source of the data. It is therefore useful to also annotate within the header:

1. purpose and characteristics of the measurement;
2. operating machine and equipment;
3. control tools;
4. measurement method.

Purpose

The purpose of data collection is not to measure everything in quantitative terms, but to provide a basis for decision making and action. The data can take any form, not necessarily numerical. Data can be divided into the following types:

- Measurement data (continuous): length, weight, etc.
- Counting data (enumerable): number of defects, etc.
- Relative merit data: judgments.
- Classification sequence data: ordinal numbers.
- Merit score data: grades.

It is not sufficient to take into account the purpose for which the data is collected: it is necessary to evaluate its validity (they must truly represent the facts, and the statistical methods used must allow for objective evaluation). The data must also be collected in a simple way through a form that is easy to complete (they must have suitable spaces for simple and concise data collection).

Designing the Data Collection Sheet

The correct design of the data collection sheet is critical, as it influences:

- the speed of collection;
- data validity (represent the facts);
- data significance (possibility of useful analysis);
- ease of analysis (the data is already in the format I need).

Data collection sheets are initially designed based on the objective of data collection, and then necessary verifications are made to facilitate recording and meet the objectives of the investigation. Proper information recording begins with the identification of the source, without which any information loses validity. The designer of the data collection sheet takes inspiration from five "base" models defined by Kaoru Ishikawa and adapts them to their own needs. These five types of data collection sheets are:

(a) analysis of the distribution of a production process;
(b) data collection sheet by type of defect;
(c) data collection sheet for physical location of defects;
(d) data collection sheet for defect causes;
(e) data collection sheet of the "checklist" type.

Each of these is described below.

Analysis of the Distribution of a Production Process

It is a very simple type of data sheet, in which data is recorded at the time of collection. The result obtained is similar to a histogram with a peak corresponding to the average of the measurements. A major disadvantage of this type of data collection sheet is the loss of temporal information: it is not possible to know which of the identified measurements was identified first compared to the others. Therefore, it does not show the variations that occur over time. The products of the production process are converted into data, which are in turn divided into classes. When the observation period ends, the spreadsheet is examined. In particular, it is desired to know whether the data is normally distributed or if there is asymmetry or bimodal distribution, if there are anomalous data, if the specification limits are respected, and how many pieces are outside the specifications. The spreadsheet allows intervention in case there is a non-normal distribution or in case most of the production is outside the specification limits, in order to bring the process back under control.

Data Collection Sheet by Type of Defect

The data collection sheet by type of defect allows for stratification of data during processing but requires that the recording methods be defined in advance. It allows for the immediate calculation of the total number of defects, the types of defects encountered, and their percentages. A list of the main types of defects is defined, but this represents a potential problem: there may be a missing category or a category that is too vague. For example, a defect of the type "imperfection" could contain completely different types of defects. In this case as well, temporal information is missing,

Project Sheet : Work 1

Process Sheet	Approved By	Part# / Part Name	Application / Number of Parts		Entered By	Date / Line	
Step	Step Name	Machine#	Basic Time		Tool Change	Processing Capacity Shift	Remarks
			Manual	Auto	Completion Change Time		
		Total					

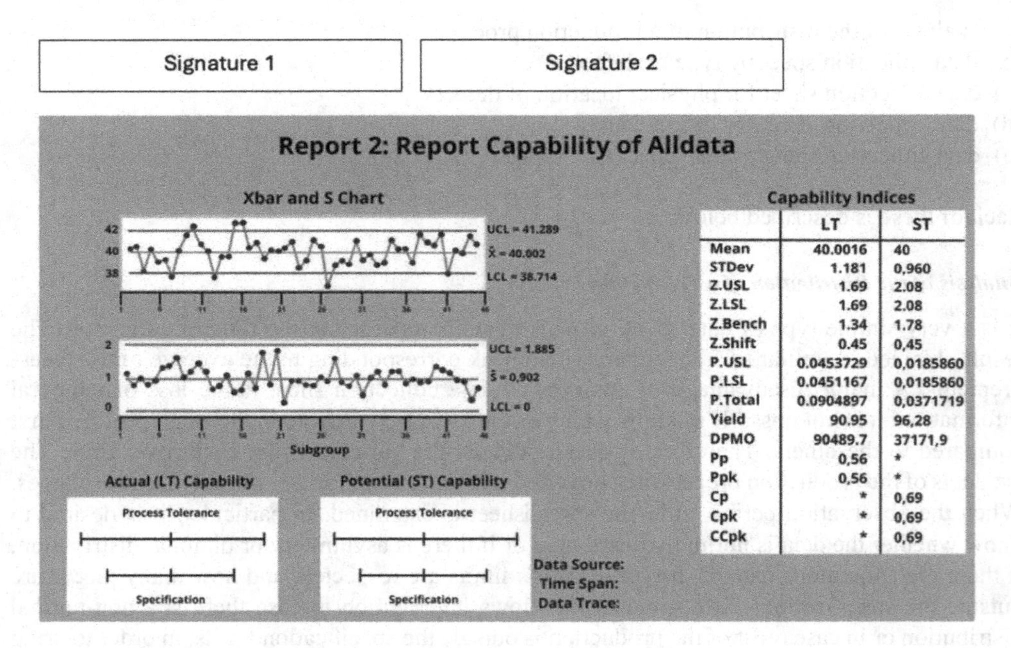

Figure 5.2 Production process sheet and example of a process capability report.

Source: authors' own elaboration

and it is therefore not possible to know the variations that have occurred over time. In addition, personnel must be trained in the correct completion of the data collection sheet.

Data Collection Sheet for Physical Location of Defects

The data collection sheet for physical location of defects is particularly important in quality management and is sometimes considered a separate tool. It shows drawings or profiles of the product, on which the location of the defect is recorded. It is very useful for making quick

Project Name: ____________________
Name of Data Recorder: ____________________
Location: ____________________
Data Collection Dates: ____________________

Defect Types / Event Occurrence	Dates							TOTAL
	Sunday	Monday	Tuesday	Wednesday	Thursday	Friday	Saturday	
Defect 1	卌 I	卌 卌	卌 卌 卌 II					
Defect 2	卌	卌 II	卌 卌					
Defect 3	卌 I	卌 I	III					
Defect 4								
Defect 5								
Defect 6								
Defect 7								
Defect 8								
Defect 9								
Defect 10								
TOTAL								

Figure 5.3 Type of defect sheet.
Source: authors' own elaboration

Location		Product	Process Name
Quality Inspector Name		Signature	Manager - Quality Control
Order No.	Job Quantity	Sample Quantity	Online Checking
Name of instrument used for Defects checking		Method used for Defect Checking	

Sr	Defect Type & Description	Category of Defect	Signature of Inspector

Defect Type - 01	Found Qty:	Category of Defect:

Reason for Defects

Defect Type - 02	Found Qty:	Category of Defect:

Reason for Defects

Figure 5.4 Physical location of defects sheet.
Source: authors' own elaboration

modifications to the production system in case a defect is frequently found in the same physical location. For example, a piece obtained by casting that frequently has a hole in the thermal centre of gravity indicates an error in sizing or positioning of the moulds. It is therefore an important tool for process analysis and facilitates the identification of causes of variability.

Defect Sheet No.	F01	Location			
Photo/Skethces:		Element			
		Component			
		Findings			
		Condition	Priority	Matrix	Colour
		Diagnosis:			
		Prognosis:			
		Possible Causes:			
		Category	/	Remedies:	
		Design		1.	
		Maintenance			
		Construction Material			
		Workmanship			
		Environment			
		Human Usage			
		Insect			
		Disaster			

Figure 5.5 Defect causes sheet.
Source: authors' own elaboration

PROJECT NAME: PROJECT MANAGER:

TASK COMPLETE (?)	PRIORITY	STATUS	TASK/DELIVERABLE	ASSIGNED TO	DATE DUE	NOTES
			PHASE 1			Running ahead of Schedule
No	Low	Overdue	Task 1		00/00/00	
No	Medium	Not Strarted	Task 2		00/00/00	
Yes	High	Complete	~~Task 3~~		00/00/00	
No	Medium	Needs Review	Task 4		00/00/00	Awaiting manager approval
No	High	Approved	Task 5		00/00/00	
No	Low	On Hold	Task 6		00/00/00	
			PHASE 2			
Yes	Medium	Complete	~~Task 1~~		00/00/00	
No	Low	In Progress	Task 2		00/00/00	
			Task 3			
			Task 4			
			Task 5			
			Task 6			
			PHASE 3			
			Task 1			
			Task 2			
			Task 3			
			Task 4			
			Task 5			
			Task 6			

Figure 5.6 Checklist sheet.
Source: authors' own elaboration

Data Collection Sheet for Defect Causes

The data collection sheet for defect causes has a structure that makes the correspondences between causes and effects clear, in order to subsequently identify which causes are the most important to address. In some respects, it is similar to the data collection sheet for defect types but, unlike the latter, it also provides temporal information. The main disadvantage lies in the difficulty of compilation and the identification of a series of symbols representing all the types of defects that may be encountered. As with the data collection sheet for defect types, there is therefore the problem of identifying the causes of the defects found, and the possibility that necessary entries may be missing or that there may be vague entries. In addition, it is assumed that the operator is able to recognize the causes.

Data Collection Sheet of the "Checklist" Type

The data collection sheet of the checklist type has the main advantage of simplicity: it is not possible for the operator to make a mistake in compiling it, as it is completely guided by the sheet itself. It is also particularly convenient for subsequent consultation: in a complex product, if a problem is found with a specific piece, it is possible to consult only the section dedicated to that piece. The checklist is compiled through each individual phase of the process. It also ensures that all checks are performed without neglecting any.

Histogram

A histogram is based on the following basic concepts:

- **Class**: the size of an interval of variability of the data that is used as a basis for representing the data itself.
- **Frequency**: the number of elements that are included in a certain class.
- **Dispersion or range**: the size of the interval between the maximum and minimum values.

The construction of a histogram involves several fundamental phases:

- Phase 1: The maximum value (M) and minimum value (m) are identified from the data table, and the range (R) is calculated as $R = M - m$.
- Phase 2: The number of classes (K) is determined based on the number of data points (N), typically following the criterion: $K \approx \sqrt{N}$.
- Phase 3: The width of each class (H) is determined by dividing the range (R) by the number of classes (K): $H = R \div K$.
- Phase 4: The class limits are defined starting from the minimum value: lower limit $= m$, upper limit $= m + H$. Subsequent class limits are identified by adding the class width value each time. Values falling on the class limits are recorded in the higher class.
- Phase 5: The frequency table is prepared, and the data are recorded.
- Phase 6: The histogram is drawn.

A histogram displays the way values are distributed for a variable. Histograms are helpful, in representing the centre, range and pattern of a dataset. They can also serve as an aid to determine if the data follows a distribution. Histograms are among the tools used in quality control. They provide a means to analyse data allowing us to gain insights into its distribution and identify

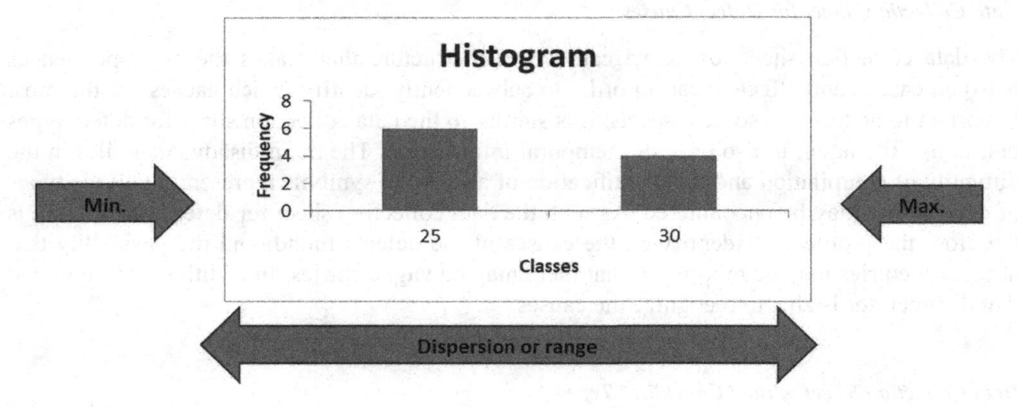

Figure 5.7 Example of histogram with the indication of the parameters adopted for its creation.

Source: authors' own elaboration

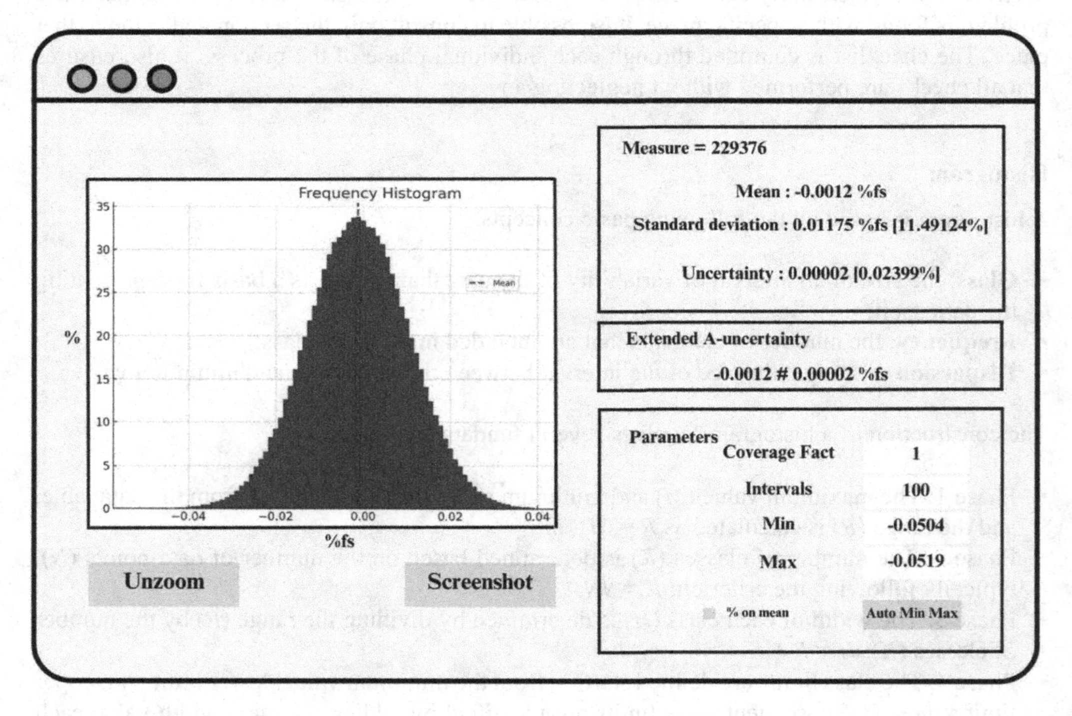

Figure 5.8 Histogram.

Source: authors' own elaboration

any outliers. Understanding how data is distributed is crucial, in selecting statistical analysis techniques. Ultimately histograms provide a representation of the data shape. The horizontal axis in both histograms and bar charts represents the data values, with each bar encompassing a specific range of values. On the other hand, the vertical axis indicates the frequency or count of data points falling within the specified range of each bar.

The primary distinction between histograms and bar charts lies in the type of data they represent. Histograms are employed when dealing with continuous data, whereas bar charts are utilized for categorical or nominal data. Notably, histograms do not have spaces between bars, as they depict the number of values falling within a given range on the horizontal axis. In contrast, bar charts may have spaces between bars, with each bar representing measured values for individual categories.

Pareto Diagram

A Pareto diagram is a kind of chart that combines a line graph and bars, with the line showing the cumulative total and the bars representing individual values arranged in descending order. The purpose of this kind of graphic is to pinpoint the main causes of a specific result or issue.

It is a graphical methodology designed to help identify the most important issues to address for improvement interventions. It is based on the assumption that typically only 20% of the problems that arise are significant, and by resolving this percentage, 80% of the previously encountered difficulties are eliminated. The graphical representation used is that of a histogram where the categories on the horizontal axis, measured with the same unit of measurement, are ordered by decreasing y-value. The graph overlays the curve of cumulative values, obtained by adding the frequency value of each category to the frequencies of all preceding categories.

A Pareto diagram is used to graphically prioritize and highlight the key elements that account for the majority of difficulties or problems. Organisations can decide where to deploy resources for most impact by identifying these critical variables. They can efficiently allocate resources by concentrating on the elements that have the greatest impact when using Pareto diagrams for issue solving. By doing so, decision-making processes become more effective and efficient overall by addressing the underlying causes of issues.

The Pareto diagram can be utilized at all stages of a process improvement and applied to all its aspects. Typically, it represents the initial step by identifying priority areas for intervention, drawing attention to these priorities, and visualizing the objectives to be achieved. Similarly, the analysis allows for evaluating the effectiveness of the improvement intervention just carried out.

The Pareto chart is attributed to Vilfredo Pareto and his principle is also known as the "80/20 rule". This rule suggests that 20% of the individuals hold 80% of the wealth; or 20% of the product line might account for 80% of the waste; or 20% of the customers could be responsible for 80% of the complaints, and so forth. Pareto mathematically established that, although numerous factors influence a specific result, only a handful have the potential to substantially alter that outcome. Approximately 80% of consequences or results stem from 20% of the causes or inputs.

Put succinctly, a small number of elements drive the majority of outcomes within a particular scenario, system, or entity.

For example, you might observe that 20% of the food you eat packs on 80% of the calories. Twenty per cent of your monthly bills consume 80% of your income.

Thus, when investigating the underlying reasons for errors, mistakes, and defects in your business operations, you'll frequently find that 80% of the issues stem from just 20% of the identified causes, as depicted in the graph below. By addressing the primary culprits responsible for the most frequent occurrences, you will observe a substantial enhancement in the business process.

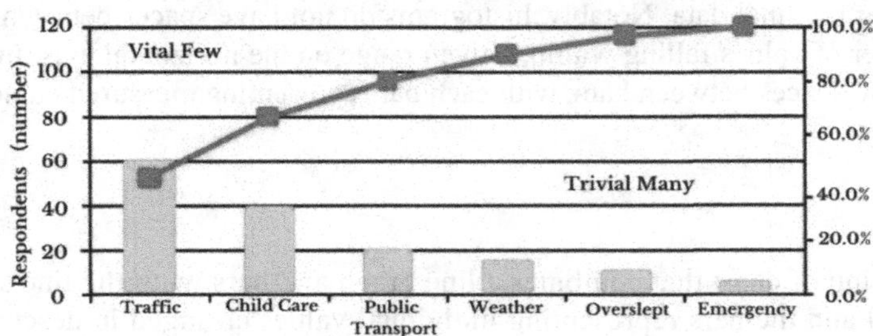

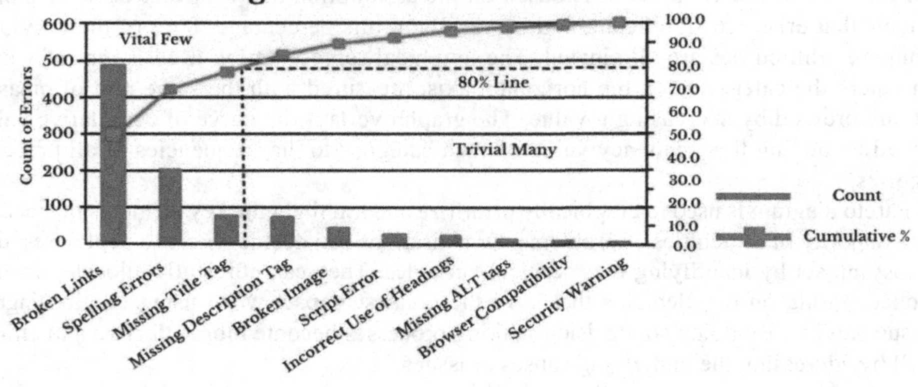

Figure 5.9 Examples of Pareto Diagram.

Source: authors' own elaboration

The procedure for making a Pareto Diagram is given below:

A. Determine which issue or problem needs to be examined.
B. Compile information on each factor's influence or frequency.
C. Determine the cumulative percentage and rank the issues in order of importance.

All things considered, Pareto diagrams are an effective tool for locating and resolving the most important problems that could seriously affect performance as a whole. Organisations can maximize efficiency and expedite time to market by concentrating on the critical few instead of the pointless many. Pareto diagrams can be used to solve problems more effectively and produce better results in a variety of industries.

Stratification

Stratification is a valuable technique used in the analysis of a phenomenon. It involves breaking down and reassembling data based on specific characteristics. The process entails classifying a dataset into different categories and subcategories, using chosen criteria. During this process, it is common for defects or non-conformities to coexist with other factors in a cause-effect diagram. However, pinpointing the exact domain where these causes occur and contribute to the resolution of the final problem can be quite challenging. To identify patterns at this stage, various data stratification methods are necessary. These methods involve grouping all items in a dataset that share common characteristics within a single layer. Subsequently, the dataset as a whole is divided into subsets, allowing for comparisons to be made between them.

Sometimes it can be useful to analyse a phenomenon not in its entirety, but in relation to multiple categories, to see how each of them influences the overall trend. Therefore, stratification is used to obtain this information. To implement it, it is necessary first of all to establish the characteristics according to which you want to disaggregate the phenomenon and then reaggregate the data according to their belonging to these characteristics; for each category, the corresponding graph will be constructed, whether it is a histogram, a correlation diagram, a Pareto diagram, or any other more appropriate one. The term stratification means to arrange something into groups. The larger the group, the more data entries will exist in the sample for that group. To find a stratified sample we need to know how many data entries are in each subgroup and the total sample size. The population is divided into smaller subgroups (strata) with the number taken from each subgroup proportional the size of the subgroup. Stratified sampling determines the number of items of data in each subgroup and so it requires a secondary sampling method to select the individual items of data. This is usually through using a simple random sampling technique (using a random number generator). This is why a stratified sample can also be called a stratified random sample. Table 5.1 summarizes stratification.

Table 5.1 Stratified sample explanation.

Sampling method	*Description*	*Example*
Stratified sampling	Smaller groups or strata within the sample are represented proportionally to the population	Finding out a favourite cake from different age categories of people in a metropolis

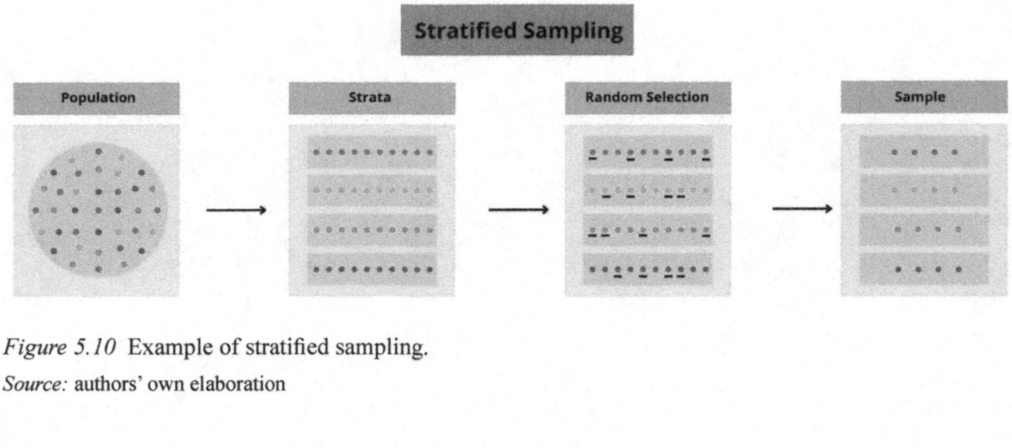

Stratified Sampling

Population → Strata → Random Selection → Sample

Figure 5.10 Example of stratified sampling.

Source: authors' own elaboration

Key Takeaways

- **Critical role of statistics**: Statistics play a pivotal role in both Total Quality Management (TQM) and Lean Thinking approaches, serving as the foundation for data-driven decision-making and continuous improvement.
- **Data collection sheet**: A structured approach to collecting data is essential for informed decision-making. Data collection sheets provide a systematic way to gather relevant information about processes, allowing for accurate analysis and identification of areas for improvement.
- **Histogram analysis**: Histograms offer a visual representation of data distribution, enabling practitioners to identify patterns, trends, and variations within a process. Understanding these distributions is crucial for pinpointing areas of inefficiency or opportunities for optimization.
- **Fishbone diagrams for root cause analysis**: Fishbone diagrams, also known as Ishikawa or cause-and-effect diagrams, are valuable tools for identifying potential causes of problems within a process. By categorizing potential influences into major branches, teams can systematically investigate root causes and develop targeted solutions.
- **Stratification techniques**: Stratification involves organizing data into meaningful subgroups to uncover underlying trends or patterns that may be obscured when analysing the data as a whole. This technique enables practitioners to identify specific factors contributing to variation and address them more effectively.

Review Questions

1. Why is statistical analysis crucial in both Total Quality Management (TQM) and Lean Thinking methodologies?
2. What is the purpose of a data collection sheet in process improvement initiatives?
3. How does using a structured data collection approach improve the accuracy and effectiveness of data analysis?
4. Describe how histograms are used in process analysis within TQM and Lean Thinking.
5. What is a fishbone diagram, and how is it utilized in root cause analysis?
6. What is the purpose of stratification in statistical analysis?

6 Management Tools in TQM and Lean Thinking

Learning Objectives

- Understand the basic characteristics of management tools used in Total Quality Management (TQM) and Lean Thinking.
- Learn how to create and utilize an Affinity Diagram to organize and categorize ideas or data effectively.
- Comprehend the concept and application of a Diagram of Relationships in visualizing connections between various elements.
- Master the use of a Tree Diagram to represent hierarchies and relationships within a system or process.
- Explore the purpose and methodology of a Matrix Diagram for analysing relationships between multiple factors.
- Learn how to construct and interpret a Data Matrix for organizing and presenting complex data sets efficiently.
- Understand the significance of a PDPC (Process Decision Program Chart) in identifying potential risks and developing contingency plans.
- Gain proficiency in creating an Arrow Diagram to illustrate the sequence of activities in a project or process.
- Master the use of PERT (Program Evaluation and Review Technique) Charts for scheduling and managing complex projects effectively.
- Develop skills in utilizing GANTT Charts for visualizing project timelines, dependencies, and progress tracking.
- Compare and contrast PERT and GANTT Charts to understand their respective strengths, limitations, and suitable applications in project management contexts.

Basic Characteristics

The work carried out with statistical tools has a reactive nature, that is, it reacts to unfavourable situations, with the aim of satisfying the customer (including the internal customer).

This improvement activity, intended as the removal of causes of problems, is insufficient for several reasons:

- Complaints and grievances are only the tip of the iceberg of customer dissatisfaction.
- Only a part of quality is addressed, which is called "negative quality" (deviation between the current situation and an ideal reference situation).

DOI: 10.4324/9781032726748-6

- The reality of many processes and business activities clearly shows that expressing them only through numbers, although necessary for an initial assessment of problems and areas of intervention, does not provide all the necessary elements for a correct and deep analysis.

Therefore, appropriate tools are needed to address problems that require a project/quality approach.

These tools are mainly aimed at the company's managers. They are the result of about five years of study by a Committee of the JUSE (Japanese Union of Scientists and Engineers), the Japanese organization that since 1946 has promoted and led the great Quality movement in this country.

The JUSE Committee, led by Professor Yoshinobu Natayami and called the Committee for Developing QC Tools, was established in 1972 and completed its work at the end of 1976. The Committee for Developing QC Tools considered some objectives essential for the problem-solving process. These objectives can be divided into two groups:

1. **Related to the problem itself:**
   a. Understand the real scope of the problem.
   b. Highlight the "processes" for solving the problems.
   c. Focus on the important points of the problem.
   d. Prevent omissions in studying the problem.
   e. Facilitate the preparation of effective programs to complete the problem.
2. **Related to those who study the problem:**
   a. Obtain creative ideas.
   b. Structure and integrate verbal expressions.
   c. Encourage systematic thinking.
   d. Enlighten the people involved and obtain their cooperation.
   e. Appeal to the users of these tools for their simplicity.

The main characteristics of these tools are as follows:

1. They are simple. With the exception of data matrix analysis, the Seven Management Tools are fundamentally simple and do not require special knowledge to be used. They can be taught in a limited number of hours, from 12 to 32. Their knowledge can thus be extended to all company managers.
2. They constitute an integrated package. The tools should be considered as an integrated package, in the sense that they lend themselves to combined use when faced with a problem. This type of use is one of the most important advantages of these tools and gives the package considerable effectiveness. Isolated and occasional use of only some tools is very limiting.
3. The application procedure is standard. The users of these tools are facilitated in their use, as the application procedure is sufficiently standardized. The use of the tool is carried out in predetermined phases, and there is therefore a track to follow. This makes group work easier because the process to follow is known to all.
4. They largely use graphic language.
5. They allow to "process" verbal information. Verbal information is translated into graphs and diagrams so that all interested parties have easy access to it. This way, even complex phenomena can be objectively framed, making it easier to process information to solve the analysed problem satisfactorily.

6. They force you to "think with your hands". With the Seven Management Tools, mental activity takes shape through the construction of charts, diagrams, and schemes. This "manual" work spontaneously activates the minds of the people involved and facilitates the search for missing elements for a more complete description of the problem and its solution.

The seven management tools can be considered:

- Semantic tools, as they help to better understand the meaning of available verbal information;
- Creative tools, in the sense that they help to restructure available verbal information (restructuring information is the mechanism underlying the creative process);
- Problem-solving tools, because they were specifically designed for problem definition and solution.

The application fields, as well as those of the seven statistical tools, are extremely wide and diverse. In particular, their contribution is essential in the following fields:

- Improvement activities in business areas;
- Analysis of information coming from the outside world of the company;
- Processes for the development of new products;
- Topics where it is necessary to operate and reason in terms of very low defect rates (parts per million).

The Seven Management Tools can be divided into three groups, corresponding to the three major phases that cover the entire process from identifying the problem to the necessary steps to arrive at its solution.

- **First group**: Analysing and formulating a problem
  - Affinity Diagram (KJ, Jiro Kawakita): clarify a complex situation through the classification, structuring, and synthesis of numerous verbal expressions;
  - Relation Diagram: identify and clarify cause-effect relationships in a complex situation.
- **Second group**: Defining measures to solve the problem
  - Tree Diagram: detail from the general to the specific in order to define actions to achieve a certain goal;
  - Matrix Diagram: correlate a series of factors to evaluate, select, decide;
  - Data Matrix Analysis: quantify and interpret relationships expressed with numerical data.
- **Third group**: Planning the activities to solve the problem
  - PDPC Diagram: identify and select alternatives in developing an action plan to define processes useful in achieving results;
  - Arrow Diagram: plan the activities to be carried out and monitor them effectively.

Affinity Diagram

Affinity mapping was developed in the 1960s by Japanese anthropologist Jiro Kawakita. Affinity diagrams are also known as the K-J method, named after Kawakita.

It allows for the collection of a large quantity of verbal expressions from multiple people (ideas, opinions, observations, etc.) and organizing them into groups based on logical and hierarchical relationships. It constitutes one of the most innovative and effective tools for applying Total Quality in a group setting, as it is able to coherently and effectively integrate two approaches that are traditionally far apart: the analytical approach and the creative approach. It is a useful tool when starting a new activity or when tackling a particularly large and complex problem, as it provides a clear and complete framework for it.

This tool works through the aggregation of homogeneous classes and the hierarchical structuring of available information, based on their affinity. The purposes and fields of application of this tool are in particular:

- identifying logical priorities: in practice, applying Pareto analysis to all those situations where it is not possible, or would be reductive, to reason only in terms of numerical priorities;
- extracting the maximum useful information from a few or scattered data and unrelated ideas;
- understanding and structuring unclear problems;
- creating new concepts.

The benefits derived from the use of this diagram are, in particular, the following:

1. it allows for the use of verbal information even in chaotic situations and for identifying problems through the synthesis of this data;
2. it allows for obtaining new ideas through an innovative way of thinking;
3. it allows for finalizing one's own and others' ideas in the form of actions to be taken and contributes to the motivation of participants;
4. it allows for obtaining an organic view of different points of view;
5. it allows for identifying relationships between different aspects of the problem, structuring them hierarchically.

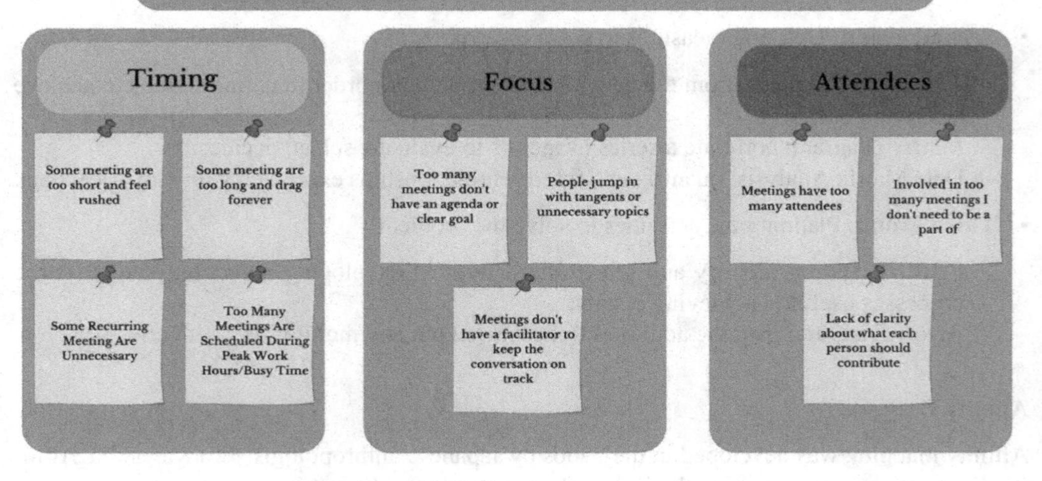

Figure 6.1 Example of affinity diagram.
Source: https://miro.com/blog/create-affinity-diagrams

Diagram of Relationships and Entity Relationship Diagram

A diagram of relationships is a useful tool for analysing a problem by breaking it down into its constituent components and identifying the fundamental causal relationships between the problem and its components, as well as between the components themselves. It shares conceptual similarities with the cause-and-effect diagram, also known as the Ishikawa diagram. In other words, a Diagram of Relationships is a graphical representation that illustrates the connections or associations between various elements, concepts, or entities. It can be used across different domains, including social networks, organizational structures, and project management. It serves as a visual aid to understand and analyse the relationships between different entities or elements within a system or context. It helps in identifying patterns, dependencies, and interactions.

The Diagram of Relationships can include nodes representing entities or elements and edges representing the connections or relationships between them. It can be hierarchical, network-based, or any other suitable structure based on the specific context. In a corporate structure analysis, a Diagram of Relationships might depict employees as nodes and their professional connections or reporting lines as edges connecting them.

This tool offers the following advantages:

1. It allows for gathering the various causes of a discussed problem, which interact closely with each other, and in particular, provides a global view of the existing relationships between all the causes.
2. It facilitates the achievement of an agreement between the participants about the relationships between the various causes.
3. It allows for developing unique and creative ideas for identifying new relationships, without being limited by constraints or restrictions.
4. It allows for isolating the few vital problems, identifying the various relationships, and ensuring that all the personnel involved have a good understanding of the problem.

An Entity Relationship Diagram (ERD) is a visual representation that illustrates how entities, such as people, objects, or concepts, relate to each other within a system. It is commonly used in database design to model and debug relational databases in various fields like software engineering, business information systems, education, and research. In an ERD, entities are represented as rectangles labelled with nouns, relationships are depicted with lines connecting entities, and attributes are shown as ovals. Cardinality in ER diagrams indicates the nature of relationships, whether they are 1-1, 1-many, or many-to-many. The ERD components include entities, relationships, attributes, and cardinality to define the structure of a database system.

Overall, an ERD serves as a crucial tool for designing relational databases by visually representing how different entities interact within the system.

Diagrams of Relationships can be used in various domains beyond databases, while ERDs are specific to database design. The former may include any type of entities or concepts, while ERDs focus specifically on entities within a database schema. In addition, Diagrams of Relationships serve to understand connections broadly, while ERDs are used for database schema design and analysis. Finally, Diagrams of Relationships can have diverse visual representations, whereas ERDs have a standardized format tailored to database design principles. In summary, while both tools visualize relationships, Diagrams of Relationships are more general and versatile, applicable across various domains, whereas ERDs are specialized tools used specifically for database design.

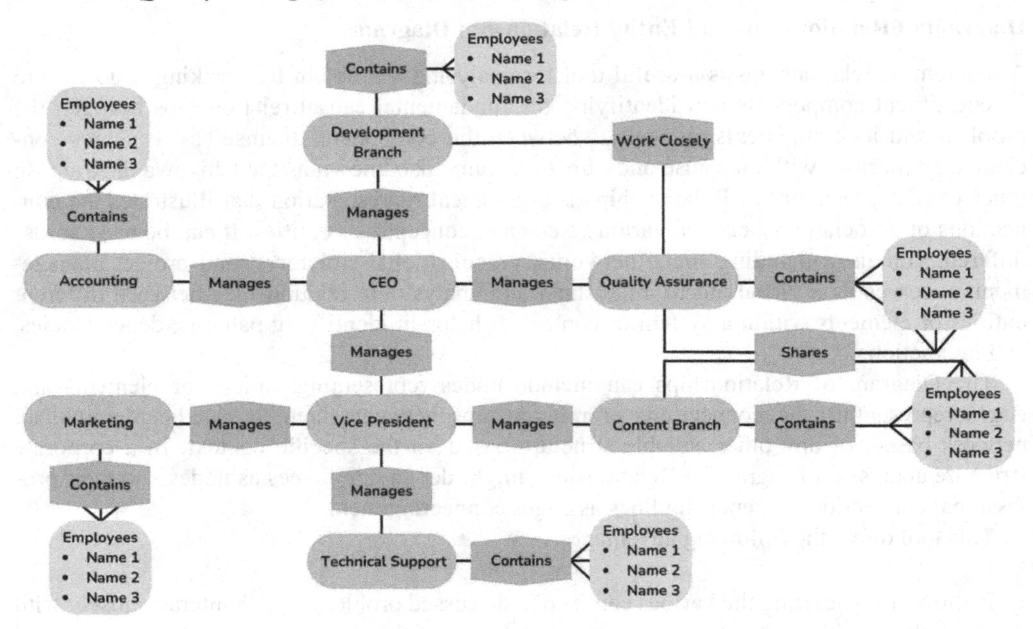

Figure 6.2 Example of diagram of relationships.

Source: authors' own elaboration

Tree Diagram

A tree diagram is a visual tool used to represent hierarchical structures, such as organizational charts, decision trees, or family trees. It consists of nodes connected by branches, with each node representing a concept, entity, or decision point, and the branches representing the relationships or connections between them. Tree diagrams are widely used in various fields, including mathematics, computer science, biology, and business, to model and analyse hierarchical relationships, processes, or decisions. In a tree diagram, the root node is positioned at the top, and branches emanate from it to connect to child nodes. Each node may have zero or more child nodes, and the arrangement of nodes and branches reflects the hierarchical relationships within the structure. For example, in a family tree diagram, the root node represents the ancestral couple, with branches extending downward to their children, grandchildren, and subsequent generations. Each node represents an individual, and the branches indicate parent-child relationships. Tree diagrams are valuable for visualizing and understanding hierarchical relationships, organizing information, and representing decision processes in a clear and intuitive manner.

They are tools used to represent, with increasing levels of detail, the set of methods, procedures, and activities most suitable for achieving a specific objective. Also called a "dendrogram" or "systematic diagram", they are used whenever a systematic analysis of a problem, concept, or objective is required. They are the cornerstone of extremely effective methodologies such as Quality Function Deployment (in the development of new products) and Policy Deployment (in the process of deploying corporate goals from the President to the lowest levels).

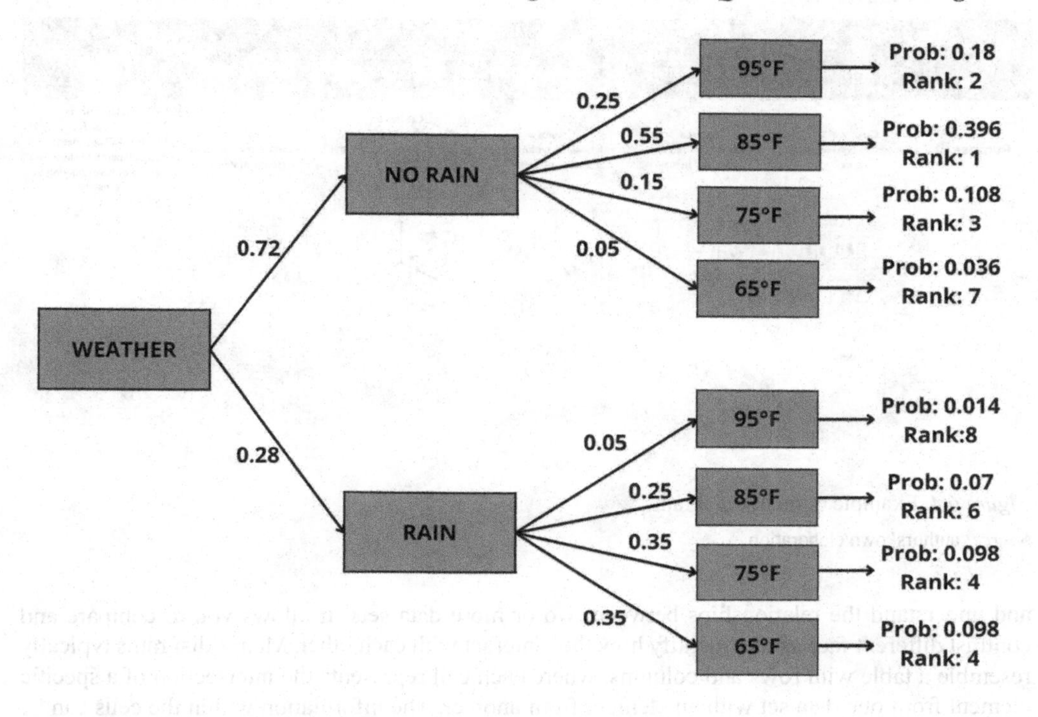

Figure 6.3 Example of a tree diagram.

Source: authors' own elaboration

The use of the tree diagram can fall into two main categories:

- Tree diagram for the development of a concept into subcomponents until the basic elements are reached. In this diagram, the fundamental question to move from one level to another is: "What is it composed of?" or "What is it made up of?"
- Tree diagram for the development of means/procedures necessary to achieve a specific result: a systematic analysis of the actions necessary to solve a problem or achieve a specific objective. Here, the fundamental question to move from one level to another is: "How?" or "What needs to be done to achieve this result?"

Overall tree diagrams are valuable tools for providing stakeholders with a comprehensive overview of complex hierarchical structures, facilitating understanding, communication, and decision-making within organizations, systems, or processes.

Matrix Diagram

A matrix diagram is a visual representation that displays the relationship between two or more sets of items. It allows identifying and expressing non-numerical relationships between two or more sets of elements (data, categories, characteristics, conditions, etc.). This diagram is the ideal tool for systematically analysing the relationships between sets of factors and highlighting the value of the connection. It is also known as a matrix chart and is a visual tool used to analyse

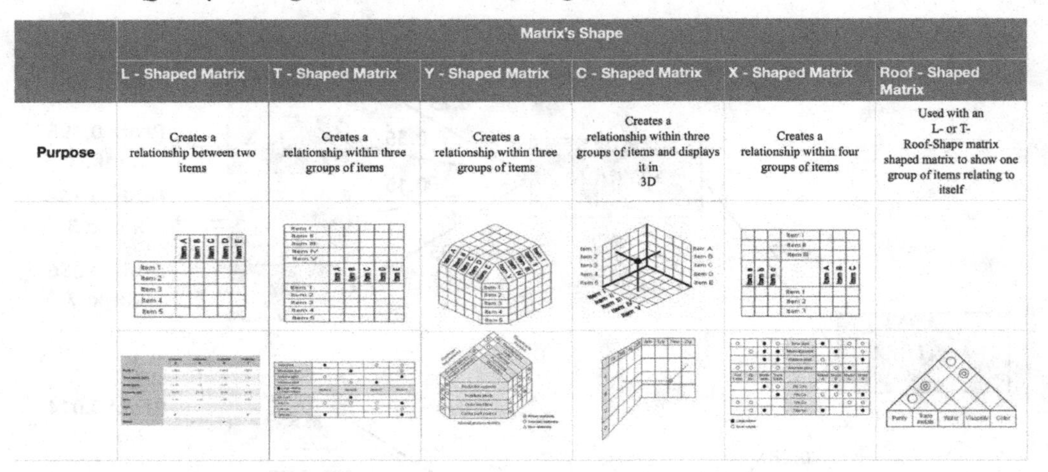

	Matrix's Shape					
	L - Shaped Matrix	T - Shaped Matrix	Y - Shaped Matrix	C - Shaped Matrix	X - Shaped Matrix	Roof - Shaped Matrix
Purpose	Creates a relationship between two items	Creates a relationship within three groups of items	Creates a relationship within three groups of items	Creates a relationship within three groups of items and displays it in 3D	Creates a relationship within four groups of items	Used with an L- or T- Roof-Shape matrix shaped matrix to show one group of items relating to itself

Figure 6.4 Example of matrix diagrams.

Source: authors' own elaboration

and understand the relationships between two or more data sets. It allows you to compare and contrast different factors and identify how they interact with each other. Matrix diagrams typically resemble a table with rows and columns, where each cell represents the intersection of a specific element from one data set with an element from another. The information within the cells can be displayed in various ways, like text descriptions, symbols, ratings, or even numbers. Depending on the number of data sets being compared, there are different types of matrix diagrams, including:

- T-shaped: compares three sets of information, often used to evaluate a single group against two different criteria.
- Y-shaped: connects three sets of items in a circular fashion, helpful for identifying dependencies and connections between them.
- C-shaped: similar to Y-shaped but in 3D, allowing for the analysis of three sets simultaneously.
- X-shaped: compares four groups of items, each connected to two others.

They provide a clear and concise picture of how different factors influence each other, making it easier to understand complex situations. In addition, by analysing the interactions within the matrix, you can uncover underlying patterns and trends that might not be readily apparent in other data representations. By visualizing the relationships between various factors, you can make informed decisions by considering the potential impact of different options.

Overall, matrix diagrams are valuable tools for various fields, including project management, quality control, business analysis, and any situation where understanding the relationships between different data sets is crucial.

Data Matrix Analysis

Data matrix analysis is the process of extracting useful and meaningful information from a set of data organized in a tabular structure. It involves a combination of statistical techniques, data visualization, and critical thinking to identify patterns, trends, and relationships between different variables.

Steps involved in data matrix analysis are the following:

1. Understanding the data matrix:
   - Identify the variables: The rows and columns of the matrix represent different variables. It is crucial to understand the meaning of each variable and the type of data it contains (numerical, categorical, etc.).
   - Assess the data distribution: Examine the distribution of values within the matrix to identify any anomalies, missing values, or non-uniform distributions.
   - Look for relationships between variables: Observe the matrix to identify relationships between different variables, such as positive or negative correlations, dependencies, or associations.

2. Statistical analysis:
   - Calculate descriptive statistics: Calculate means, medians, standard deviations, and other statistics for each variable to summarize their main characteristics.
   - Apply statistical tests: Depending on the analysis goals, apply appropriate statistical tests to confirm or refute hypotheses formulated about the relationships between the variables.
   - Visualize the data: Create graphs and tables to visualize the data and relationships between variables clearly and understandably.

3. Interpreting the results:
   - Summarize the findings: Summarize the results of the statistical analysis and translate them into meaningful conclusions based on the problem context.
   - Identify implications: Assess the implications of the results for decision-making, strategic planning, or solving specific problems.
   - Communicate the results: Communicate the analysis results clearly and concisely to an appropriate audience, using understandable language appropriate to the context.

There are various tools and techniques for data matrix analysis, including:

- Statistical software: Software like R, SPSS, or SAS offers a wide range of functions for statistical data analysis.
- Spreadsheet software: Programs like Microsoft Excel or Google Sheets can be used to calculate descriptive statistics and create simple graphs.
- Multivariate analysis techniques: Techniques like factor analysis or multiple regression can be used to identify complex relationships between multiple variables.

Data matrix analysis requires a combination of statistical knowledge, computing skills, and critical thinking abilities. The ultimate goal is to extract useful and actionable information from data to support informed decisions and improve the understanding of a problem or complex system.

For example, suppose we have a data matrix containing information about online store customers, such as age, gender, products purchased, and amount spent. By analysing this matrix, we could identify relationships between different variables, such as:

- Older customers tend to buy more expensive products.
- Women buy products from a particular category more frequently than men.
- There is a positive correlation between customer age and the amount spent.

These data are exploited to improve marketing strategies, optimize the customer experience, or develop new products and services.

Data matrix analysis is a versatile and powerful process applicable to various disciplines and contexts. With a systematic approach and the use of appropriate tools, it is possible to transform raw data into valuable information for the growth and success of any organization.

It allows for a clear interpretation of large amounts of numerical data by identifying the main variables; it consists of searching for the principal components through multivariate analysis techniques (Principal Component Analysis).

Its use is therefore advantageous:

- in the study of parameters in production processes;
- in the analysis of market information;
- in identifying relationships between numerical and non-numerical variables.

Diagram PDPC (Process Decision Program Chart)

A PDPC is typically represented by a flowchart-like diagram with several interconnected elements. Its main components are the following:

1. **Start/End Diamond:**
   - Symbol: A diamond shape.
   - Function: Represents the beginning and/or end of the process or specific tasks within it.
   - Label: Can be labelled "Start" or "End" depending on its function.
2. **Process Rectangle:**
   - Symbol: A rectangle with rounded corners.
   - Function: Represents a specific step or task in the process.
   - Label: Briefly describes the action or decision being made.
3. **Decision Diamond:**
   - Symbol: A diamond shape.
   - Function: Represents a point where a decision needs to be made based on a specific condition.
   - Label: Contains the question or condition for the decision (e.g., "Is there an error?").
4. **Arrows:**
   - Function: Connect different elements in the diagram and show the flow of the process.
   - Types: solid lines indicate the normal flow of the process, dashed lines indicate alternative paths or potential deviations from the standard flow.
5. **Text Boxes (Optional):**
   - Function: Can be added to provide additional information or clarification about specific steps, decisions, or outcomes.

Figure 6.5 shows a simple example of a PDPC diagram for processing an online order. It is a tool to support the establishment of the most appropriate procedures and actions for carrying out a project. Its purpose is to foresee, from the planning stage, all the unexpected events that may arise during the project and to indicate the corresponding prevention measures or countermeasures, evaluating various alternatives.

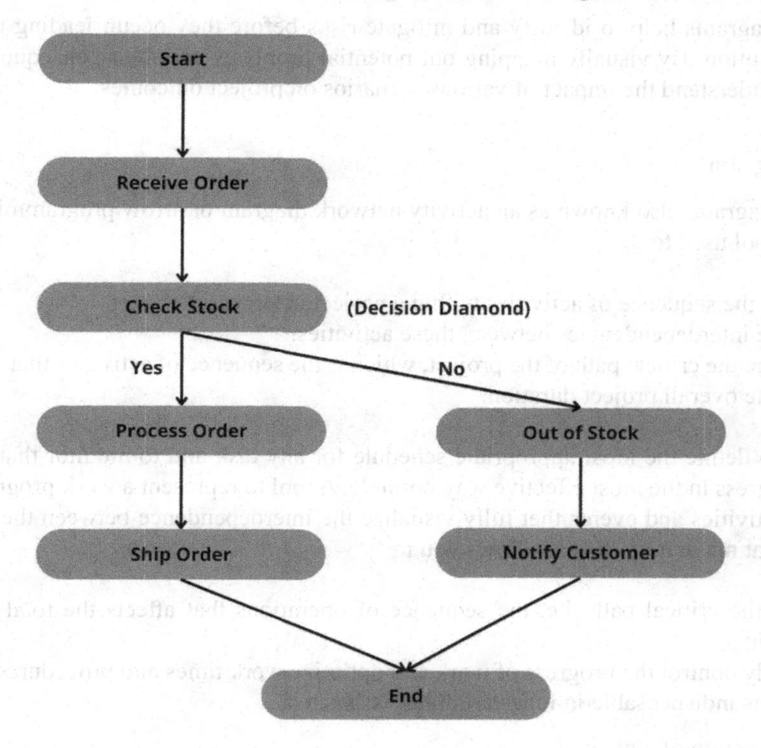

Figure 6.5 PDPC diagram with symbols.

Source: authors' own elaboration

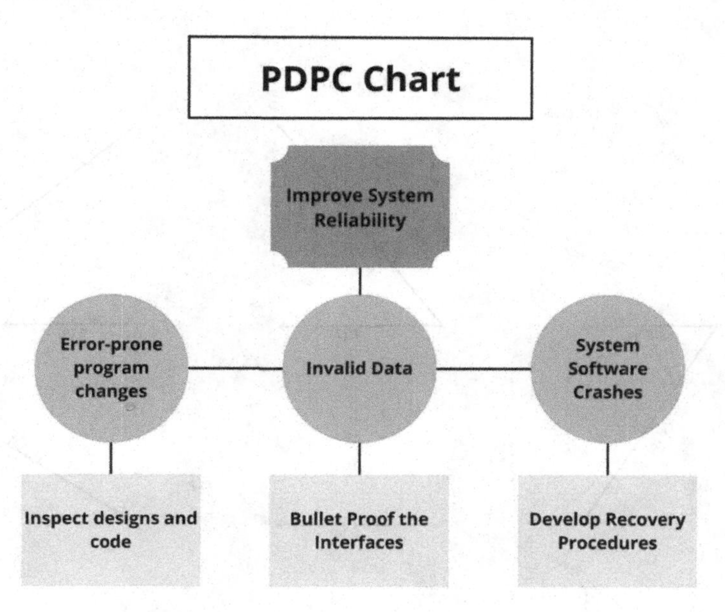

Figure 6.6 Example of a Diagram PDPC based on a real case study.

Source: authors' own elaboration

PDPC diagrams help to identify and mitigate risks before they occur, leading to smoother project execution. By visually mapping out potential problems and their consequences, teams can better understand the impact of various scenarios on project outcomes.

Arrow Diagram

An arrow diagram, also known as an activity network diagram or arrow programming method, is a visual tool used to:

- Illustrate the sequence of activities within a project or process.
- Show the interdependencies between these activities.
- Determine the critical path of the project, which is the sequence of activities that will directly impact the overall project duration.

It is used to define the most appropriate schedule for any task and to monitor that the task or actions progress in the most effective way possible. A tool to represent a work program through a grid of activities and events that fully visualize the interdependence between the elementary activities that make it up. Its use allows you to:

- identify the critical path, i.e. the sequence of operations that affects the total duration of execution;
- effectively control the progress of work and optimize work times and procedures. The arrow diagram is indispensable in long-term projects such as:
  - construction of a plant;
  - development of new products;
  - preparation of events that require a large number of participants.

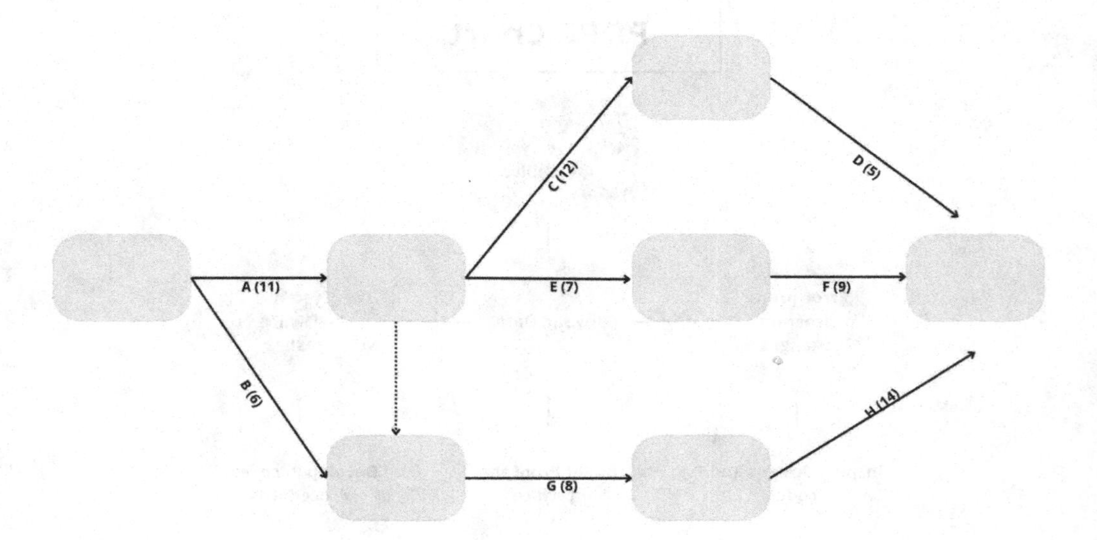

Figure 6.7 Example of arrow diagram.

Source: authors' own elaboration

PERT (Program Evaluation and Review Technique chart)

A PERT chart (Program Evaluation and Review Technique chart) is a project management tool used to plan, schedule, and coordinate tasks within a project. PERT charts use a graphical representation to show the sequence of activities in a project, the dependencies between them, and the critical path that determines the minimum time required to complete the project. It uses a powerful graphical representation in which phases are represented by circles and the sequentially/interdependence between phases is highlighted by arrows.

In this example, activities A, B, C, and D are the main tasks in the project, and activities E through L are sub-tasks or sub-components of those tasks. The arrows between the tasks show the dependencies, meaning that certain tasks must be completed before others can start.

Here an example of a PERT chart:

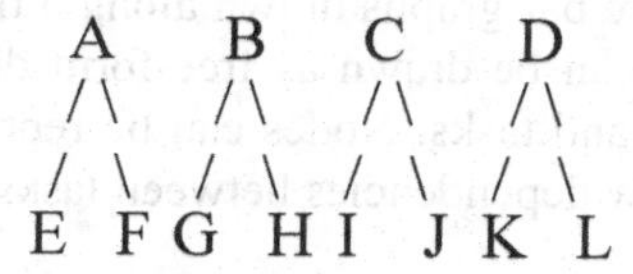

For example, in the above chart, activity A must be completed before activities E and F can begin. Once E and F are completed, activities G, H, I, J, K, and L can start, but activities G and H must be completed before activity D can begin.

By using a PERT chart, project managers can identify the critical path (the sequence of activities with the longest duration) and adjust the project schedule or allocate resources as needed to ensure that the project is completed on time.

GANTT

At the beginning of the twentieth century, Henry Gantt created charts that recorded workers' progress in completing tasks. Supervisors could quickly see if production schedules were behind, ahead, or on time. Gantt charts revolutionized project management, contributing to the construction of major works such as the Hoover Dam and the US highway network. Initially, Gantt charts were written on paper. With the advent of computers in the 1980s, Gantt charts became increasingly complex and sophisticated and, to this day, remain one of the most widely used project management tools.

A Gantt chart is a type of bar chart that is commonly used in project management to illustrate a project schedule. The chart shows the start and end dates of each task in a project and the dependencies between them.

Here an example of a Gantt chart:

In this example, each horizontal bar represents a task, and the length of the bar indicates the duration of the task. The start and end points of each bar show the start and end dates of the task, respectively. The chart also shows the dependencies between tasks, which are indicated by the overlap of the bars.

For example, in the above chart, Task 1 must be completed before Task 2 can begin, and Task 2 must be completed before Task 3 can begin. Task 4 can start as soon as Task 3 is completed, but it does not have any dependencies on the other tasks.

Gantt charts can be useful for project managers to plan and monitor project progress. They can help to identify tasks that are falling behind schedule, and can be used to adjust the project timeline or allocate resources as needed to ensure that the project is completed on time.

Comparing PERT and GANTT Charts

PERT charts and Gantt charts share similarities in that they provide visual representations of a project's tasks, schedules, and timelines, but there are significant differences between these two project management tools.

Unlike Gantt charts, which are bar graphs drawn along a timeline to represent the project's tasks and phases, PERT charts can be drawn as free-form diagrams, with nodes and arrows to represent events, milestones, and tasks. Nodes can be represented by boxes or circles, and arrows connect the nodes to show dependencies between tasks and the estimated time for completing each task.

Another significant difference is that PERT charts illustrate task dependencies, while Gantt charts do not. PERT charts use directional arrows to show which tasks must be completed sequentially due to interdependencies, while diverging arrows indicate functions that can be completed in parallel or out of order because they do not have dependencies. On the other hand, each bar in a Gantt chart stands alone, making it difficult to understand how a missed deadline could affect other chart tasks.

Key Takeaways

- **Basic characteristics**: This section likely covers fundamental principles and attributes of management tools in TQM and Lean Thinking, emphasizing their importance in achieving organizational goals through continuous improvement and waste reduction.
- **Affinity diagram**: Affinity diagrams facilitate the organization and categorization of ideas, data, or issues by grouping them based on their natural relationships or similarities. The key takeaway here is the ability to identify patterns and themes within complex information, aiding in problem-solving and decision-making processes.
- **Diagram of relationships**: This tool visually represents the connections and interactions between various elements or components within a system or process. The key takeaway is the enhanced understanding of how different factors influence each other, helping in analysing root causes and devising effective solutions.
- **Tree diagram**: Tree diagrams illustrate hierarchical structures or breakdowns of a problem, process, or system. The key takeaway is the ability to break down complex concepts into manageable components, facilitating better organization and comprehension.
- **Matrix diagram**: Matrix diagrams provide a systematic way to analyse relationships between different sets of factors or variables. The key takeaway is the ability to identify correlations, dependencies, and strengths of relationships, aiding in strategic decision-making and resource allocation.
- **Data matrix analysis**: This tool is likely used to organize and analyse data in a structured format, allowing for easy comparison and interpretation. The key takeaway is the ability to derive meaningful insights from data sets, supporting evidence-based decision-making and performance improvement efforts.

- **Diagram PDPC (Process Decision Program Chart)**: PDPC charts visualize the potential outcomes and contingency plans associated with different decision paths within a process or project. The key takeaway is the proactive identification and mitigation of risks, ensuring smoother project execution and quality outcomes.
- **Arrow diagram**: Arrow diagrams depict the sequence and dependencies of activities within a project or process, often used in project management. The key takeaway is the visual representation of critical paths and interdependencies, aiding in project planning, scheduling, and resource allocation.
- **PERT (Program Evaluation and Review Technique Chart)**: PERT charts are network diagrams used to map out and schedule the tasks and milestones of a project. The key takeaway is the ability to visualize project timelines, identify critical path activities, and estimate project completion dates.
- **GANTT**: Gantt charts provide a visual representation of project schedules, showing tasks, timelines, and dependencies over time. The key takeaway is the clear visualization of project progress, facilitating effective project management, and resource allocation.
- **Comparing PERT and GANTT charts**: This section likely highlights the differences between PERT and Gantt charts in terms of their features, applications, and advantages. The key takeaway is understanding when each chart is most appropriate to use based on the project's characteristics and management needs.

Review Questions

1. What are the basic characteristics of management tools in Total Quality Management (TQM) and Lean Thinking? How do these characteristics contribute to organizational improvement?
2. Explain the purpose and process of creating an affinity diagram. Provide examples of situations where affinity diagrams would be useful in problem-solving or decision-making processes.
3. Describe the significance of a diagram of relationships in analysing complex systems or processes. How does it help in identifying key connections and dependencies?
4. How can a tree diagram be utilized to break down complex concepts or problems into manageable components? Provide an example scenario where a tree diagram would be beneficial.
5. Discuss the importance of matrix diagrams in understanding relationships between different factors or variables. How can matrix diagrams assist in strategic decision-making?
6. Explain the function of a data matrix and its relevance in organizing and analysing data sets. What are the benefits of using a data matrix in data analysis?
7. Describe the purpose of a Process Decision Program Chart (PDPC) and how it helps in risk management and contingency planning.
8. Compare and contrast arrow diagrams, PERT charts, and Gantt charts in terms of their features, applications, and advantages. When would each type of chart be most appropriate to use?
9. How do PERT charts aid in project planning and scheduling? What are the key components of a PERT chart, and how are critical path activities identified?
10. Explain the significance of Gantt charts in project management. How do Gantt charts visualize project schedules, and what information do they provide to project managers?
11. Reflect on the differences between PERT and Gantt charts. In what scenarios would you prefer to use a PERT chart over a Gantt chart, and vice versa?

7 QFD Methodology and Risk Management in TQM and Lean Thinking

Learning Objectives

- Understand the concept and purpose of Quality Function Deployment (QFD).
- Learn the steps involved in implementing QFD.
- Explore the various tools and techniques used in QFD.
- Define the concept of risk and its importance in decision-making processes.
- Understand how risk is assessed and categorized.
- Familiarize with the ISO 31000:2018 standard and its significance in risk management.
- Learn the principles and framework outlined in ISO 31000:2018.
- Understand the integration of risk management into the ISO 9001:2015 quality management system.
- Identify the key concepts and requirements related to risk management in ISO 9001:2015.
- Gain familiarity with the essential keywords and terminology used in ISO 9001:2015.
- Gain an overview of the contents and structure of ISO 9001:2015.
- Identify the key clauses and sections within the standard.
- Understand the concept of Poka-Yoke (mistake-proofing) and its importance in quality management.
- Learn different types of Poka-Yoke techniques and their applications.

Quality Function Deployment (QFD)

Quality Function Deployment (QFD) is a concrete method for ensuring quality in new products from the design and development stages. The QFD methodology was developed by Akao, Mizuno, and Furukawa of the Union of Japanese Scientists and Engineers (JUSE) in 1970. It was created following a trial of the model at Mitsubishi Heavy Industries, which aimed to obtain measurable quality parameters in the design field. The QFD system is part of Lean Production (developed by Taiichi Ohno), along with other methodologies developed in Japan during the same period, such as the Ishikawa methodology, the 5S Method, and other analysis methodologies to increase corporate production efficiency.

It constitutes an extremely linear and structured thread to move from the often vague and unexpressed needs of customers to:

- characteristics and specifications of new products/services;
- detailed design;

DOI: 10.4324/9781032726748-7

- industrialization;
- production;
- distribution;
- installation, without the risk of misunderstandings, distortions, and inconsistencies.

During the design phase, to follow the Quality Function Deployment (QFD) methodology, it is necessary to lead market research to collect information on the needs of potential users of this product, to create the necessary documentation that describes exactly what the customer wants, their needs, and their preferences (Voice of Customer – VOC). Once the customers' requests are mapped and the areas of intervention to meet them are identified, development of the product can begin according to these specifications and with an approach of continuous verification according to the most well-known methodologies.

This phase serves the company to establish which product characteristics are considered important by the customer and how he evaluates the offer compared to that of competitors. In this way, it is possible to clearly identify the areas for improvement for an existing product or the key attributes on which to base the design in the case of a new product. The second step involves the involvement in brainstorming analysis to collect the ideas of company personnel and bring

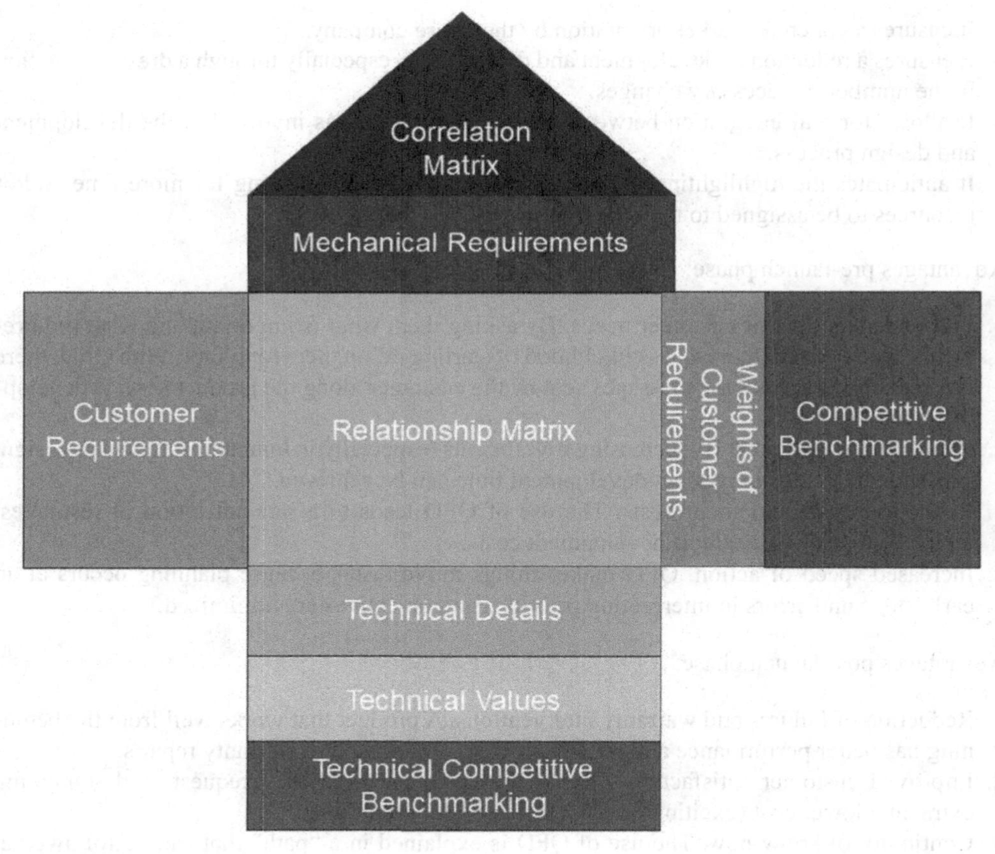

Figure 7.1 QFD method.

Source: authors' own elaboration

a list of characteristics that the product must have to meet customer requirements. Information on customer preferences forms the basis for building the "House of Quality" matrix (HOQ). These needs are translated into corresponding process characteristics (materials, machinery, etc.), expressed in terms that can be quantified and qualified, grouped into first, second, and third levels, and reported in the first row of the matrix.

It is established whether there is a relationship between each requirement and each process characteristic, classifying it as strong, medium, or weak, indicating it with a specific symbol. The units of measure chosen for each process characteristic make it possible to set objective values. The objective figures can be increased or decreased with the improvement process, and for each of them, an arrow is shown on the table, the direction of which indicates whether the improvement acts by increasing or decreasing the objective value. The triangular correlation matrix created to indicate the intersections between process characteristics serves to see if there is a relationship between them or not; it is only reported as positive or negative, and in particular, it is highlighted if there is a "strong" relationship. By assigning weights to each requirement/characteristic relationship, a general evaluation of the relative importance of each process characteristic can be obtained; each of them is obtained by taking into account the weight previously attributed by the market to each product characteristic.

It is important for companies because:

- It ensures a concrete market orientation by the entire company.
- It ensures a reduction in development and design costs, especially through a drastic reduction in the number of necessary changes.
- It allows for real integration between the company functions involved in the development and design process.
- It anticipates the highlighting of problems and difficulties, allowing for more time and/or resources to be assigned to their resolution.

Advantages pre-launch phase:

1. Full understanding of customer needs. By asking them what is important and what requirements are expected, there is less likelihood of starting off on the wrong foot. With QFD, there is a chance not to miss the messages sent by the customer along the product/service development path.
2. Reduction of lead time. By increasing investments (especially in human resources) in design, subsequent savings in product development time can be achieved.
3. Reduction of development costs. The use of QFD leads to a rationalization of resources, which also positively affects development costs.
4. Increased speed of action. QFD makes things move faster because planning occurs at an early stage and errors in interpreting priorities and objectives are minimized.

Advantages post-launch phase:

1. Reduction of failures and warranty interventions. A product that works well from the beginning has better performance and requires lower costs related to warranty repairs.
2. Improved customer satisfaction. Giving the customer what they request, and something extra, at a lower cost (exciting quality) is a very important plus.
3. Continuity of know-how. The use of QFD is explained in a "path" that can be followed at later times. It also becomes a valid model for new resources entering the company and prevents know-how from being lost when others leave the company.

In conclusion and to summarize, the House of Quality system is a visual methodology for developing all aspects of technological and quality iteration that intersect for the production and design of a new product in the same layout. It is built by the QFD team, starting from feedback from customers to evaluate the best design and product design choices.

The strongly visual aspect of the matrix allows the working group to translate customer requirements into specific design technical specifications, immediately highlighting the strengths and weaknesses that the technical departments will have to address. In fact, once the main characteristics and areas for product improvement have been established, they are graphically formalized in the matrix. Each prototype or sample product should be accompanied by an analysis based on the House of Quality in order to provide immediate insight into design difficulties and engineering compromises to be adopted. The process encourages different company departments to work together, listening to the voice of the customer, with the aim of creating a product suitable for meeting their needs.

Operations to be carried out to develop a House of Quality diagram are as follows.

1. **Collect customer requirements (VOC) and develop the technical concept**. The first step requires an understanding of customer requirements, including product usage, design aspects, materials used, commercial and market aspects, as well as a hypothesis of the cost that the customer is willing to bear.

   1.1 **Prioritize customer requests**. All the elements collected from the customer will now be prioritized. One advice at this stage could be to avoid having all elements with the same level of importance. It would be better to avoid having identical numbers and that the sum of all values gives a value of, for example, 100, but as in our example, repeated numbers with any sum are also acceptable. Finally, we have the levels of priority and importance perceived by the customer.

2. **Develop technical aspects,** assuming that the product to be developed has been well understood by the team, now we need to start the Brainstorming phase in order to create a technological idea that can assess all the technical processes to develop the project. Each team operator will provide elements to better structure the production process, tackle technological problems, and any solutions to be pursued. Obviously, it is necessary to keep the customer's needs in focus and any cost or production budget "overruns". In the subsequent phase, the topmost part of the matrix integrates an exhaustive compilation of technical attributes. These attributes embody the HOWs or design elements crucial for addressing the WHATs, which represent the fundamental customer requirements or needs that the product or service is intended to fulfil. These encompass a range of aspects such as functionality, performance, usability, reliability, and other features that customers prioritize when evaluating a solution. Employing the Quality Function Deployment (QFD) methodology, the technical specifications undergo meticulous assessment vis-à-vis each other and in conjunction with the WHATs to determine their effectiveness in fulfilling the customer's requirements (evaluating HOW well they address the customer's needs).

Prioritize technical priorities also for the technical part, it will be important to understand which priorities are more important than others. At this stage, we recommend giving an increasing value from 1 to 5, or if preferred, adopting the same set of different numbers for a total of 100. This phase has only an indicative aspect, however, it is also possible to introduce a calculation

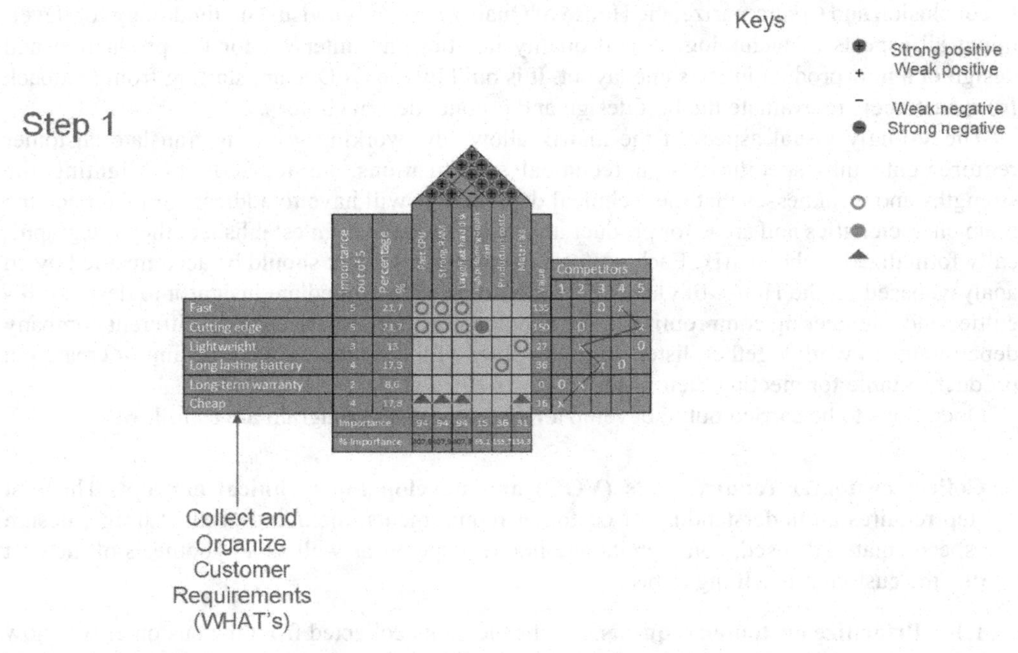

Figure 7.2 QFD: Step 1 – voice of customer.

Source: authors' own elaboration

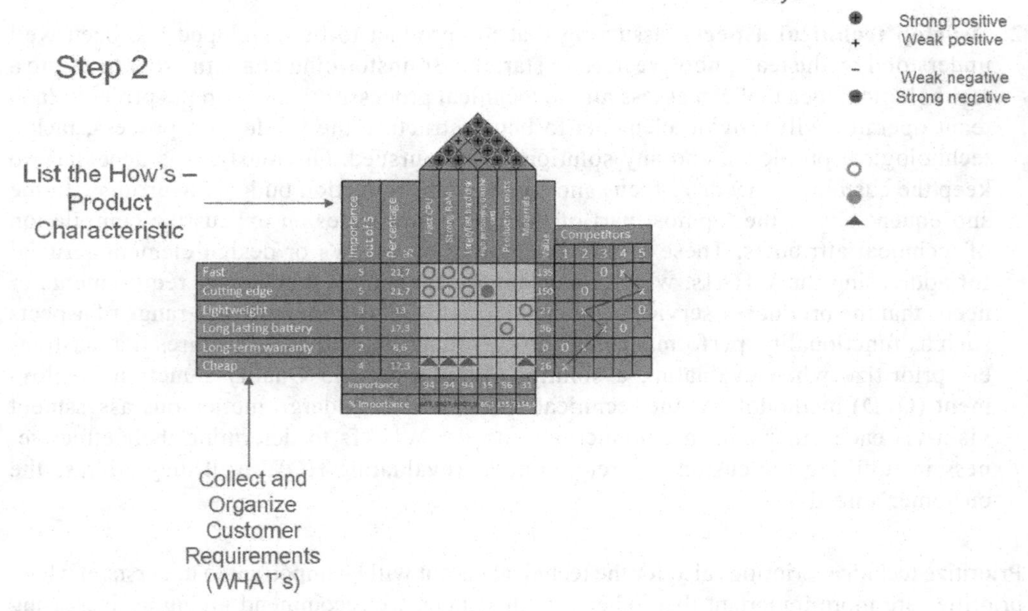

Figure 7.3 QFD: Step 2 – HOWs to VOC.

Source: authors' own elaboration

so that the priority gives greater weight to the final calculation (e.g., adding this value to the overall development of priority values that we will see in point 4).

3. **Analysis of technical importance links to requests**. This is one of the most delicate phases; now we need to start giving values of links in the developed matrix. For each intersection cell, we will give a link value that corresponds to how well that entry technically fully satisfies the customer's requests. We also recommend using a small scale (e.g., from 1 to 10), and each of these values will then be multiplied by the customer's priority level (point 1.1).

One of the main goals of Quality Function Deployment (QFD) is to pinpoint areas where trade-offs are necessary to achieve a technological breakthrough that meets customer requirements. Numerical values, ranging from –9 (indicating a strong negative interaction) to 9 (indicating a strong positive interaction), can be utilized. Different QFD models may use various symbols and may not rely on numerical values.

For example, if the customer requests a particular surface treatment, it follows that a phase will be developed in the process for that specific treatment. Consequently, the bond will be the maximum. Conversely, for the same surface treatment, the bond with the development of minimum batch sizes of 1000 pieces could have a smaller impact, with a value that is even inconsequential (i.e., 0).

4. **Relating the technical aspects**. A critical element could also be given by the fact that two technical aspects conflict with each other. Often, it is necessary to give in to what are called "engineering compromises", that is, having to downsize certain technical functions in favour of a globally more acceptable result and in function of other contrasting characteristics. This phase highlights the fragilities or criticalities of the project. The advice in this case is to use symbols that quickly make it clear where there is affinity and where there are incompatibilities. The scale we propose is developed on 5 levels, 0 is neutral, we then have 2 positive values indicating affinity (+ and + +) and 2 negative values indicating incompatibility (– and – –). It is precisely on the values of incompatibility that it is necessary to pay more attention in order to find the best solution or compromise for the success of the project.

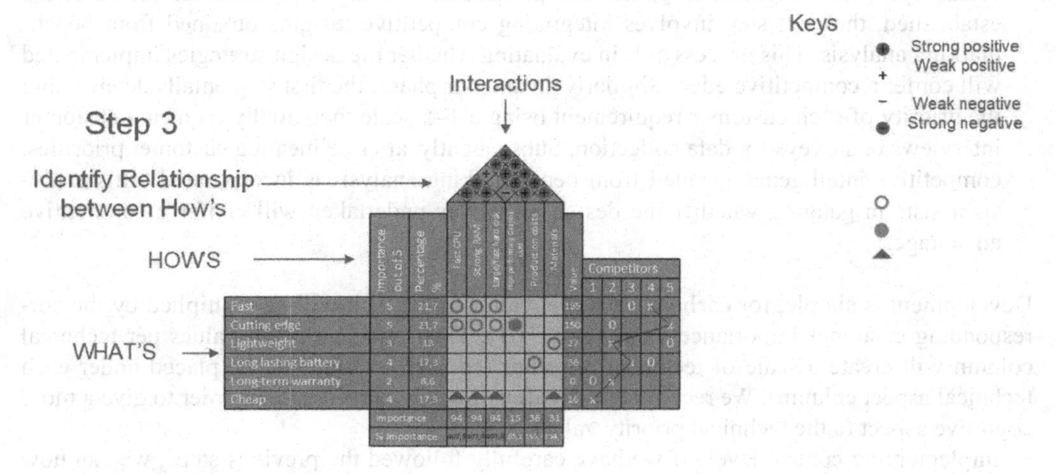

Figure 7.4 QFD: Step 3 – HOWs Interaction.

Source: authors' own elaboration

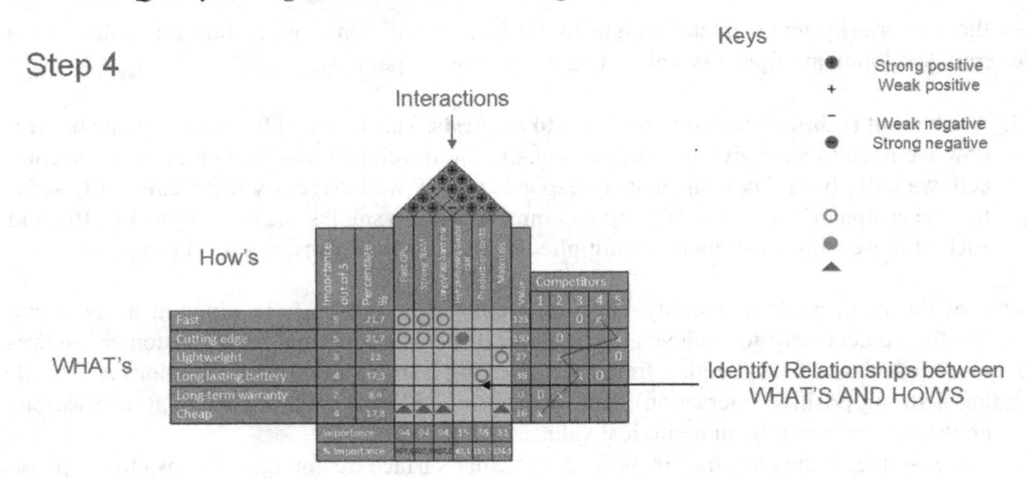

Figure 7.5 QFD: Step 4 – main interaction.

Source: authors' own elaboration

During this stage, it is essential to identify and assess the primary connections between the WHATs and HOWs. Utilize a numerical and color-coded system to determine the intensity of the relationship between characteristics. The HOWs that exhibit the strongest interaction with the most critical WHATs should be prioritized as the top solutions for meeting customer needs.

5. **Development of priority and control values.** We have finally put on paper all the customer diagnosis values and the technical aspects involved. We have understood the technical criticalities to be addressed and developed the process aspects. Finally, we can begin to give priority and importance values to the project and the implementation process. During this stage, the initial task involves assessing the importance of each customer requirement. We utilize a scale ranging from 1 to 5 for this purpose. It may be necessary to conduct interviews or surveys with customers to gather this prioritization data. Once customer priorities are established, the next step involves integrating competitive insights obtained from benchmarking analysis. This process aids in evaluating whether the design strategies implemented will confer a competitive edge. Similarly, in another phase, the first step entails determining the priority of each customer requirement using a 1–5 scale, potentially requiring customer interviews or surveys for data collection. Subsequently, after delineating customer priorities, competitive intelligence gleaned from benchmarking analysis is incorporated. This analysis assists in gauging whether the design initiatives undertaken will confer a competitive advantage.

Development is simple, for each value in the matrix (point 2) it will be multiplied by the corresponding customer importance value (point 1.1). Then, the sum of all values per technical column will create a scale of technical priorities (so such total should be placed under each technical aspect column). We recommend dividing them by percentage, in order to give a more cognitive aspect to the technical priority values.

Implementing control levels If we have carefully followed the previous steps, we can now address the process control elements. This includes technical problems, so it will be important to carefully define the verification parameters for Quality control. Minimum and maximum tolerance parameters (LSL and USL) should be established to ensure that the process is on target with the customer's expectations.

Adding any technical specifications if there are technical process details, it is a good practice to indicate them in the control section, such as the type of verification, number of development batches, number of checks per unit, etc.

In Step 6, it is essential to calculate the deployment priorities for each Technical Feature. This involves summing the product of each interaction value, derived from Step 4, and its corresponding Customer Priority value. By ranking these scores, one can gain insight into which Technical Features are of utmost importance in meeting the diverse array of Customer Requirements.

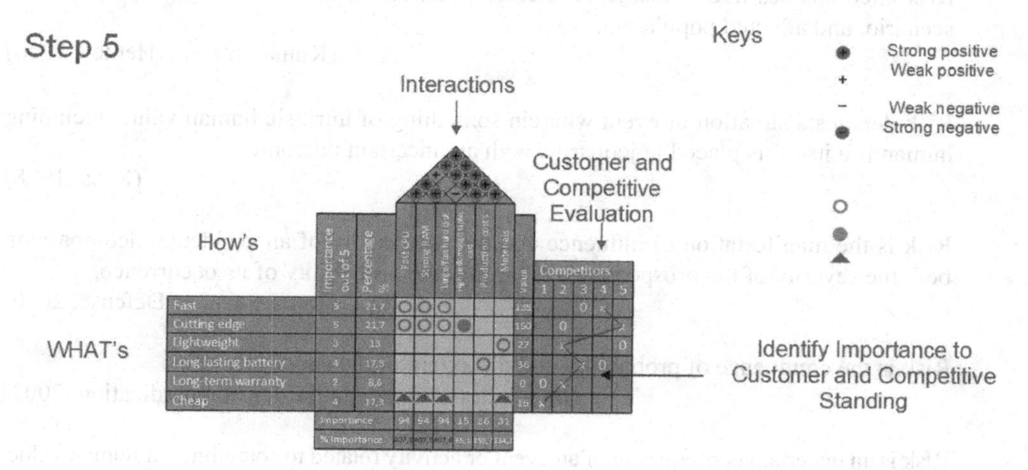

Figure 7.6 QFD: Step 5 – competitive analysis.

Source: authors' own elaboration

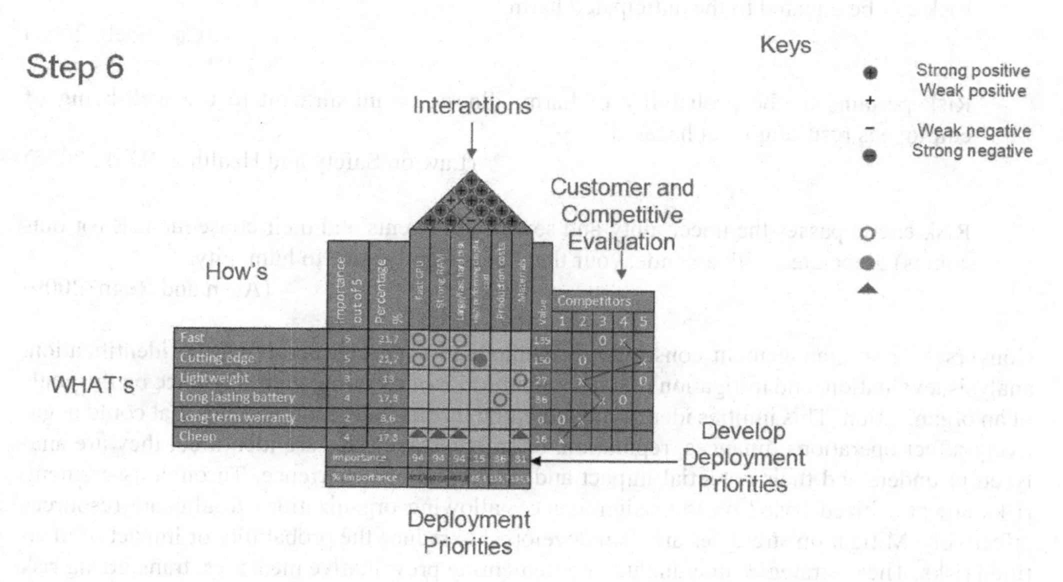

Figure 7.7 QFD: Step 6 – deployment priority.

Source: authors' own elaboration

Risk Management and Risk Definition

Risk can be defined as the amalgamation of the probability of an event occurring and the subsequent ramifications it entails. International technical literature provides various definitions of risk:

> Risk is the product of probability and severity.
>
> (Crouch and Wilson, 1982)

> Risk encompasses five fundamental elements: outcome, likelihood, significance, causal scenario, and affected population.
>
> (Kumamoto and Henley, 1996)

> Risk denotes a situation or event wherein something of intrinsic human value, including human life itself, is placed in jeopardy, with an uncertain outcome.
>
> (Rosa, 1998)

> Risk is the manifestation of influence and the potentiality of an accident, encompassing both the severity of the prospective mishap and the probability of its occurrence.
>
> (Department of Defense, 2000)

> Risk is the confluence of probability and the extent of the consequences.
>
> (International Organization for Standardization, 2002)

> Risk is an uncertain consequence of an event or activity related to something of human value.
>
> (IRGC, 2005)

> Risk can be equated to the anticipated harm.
>
> (Campbell, 2005)

> Risk pertains to the probability of harm, illness, or impairment to the well-being of employees resulting from hazards.
>
> (Law on Safety and Health at Work, 2005)

> Risk encompasses the uncertainty and severity of events and their consequences (or outcomes) associated with an endeavour that holds significance to humanity.
>
> (Aven and Renn, 2009)

Conversely, risk management constitutes a methodical approach involving the identification, analysis, evaluation, and mitigation of risks with the aim of reducing their influence on the goals of an organization. This implies identifying potential threats and vulnerabilities that could negatively affect operations, finances, reputation, or safety. Once risks are identified, they are analysed to understand their potential impact and likelihood of occurrence. Through assessment, risks are prioritized based on their significance, allowing organizations to allocate resources effectively. Mitigation strategies are then developed to reduce the probability or impact of identified risks. These strategies may include implementing preventative measures, transferring risk through insurance or contracts, or accepting certain risks as unavoidable. Consistent monitoring and periodic reassessment of risk management procedures guarantee the continued efficacy and adaptability of strategies in addressing emerging threats and adapting to shifts within the business landscape. Effective risk management enhances organizational resilience, improves decision-making, and ultimately contributes to long-term success.

Risk management typically involves several key steps:

- **Identification of risks**: During this stage, the systematic identification of potential risks impacting the organization's goals takes place. These risks can emanate from diverse origins such as internal operations, external occurrences, human elements, technological aspects, and regulatory frameworks.
- **Risk analysis**: After identifying risks, it becomes imperative to analyse them to comprehend their characteristics, potential consequences, and probability of occurrence. This analysis aids in prioritizing risks according to their importance to the organization.
- **Risk assessment**: In this step, the identified risks are assessed to determine their overall risk level, considering both their potential impact and likelihood of occurrence. This assessment helps in prioritizing risks and allocating resources effectively for their management.
- **Risk mitigation**: After assessing the risks, strategies are developed to mitigate or manage them effectively. This could entail putting in place precautionary measures to decrease the probability of risks occurring, alongside actions to mitigate their effects should they manifest. Risk mitigation approaches might involve transferring risk via insurance, avoiding risk altogether, reducing risk exposure, or acknowledging and managing risk consequences.
- **Implementation of risk controls**: Once mitigation strategies are identified, they need to be implemented across the organization. This entails setting up controls, protocols, and systems to diminish or eradicate the repercussions of identified risks.
- **Monitoring and review**: Risk management is a continual endeavour demanding consistent surveillance and evaluation. In this phase, it encompasses evaluating the efficacy of risk mitigation measures, identifying any novel or emerging risks, and adapting strategies accordingly. Routine assessment guarantees that the organization's risk management procedures stay relevant and proficient in addressing shifting threats and alterations within the business milieu.
- **Communication and reporting**: Effective communication of risks and risk management activities is crucial for stakeholders to understand the organization's risk profile and the measures in place to manage them. Reporting on risk management activities, including any changes in risk exposure or the effectiveness of mitigation measures, facilitates the preservation of transparency and accountability in the company.

By adhering to these procedures, organizations can methodically recognize, evaluate, and address risks to accomplish their goals while mitigating potential negative consequences.

ISO 31000:2018

In the year 2009, the International Organization for Standardization (ISO) released the ISO 31000 standard titled "Risk Management – Principles and Guidelines" (subsequently updated in 2018), along with the ISO 31010 standard on "Risk Management – Techniques for Risk Assessment". ISO 31000 aims to harmonize the concepts of risk management that had emerged until that moment in various scientific fields, in order to offer itself as a "reference standard" for the present and the future in this rapidly developing sector. In this sense, ISO 31000 is not to be considered a standard to be applied to obtain a "conformity certification", i.e., a quality stamp issued by third parties, as in other cases of ISO standards. ISO 31000 is only a principal

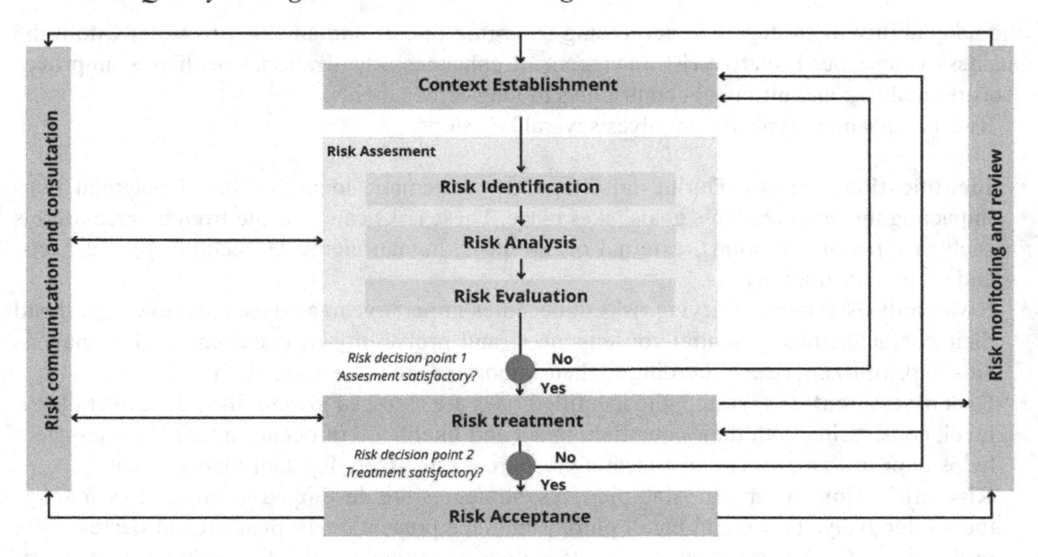

Figure 7.8 Graphical representation of the risk management process.

Source: authors' own elaboration

approach that is valid for all sectors, concerning the issue of risk management, with the aim of standardizing vocabulary, language usage, and management process phases that were rather varied in different sectors until then.

The main differences compared to the previous version of ISO 31000 are as follows:

- **Leadership**: Significant importance is attributed to the role of top management in leading risk management efforts. The updated ISO 31000 standard promotes a "holistic" method to risk management throughout the organization and across all operations. The primary initiation of risk management stems from the organization's governance structure.
- **Continuous interaction with the external environment**: The updated version advocates for continual engagement with the external environment of the organization to facilitate continuous enhancement over time, aiming to "keep pace". An open system that fosters ongoing interaction with the external environment facilitates seamless and organic adjustments to changes and the prevailing circumstances.
- **"Interactive" risk management**: As highlighted in the preceding statement, this results in an ongoing examination of data derived from continual interactions with the external environment. It is imperative to examine this data to establish measures and procedural controls geared towards effectively managing the associated risks. In ISO 31000, risk management is defined as the coordinated activities to direct and control an organization with respect to the risks it faces. In other words, risk management can be considered as the systematic process that enables the identification, analysis, evaluation, and mitigation of risk. Within the scope of standardization, risk is defined in a thorough and inclusive manner as "the impact of uncertainty on objectives". This definition explicitly recognizes that an impact can manifest as a departure from the expected result, whether favourable or unfavourable. Furthermore, it recognizes that objectives can span various domains, such as financial, health and safety, and environmental, and can be applied at different hierarchical levels, including strategic, organizational, production, and operational.

Risk, as commonly understood, is characterized by the occurrence of events and the subsequent consequences that ensue. Its evaluation hinges upon the assessment of both the value of the consequences resulting from a specific occurrence and the probability of said event materializing. It is important to note that uncertainty, an integral component of risk, represents the unknown element associated with the probability of an event and the resulting consequences in practical terms. It should also be noted that the concept of risk is applied in various areas of society, from the public sphere (health risk, natural disaster risk, national security risk) to the private and productive sector (business risk, corporate risk, investment risk . . .).

Public institutions are typically interested in reducing the impact of catastrophic events in terms of:

- reducing deaths and serious injuries;
- damage to the economy;
- maintaining the quality of life of the population;
- damage to national interests;
- damage to image;
- identifying assets to be protected;
- prioritizing activities to implement risk reduction;
- carrying out prevention and awareness-raising actions;
- coordinating any rescue interventions.

In this case, therefore, the approach is oriented towards reducing the consequences of an event on the quality of life of citizens. Instead, private operators and, in general, investors, are interested in reducing the impact of losses in terms of:

- economic damage to productive infrastructure;
- turnover;
- image;
- employee health;
- prioritizing risk reduction activities;
- protecting the core part of their business;
- transferring the most relevant and not directly manageable risks to third parties, with the intervention of insurance coverage.

These different, sometimes completely complementary, viewpoints have imposed in the last decade an increasingly greater need for coordination between the Public and Private sectors in order to counteract natural hazards and anthropogenic threats of various kinds, to reduce the cumulative risk that affects the population and productive activities in the complex modern society.

The ISO 31000 standard, which describes the Risk Management process, highlights how it adopts the PDCA cycle methodology (Plan/Do/Check/Act). The PDCA cycle, developed in the 1920s by Walter Shewhart and subsequently popularized by W. Edwards Deming, consists, as seen previously, of four phases: Plan – what to do and how to do it (planning) to meet policy and objectives that have been determined. Do – implement what has been planned. Check – verify if what has been planned has been done and if what has been done is effective in achieving the objectives. Act – how and what to improve?

Keeping this methodological approach in mind, it can be noted that Figure 7.8 represents the details of the actions to be carried out in a Deming cycle specific to risk management. In particular, the following phases/actions are highlighted:

- Definition of the scope, context, and analysis criteria, precisely delimiting the process/system under examination.
- Conducting risk assessment, which can be divided into three distinct sub-phases: (i) risk identification, (ii) risk analysis, and (iii) risk evaluation. The purpose of risk assessment is to determine as objectively as possible whether the risk is tolerable and therefore acceptable, or if it is too high and therefore requires mitigation actions to reduce its value.
- Risk treatment, typically divided into two sub-phases: risk mitigation through appropriate countermeasures, acceptance of residual risk.

As shown in Figure 7.8, the central phases of the risk management process examined so far are accompanied by transversal phases of:

- Communication and consultation with the subjects/experts involved in the management/use of the process;
- Monitoring and review of the results, methodologies, and actions implemented in the different phases of the management process.

Returning to the definition of risk, in accordance with ISO 31000, we can consider risk as a combination of the probability of a particular event occurring and the consequences generated by the event itself. In this more applied vision, the consequences are to be considered as damages, and therefore negative effects, produced by the occurrence of the event and can be expressed in terms of different perspectives: impacts on people's health and lives, economic impacts, environmental impacts, and also socio/political impacts.

The risk assessment can be carried out using one of the following three distinct conceptual modes:

- Qualitative terms (e.g., dividing the risk into three levels: High, Medium, Low), by analysing the combination of damages incurred and the probability of the event occurring, also evaluating these two variables with qualitative criteria.
- Semi-quantitative terms, using numerical scales that do not have an absolute value but allow the translation of the probability of the event, damage, and risk into relative numbers. In this case, numerical scales can be linear or logarithmic, depending on the specific cases considered and the magnitude variation of the two variables, probability of the event and damage, on which the risk assessment is based.
- Quantitative terms, using numerical rating scales with an absolute meaning (probability of the event expressed in mathematical terms, such as 1/100; 1/1000, etc.; damages expressed, for example, in terms of the number of deaths and severe injuries requiring hospitalization or measured economically in thousands/millions of euros/dollars, and so on).

It should be noted that quantitative analysis is often impossible due to the lack of specific and reliable data and, more generally, due to the limited information available to those carrying out the analysis. In these cases, when a risk assessment is necessary, it is appropriate to refer to the first two modes, choosing the qualitative method when information is scarce or when an approximate risk assessment is sufficient.

It is also important to specify the concept of consequence analysis, often referred to as impact analysis. This type of analysis evaluates the nature and extent of damage (synonymous in this case with impact and consequence) once the event has actually occurred. As previously mentioned, the consequences of an event can be observed from different perspectives, such as the health status of people, the number of deaths, or the economic and environmental damage, including socio/political impacts that occur after the event.

Risk Management in ISO 9001:2015

In earlier editions of ISO 9001, risks were solely dealt with through preventive actions. However, in the updated 2015 version, risk is comprehensively integrated throughout the standard to underscore the significance of continuous evaluation and analysis.

The new standard is primarily concerned with the fact that everyone finally understands that setting up a Quality System does not mean adding something artificial to daily work structured by ISO 9001 procedures that no one will ever read because our everyday way of working and the Quality System must be the same thing. Therefore, do not think that this new requirement inserted in ISO 9001:2015 means producing more paperwork to accompany what you have already produced to describe how your system works. It is true, in fact, that risk management can be considered the backbone of the new document, but is it not also true that the most savvy organizations evaluate and analyse risks on a daily basis? How many times, in fact, do they wonder whether or not to carry out a project (by simply analysing costs and benefits) or what the risks are associated with the purchase of a new tool or machine, especially in the case of inadequate training for the operators who will use them? And who, when changing their software, has not stopped to think about how to deal with the introduction period in order to best manage the period of time when not everyone is yet able to master the new program? And are the risks associated with the delivery of a particular product not considered? What about Salespeople who must decide whether to try to enter a new market or not? Do they not carry out a risk analysis? Finally, even a supplier who delivers materials late can pose risks that need to be evaluated before deciding whether to buy from them.

In short, we are certain that each of you has a clear understanding of what we are talking about and is aware that within your organization, risk analysis is regularly conducted multiple times a day. If your organization is doing well in the market, it is certain that you are already conducting a serious analysis of the risks that it may encounter.

Risk management in ISO 9001:2015 simply requires evidence of how and when risks have been evaluated and plans put in place to address them.

The global economy of the twenty-first century, dynamic and interconnected, represents a challenge for organizations that want to create and provide value to their customers. Performance today is linked to various factors such as the external global environment, the sector in which they operate, each organization's strategy, and more or less fortunate random and unpredictable events, depending on the case. Trying to navigate these extraordinary times by achieving excellent performance and good results, many organizations worldwide have implemented the ISO standards' requirements related to management systems to demonstrate their ability to adhere to them, which is certified through a third-party audit. The most popular and widespread management system standard in the world is ISO 9001, which deals with quality systems. "Quality" can be characterized not only by an organization's capacity to fulfil customer demands but also by its ability to consistently achieve anticipated outcomes and execute tasks correctly upon the initial execution of an activity.

In September 2015, the ISO Technical Committee responsible for revisions and new issues of ISO 9001 completed the major changes that characterize the current ISO 9001 standard and published its latest revision: ISO 9001:2015. All changes compared to ISO 9001:2008 were designed to ensure that the standard continued to be used by top management as a management tool, adapting over time to changes in the business world. Hence, ISO 9001:2015's attention to the context in which organizations operate, documented information that, in a changing world, may have new and ever-different formats and supports, and awareness of risks and opportunities.

The Keywords of the ISO 9001:2015

Starting with the context of the organization, it is nothing more than the environment in which a company operates. It includes internal and external factors that can have an effect on products and services, influence the quality system, or be relevant to business strategies.

Another definition to focus on is that of documented information, which refers to information that must be controlled by the organization along with the support that contains it. Documented information includes information related to the quality system and its processes, as well as all information necessary for the organization to function as intended and to document the results achieved.

The definition of stakeholders is also interesting because it identifies anyone who may have an effect on a business decision or activity, and anyone who may be affected by any decision or activity of an organization.

Knowledge, on the other hand, is a set of information that is believed to be true with a reasonable degree of certainty, as well as familiarity, awareness, and understanding of something such as facts, information, descriptions, skills, etc. Knowledge is usually acquired through experience or education. An interesting aspect of ISO 9001:2015 is the concept of knowledge as a resource because, once again, we can see that in 2015 they were trying to adapt to changing times. Knowledge has become a key element and a fundamental resource to ensure successful project management and to develop new business opportunities.

Risks, on the other hand, are the effect of uncertainty or an unexpected outcome, a positive or negative deviation from what is expected. Risk-based thinking refers to a whole range of coordinated activities and methodologies that organizations carry out to manage and control most of the risks that threaten their ability to achieve the objectives they have set.

Also, in the context of a changing world that is changing faster and faster, the concept of improvement has been introduced by means of a radical change that goes alongside the incremental improvement already present in the previous version of the standard. In this way, ISO 9001 has come closer to excellence models such as the EFQM (European Foundation for Quality Management) model, which emphasize that flexibility and the ability to change quickly are essential for long-term success. Moreover, experts have repeatedly noted a strong correlation between improvement and change management: organizations that manage change well by planning, designing, implementing, and controlling it effectively demonstrate greater improvement. It is perfectly logical that companies that strive to identify and respond effectively to risks and opportunities, changing and improving continuously, are those that are most able to satisfy the needs of their stakeholders over time. ISO 9001 also allows for the management of change through a systemic approach, as suggested by the model itself.

The first ISO 9001, which was released in 1987, was prepared over a period of seven years. Seven years later, the ISO 9001:1994 was published, and for the overhaul of the famous ISO 9001:2000, also called Vision 2000, it took only six years. Another eight years passed, and in 2008, the penultimate version of the document was published, which preceded the current standard.

For the ISO 9001:2015, it took another seven years, but this time they were really well spent, as the new document completely redesigns our approach to quality.

In order to gain deeper insights into the present and anticipated requirements of ISO 9001 and ISO 9004:2018 customers and to guarantee their contentment, an online survey was initiated among users of the standards in October 2010. The findings from this survey were then incorporated into the strategic planning process for the existing standard.

The survey provided an opportunity to survey users about requirements and how the quality management system standard should change. Based on this feedback, the ISO released a document elucidating the rationale behind the necessity for changes in ISO 9001 and the significance of the 2015 version. These reasons include:

- adjusting to evolving global dynamics;
- acknowledging the growing complexities within organizational environments;
- establishing a steadfast foundation for forthcoming endeavours;
- ensuring that the new standard addresses the requirements of all involved parties;
- guaranteeing harmonization with other management system standards.

Two of the most important goals in revising the ISO 9000 family of standards declared by the TC 176 technical committee were:

- to develop a simplified set of standards that could be equally applicable to small, medium, and large organizations;
- to ensure that the amount and detail of required documentation, documented information, was more relevant to the results of the organization's process activities.

Contents of ISO 9001:2015

In ISO 9001:2015, compared to its predecessors, notable changes have been implemented regarding its structure, terminology, requirements, and emphasis on certain concepts. However, the fundamental intent, objective, scope, and applicability of ISO 9001 have remained unchanged. The Technical Committee (TC) 176 outlines the primary differences in content between the old and new versions of ISO 9001, which include:

- Adoption of the high-level structure outlined in Annex SL of ISO/IEC Directives Part 1, known as the HLS.
- Explicit requirement for risk management to enhance understanding and application of the process approach.
- Reduction in prescriptive requirements.
- Increased flexibility concerning documentation.
- Enhanced suitability for the service sector.
- Mandate to define the quality system's scope.
- Greater emphasis on the organizational context.
- Heightened requirements regarding leadership.
- Increased focus on achieving desired process outcomes to enhance customer satisfaction.

Approximately 19% of the requirements in ISO 9001:2015 are new, meaning that roughly 80% of the requirements from the previous 2008 version have remained largely unchanged. Some requirements are expressed similarly to before, while others have been rephrased but maintain the same underlying intent. Additionally, some modifications alter the scope or applicability of the requirement without constituting an entirely new requirement.

The degree of prescription has been reduced in the 2015 version of ISO 9001. Many instances now require organizations to consider various factors and apply risk-based thinking to assess the potential benefits and drawbacks of different choices. The standard's approach has shifted from a prescriptive definition dictating what organizations must do to demonstrate compliance to one placing greater responsibility on top management of each organization.

ISO 9001:2015 stands out from its 2008 counterpart primarily due to the structural framework imposed on all management system standards, known as the HLS. One of the major difficulties that the ISO Technical Committee faced in drafting it was precisely to align the ISO 9001:2015 as much as possible with the standards that regulate other management systems such as ISO 14001, which deals with environmental management. All of this, of course, with a view to promoting more and more integrated systems.

The strong emphasis on risk-based thinking allows management to decide the level at which requirements apply to provide products and services that improve customer satisfaction. One of the main changes, in fact, is certainly the great attention given to risk management, a topic that until a few years ago seemed to have little chance of crossing the threshold of the Quality Manager's office.

We have extensively discussed risk management and ISO 9001:2015, as well as all the other novelties of the standard, explaining how processes, individual documents, daily activities, and ultimately the entire way of working need to be rethought. It can no longer be limited to preaching a process-based approach while continuing to think in terms of functions. It must really be articulated in a collaborative project capable of revolutionizing every organization in its essence to make all stakeholders deeply satisfied and to emerge unscathed from a crisis that seems to be endless also due to the inability of companies to use the quality tool as a competitive lever.

Poka-Yoke

The poka-yoke, or foolproof, methodology is designed to preclude errors by establishing operational conditions where operators are unable to make mistakes. Shigeo Shingo, a Toyota engineer, championed the concept of Zero Quality Control, which extensively employs poka-yoke principles. These mechanisms are employed to avert specific error causes and ensure, in a cost-effective manner, that each produced item is devoid of defects. A poka-yoke technique refers to any mechanism that either prevents an error from occurring or immediately highlights the error. The ability to swiftly identify errors is critical because, according to Shingo, product defects stem from worker errors, necessitating careful identification and analysis of these shortcomings. Therefore, if operator errors are identified and eliminated in advance, they will not result in defects. An example cited by Shingo shows how spotting errors at a glance allows defects to be prevented. Suppose a worker must assemble a piece consisting of two buttons, with a spring to be placed under each button. Sometimes, the worker may forget to place the spring, resulting in an error. A simple poka-yoke application was therefore developed. The worker must remove from the container exactly the number of springs needed for the operation and place them on a disc. If, after assembly is complete, a spring is left on the disc, an error has been made. The operator immediately realizes that a spring has been omitted and corrects the error. The cost of inspection is minimal, as is the cost of eliminating imperfections. This example demonstrates that the poka-yoke approach is particularly suitable for eliminating oversights and omissions. The poka-yoke approach aims to:

- Incorporate quality into design and processes.
- Make it impossible to produce defective products.

- Use poka-yoke protections incorporated into products and processes.
- All unintentional errors and defects can be eliminated. One must firmly believe that errors can be eliminated.
- Errors and defects can be eliminated when everyone works together.
- Search for root causes: one must understand the actual cause of the problem and apply countermeasures.

Shingo has identified three different types of inspections: "judgment inspection", which involves separating defective products from those considered acceptable and is also called "quality control"; "informative inspection", which involves using the data obtained from inspections to control processes and prevent defects. The more frequent the feedback, the faster the quality improvements that can be achieved. It is important to conduct inspections throughout all phases of the process to intervene as quickly as possible. If the operator is also responsible for checking their own work, the benefits will be even more substantial. In fact, since workers check each individual unit produced, they are able to immediately identify the cause of the defect.

Such inspections and checks allow for obtaining information "after the fact". "Source inspection", on the other hand, involves determining "before the fact" whether the necessary conditions exist for a high level of quality. Poka-yoke mechanisms are used in this type of inspection, for example, to prevent production until the appropriate operating conditions are met. Poka-Yoke was developed by Japanese engineer Shigeo Shingo in the Toyota Production System (TPS). Shingo also developed and described Zero Quality Control, a combination of Poka-Yoke techniques to correct potential defects and source inspection to prevent them. Poka-yoke systems are introduced in the two phases of value creation: Design Poka-yoke and Process Poka-yoke. In both cases, the goal is to eliminate the possibility of making errors, which could result in defects in the final product. Simplicity of execution should guide advocates of these systems, as laboriousness would generate non-value-adding activities. The Poka-Yoke approach comprises five steps:

1. Identify and describe the defect.
2. Locate the source of the defect by visiting the workplace, known as Gemba. This is where the underlying causes of issues can be uncovered.
3. Analyse the process where the defects were detected.
4. Investigate the root cause of the defect.
5. Develop and implement a Poka-Yoke solution to prevent the recurrence of the defect in the future.

Poka-Yoke is used to indicate a design choice or equipment that, by limiting the way an operation can be performed, forces the user to execute it correctly. The word literally means "avoid (yokeru) inadvertent errors (poka)". Common examples of Poka-Yoke are found in the computer and audio/video sectors, such as in the case of many electronic connectors. Three types of Poka-Yoke can be distinguished: the contact method, where the physical characteristics of an object (shape, color, etc.) allow for distinguishing the correct position or prevent malfunctions caused by incorrect contact; the fixed-value method, which checks whether a certain number of operations have been performed; and the motion-step method, which checks whether all the phases of a particular process have been performed in the correct order.

The starting point for Poka-Yoke is the awareness that no person or system is capable of completely avoiding unintentional errors. In the case of industrial systems, such as machines and plants, information on errors can normally be provided using the Mean Time Between Failure (MTBF) parameter or the average time between failures.

Wrong actions such as carelessness, omission, confusion, forgetfulness, misreading, misinterpretation can also be exacerbated by stress, harmful environmental factors, poor working conditions. All these aspects are inherent in human nature and, despite all efforts, cannot be completely eliminated with certainty.

With the help of Poka-Yoke, mostly simple but effective systems are used to ensure that such errors in the production process do not lead to errors in the final product or go unnoticed. It is also used in combination with a respective inspection method in order to preclude defects that have already occurred in the past and have been corrected. Source Inspection, also developed by Shigeo Shingo, has been particularly effective. Since the entire causal chain between incorrect manipulation in the process and the error in the product is considered, the causes of errors will be highlighted and eliminated. In this way, the potential repetition of an error will be effectively prevented, leading to the concept of zero errors.

In practical use, a Poka-Yoke system consists fundamentally of two basic elements: a detection mechanism and a regulation mechanism. In addition, there are special precautions in the form of design measures that preclude possible irregularities from the outset.

Poka-yoke techniques can be broadly categorized into two types based on their approach to error prevention or detection:

1. **Preventive Poka-Yoke**: Preventive poka-yoke techniques aim to eliminate errors or mistakes from occurring in the first place. These techniques focus on making it impossible or difficult for errors to happen by elaborating processes, tools, or systems in a way that guides users towards correct actions. Some examples of preventive poka-yoke include:
    - Physical constraints: Designing equipment or tools with physical features that prevent incorrect assembly or operation.
    - Color coding: Using color-coding schemes to differentiate between components or steps in a process, reducing the likelihood of confusion or errors.
    - Shape coding: Designing components or interfaces with unique shapes that only fit together in the correct orientation.
    - Checklists: Implementing checklists or standardized procedures to guide users through tasks and ensure all necessary steps are completed correctly.

2. **Detective Poka-Yoke**: Detective poka-yoke techniques focus on quickly detecting errors or defects when they occur, allowing for immediate correction or intervention before they lead to further problems. These techniques often rely on sensors, alarms, or visual indicators to signal when an error has occurred. Examples of detective poka-yoke include:
    - Error alarms: Installing sensors or detectors that trigger alarms or alerts when abnormalities or errors are detected in a process.
    - Visual controls: Using visual indicators such as lights or flags to signal when a step in a process has been completed or if there is a deviation from the standard.
    - Error-proofing software: Implementing software solutions that automatically validate data or inputs, flagging potential errors for review before they cause issues downstream.
    - Automated inspections: Incorporating automated inspection systems or quality control checkpoints to detect defects or deviations from specifications during production.

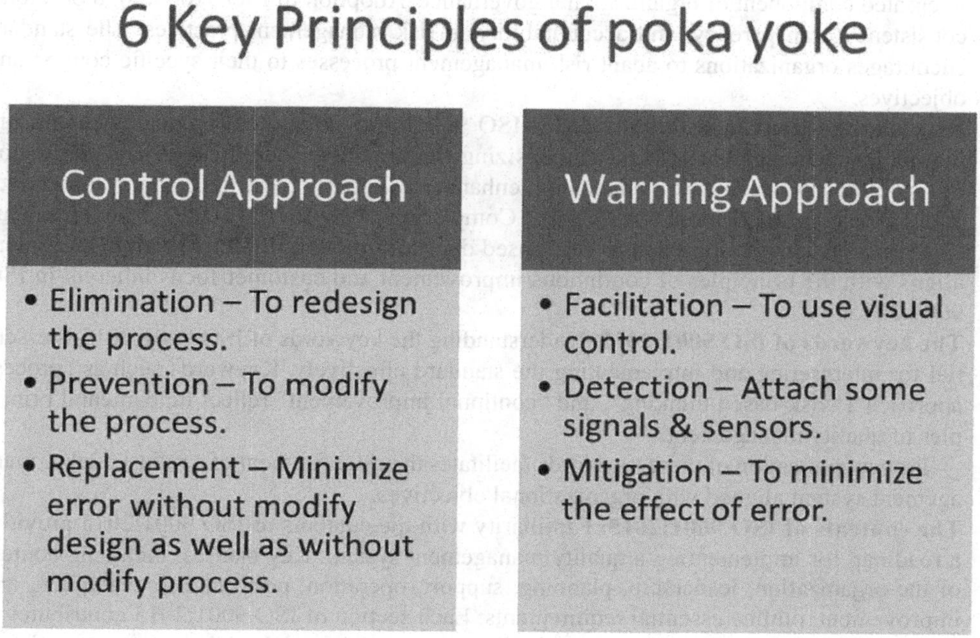

6 Key Principles of poka yoke

Control Approach	Warning Approach
• Elimination – To redesign the process. • Prevention – To modify the process. • Replacement – Minimize error without modify design as well as without modify process.	• Facilitation – To use visual control. • Detection – Attach some signals & sensors. • Mitigation – To minimize the effect of error.

Figure 7.9 Poka-Yoke principles.

Source: authors' own elaboration

Both preventive and detective poka-yoke techniques play complementary roles in error-proofing processes. Preventive techniques aim to minimize the occurrence of errors upfront, while detective techniques provide an additional layer of assurance by quickly identifying and addressing errors as they arise. Effective implementation of both types of poka-yoke can significantly improve quality, productivity, and efficiency in manufacturing and other operational contexts.

Key Takeaways

- **Quality function deployment (QFD)**: QFD facilitates translating customer requirements into specific product or service characteristics. It promotes cross-functional collaboration and customer-centric design. QFD aids in prioritizing design features based on customer needs and technical requirements. Implementation of QFD leads to improved product quality, reduced development time, and increased customer satisfaction.
- **Risk management and risk definition**: Risk definition is crucial for identifying, assessing, and managing potential threats and opportunities. Understanding and articulating risks accurately enhances decision-making processes. Effective risk definition enables organizations to proactively address uncertainties and mitigate negative impacts. Clear risk definitions foster a common understanding among stakeholders, facilitating alignment of risk management efforts.
- **ISO 31000:2018**: ISO 31000:2018 offers a structured framework for managing risks at every level of an organization. It underscores the significance of risk management as a vital and

integrated component of organizational governance. Adoption of ISO 31000:2018 promotes consistency, transparency, and accountability in risk management practices. The standard encourages organizations to adapt risk management processes to their specific context and objectives.

- **Risk management in ISO 9001:2015**: ISO 9001:2015 integrates risk management into the quality management system, emphasizing the proactive identification and mitigation of risks. Incorporating risk management enhances the ability to achieve quality objectives and improve organizational resilience. Compliance with ISO 9001:2015 requirements ensures a systematic approach to risk-based decision-making. Effective risk management aligns with the principles of continuous improvement and customer focus inherent in ISO 9001:2015.
- **The keywords of ISO 9001:2015**: Understanding the keywords of ISO 9001:2015 is essential for interpreting and implementing the standard effectively. Keywords such as "process approach", "risk-based thinking", and "continual improvement" reflect fundamental principles of quality management.

  Proper application of these keywords facilitates the establishment of a robust quality management system aligned with organizational objectives.
- **The contents of ISO 9001:2015**: Familiarity with the contents of ISO 9001:2015 provides a roadmap for implementing a quality management system. Key clauses, including context of the organization, leadership, planning, support, operation, performance evaluation, and improvement, outline essential requirements. Each section of ISO 9001:2015 contributes to establishing a systematic approach to quality management, emphasizing customer satisfaction and organizational effectiveness.
- **Poka-Yoke**: Poka-Yoke techniques aim to prevent errors or defects at the source, eliminating the need for rework or correction. Implementation of Poka-Yoke contributes to improve product quality, increase productivity, and reduce costs. Poka-Yoke emphasizes simplicity and practicality, making it accessible for application across various industries and processes. Adoption of Poka-Yoke fosters a culture of continuous improvement and error prevention within organizations.

Review Questions

1. What is Quality Function Deployment (QFD) and why is it important in product/service development?
2. Describe the main steps involved in implementing QFD.
3. How does QFD facilitate alignment between customer requirements and product/service characteristics?
4. Define risk and explain its significance in organizational decision-making processes.
5. How does clear risk definition contribute to effective risk management?
6. Discuss the differences between potential risks and opportunities within an organization.
7. What is ISO 31000:2018 and what is its purpose in risk management?
8. Describe the principles outlined in ISO 31000:2018 for managing risks.
9. How does ISO 31000:2018 provide a framework for integrating risk management into organizational processes?
10. How does ISO 9001:2015 integrate risk management into the quality management system?
11. Discuss the relationship between risk management and continuous improvement in ISO 9001:2015.

12. What are the key requirements related to risk management in ISO 9001:2015?
13. Identify and define the key keywords used in ISO 9001:2015.
14. Explain the purpose and objectives of each section within ISO 9001:2015.
15. What is Poka-Yoke and how does it contribute to quality improvement?
16. Describe common Poka-Yoke techniques used in error prevention.
17. How does Poka-Yoke foster a culture of continuous improvement within organizations?

8 Customer Analysis Techniques in TQM and Lean Thinking

Learning Objectives

- Understand the different types of quality and their significance in various industries.
- Explore the concept of customer satisfaction within quality management frameworks.
- Analyse the Kano Model and its application in understanding customer preferences and satisfaction levels.
- Evaluate the SERVQUAL model as a tool for assessing service quality and identifying areas for improvement.
- Examine various tools used in the evaluation of SERVQUAL and their practical applications.
- Develop skills in analysing SERVQUAL data and utilizing findings for enhancing service quality.
- Learn negotiation techniques and problem-solving strategies applicable in diverse contexts.
- Identify the four negotiation strategies and their respective applications in different negotiation scenarios.
- Explore the six pillars of win-win negotiation and their role in achieving mutually beneficial outcomes.
- Understand the principles of Prospect Theory and its implications in decision-making processes.
- Analyse the two phases of decision-making processes within Prospect Theory and their influence on outcomes.
- Learn effective mediation strategies for resolving conflicts and reaching agreements in contentious situations.
- Develop problem-solving skills utilizing techniques such as TRIZ (Theory of Inventive Problem Solving) to overcome challenges effectively.

Quality Types

The quality of products and services is measured by the ability to meet the needs of customers. Detecting customer satisfaction is not just a slogan. In other words, simply wanting to be on the side of the customer is not enough to collect, understand, and interpret their feedback on the organization's performance.

DOI: 10.4324/9781032726748-8

Quality of service from both the customer and provider perspectives can be:

- **Expected quality**: The focus is on the customer, and the primary goal is to identify what they want, their implicit, explicit, and latent needs, evaluation criteria, and judgments on the quality of service.
- **Designed quality and promised quality**: The focus shifts within the organization. The goal is to identify what is to be delivered to the customer and how. This involves identifying customer types, targets, operational standards to be ensured, designing the characteristics of the organizational system and service delivery system. This subsystem also includes the promised quality that affects the level of expectations. Therefore, attention is needed not to promise what cannot be delivered. In many cases, the standards defined in this phase are included in service charters and define the commitments the company makes to customers.
- **Delivered quality**: This examines internal operations, focusing on processes. Its primary objective is to methodically oversee the performance of the service delivery system, ensuring alignment with defined standards and assessing what the organization truly delivers.
- **Perceived quality**: The focus lies on the customer – their perceptions, evaluations of the received service, overall satisfaction level, and the quality factors or individual elements of the service delivery system. This finalizes the assessment of Delivered Quality by scrutinizing the non-standardizable and challenging-to-evaluate aspects of the service.
- **Compared quality**: The reference is directed towards comparable structures offering similar services or towards other entities, aiming to ascertain the distinctions in quality and their respective sources.

Customer Satisfaction in Quality Management

Customer satisfaction can represent and highlight the needs and expectations of customers, which, once well identified, are prioritized and defined in terms of minimum acceptable performance and ideal performance. Perception is also compared with indications related to expectations, which allows identifying where to focus efforts.

It can facilitate the understanding of latent needs by developing sensitivity and the ability to capture weak signals, anticipate needs, and discover latent needs. The ability to understand latent needs constitutes a strong stimulus for innovation and the definition of new responses to needs.

It can help capture ideas, suggestions, and recommendations since attentive listening is an inexhaustible source of proposals, suggestions, and stimuli for the definition of increasingly effective interventions. Customer satisfaction can combine the flow of information that comes from the outside with that which comes from the inside. It can facilitate the overcoming of internal constraints given by the repetitive and routine actions of the organization and support the verification and understanding of the effectiveness of policies through systematic monitoring over time of the level of customer satisfaction. It can also help strategically define new service packages or interventions to improve existing ones.

Kano Model

The Kano Model is a theory developed in 1980 by Professor Noriaki Kano that classifies customer preferences into four categories:

- **Basic factors**: These are the characteristics that customers consider implicit and taken for granted, and that generate dissatisfaction when they are missing. In the Kano Model, these factors have the highest weight because they must always be respected by the company. Although they do not lead to real satisfaction, their absence in the product or service generates dissatisfaction, often severe.

- **Excitement factors**: These are attractive features that cause satisfaction when provided but do not generate dissatisfaction when not provided. These are the factors that surprise the customer, generating pleasure and allowing the company that provides them to distinguish themselves from competitors.
- **Performance factors**: These are factors related to a single performance and generate satisfaction or dissatisfaction depending on whether this performance is high or low. Usually, these factors are related to specific needs or desires of customers that are clearly expressed. Their determination mainly occurs through appropriately conducted market research.
- **Indifferent attributes**: These are characteristics that customers do not evaluate because they are not interested.

SERVQUAL

According to Baldwin (2006), "The patient who undergoes dental treatment cannot see the result before the purchase, and therefore seeks quality signs to reduce uncertainty". Thus, the most important elements (the "quality signs") that patients pay attention to, and on which they base their judgment, are:

- **Factor 1** – *Ability and punctuality*. This factor represents the "expertise", the "skill" that the dentist has acquired through study and practice, and is associated with the timeliness and punctuality of service delivery, with minimal waiting times and without delays. Equally important is their ability to treat the patient as an individual, to communicate their intentions regarding the treatment plan and to convince them to adhere to it and follow the instructions.
- **Factor 2** – *Responsiveness and reliability*. This parameter includes safety, reliability, and technical competence. At the base of this factor, we find the manifest willingness to promptly provide assistance to the patient, the willingness to answer their questions, the ability to inspire confidence and to inform the patient of their rights and duties without delay. In other words, the patient wants to "feel" that the dentist understands and recognizes their need to be treated in times and ways that respect their life rhythms, and in an environment that encourages them to feel less anxious. On the technical competence side, in qualitative terms, patients have indicated that a low percentage of treatment failure and the need for retreatment, high levels of quality control of processes and materials, and an attentive and enthusiastic character – combined with the ability to effectively perform agreed-upon treatments – all contribute to generating the impression that the dentist is responsible and trustworthy.
- **Factor 3** – *Tangible elements*. The equipment in the dental office, the furnishings, the waiting room, the appearance of the staff uniforms, the attractiveness of the informational material are tangible elements that contribute to the success of this factor as a predictive element of service quality. Equally important are the cleanliness and overall appearance of the office, the dentist/s who work there, and the rest of the staff.

Tools for Evaluating SERVQUAL

SERVQUAL, which has undergone some adjustments over the years, is an established method (among many) for improving service quality. Based on an understanding of customer needs,

Table 8.1 Analysis of the discrepancies between expectations and effective evaluation.

Expectations < effective evaluation	Performance exceeding expectations	Customer loyalty	Satisfaction
Expectations > effective evaluation	Expectations betrayed	Loss of customers	Dissatisfaction
Expectations = effective evaluation	Indecision in judgment	Acquirable customers	Uncertain satisfaction

identified through a specific questionnaire, SERVQUAL compares perceptions of service quality with corresponding characteristics of an "excellent" organization. The analysis of the discrepancies between perceived and ideal characteristics can be effectively used to improve service quality.

Thus, SERVQUAL is a highly standardized quantitative methodology specifically designed to measure customers' judgments on service quality, allowing for the comparison of user expectations and perceptions regarding a specific service which leads to the development of the company.

It consists of a series of 22 predefined questions divided into two repeated groups concerning user expectations and judgments on various aspects of the service, or distributed more compactly in a single series of questions. The 22 questions, valid for any type of service, allow for the separate measurement of perceived quality and expectations across 5 dimensions deemed essential by the author for judging service quality. These dimensions are as follows:

- Tangible elements (appearance of physical structures, equipment, and personnel).
- Reliability (ability to deliver the promised service reliably and accurately).
- Responsiveness (willingness to help customers and provide prompt service).
- Assurance (competence and courtesy of employees and their ability to inspire confidence and security).
- Empathy (caring and personalized assistance reserved for customers and users).

Apart from tangible aspects and service accessibility, the identified elements encompass a range of characteristics including communication, safety, competence, courtesy, understanding

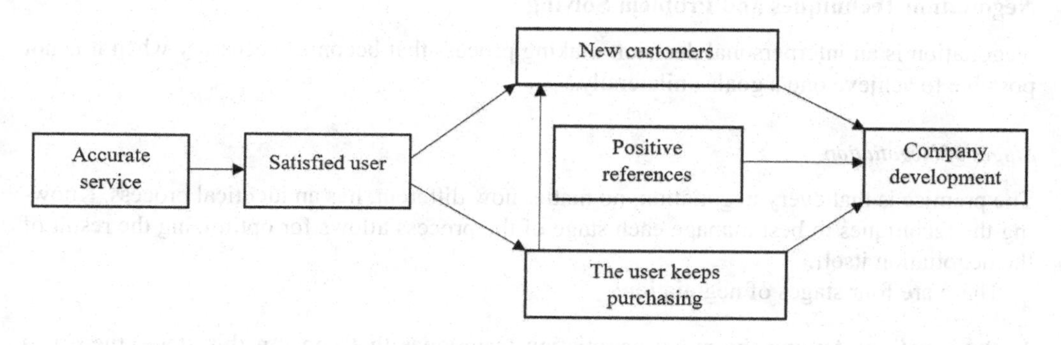

Figure 8.1 SERVQUAL concept.
Source: authors' own elaboration

of customer needs, and service accessibility, all of which underscore the importance of the relationship – a fundamental component of every service interaction.

Analysis and Use

The standard questionnaire uses a numerical scale of 1–7 for each question, which makes the tool particularly easy to use. This ease of use is both the strength and weakness of the model; standardization makes comparisons between measurements obtained in different organizations and/or times very easy. The simplicity and ease of use of the questionnaire, which uses a numerical scale of 1 to 7 to code responses, make SERVQUAL a particularly suitable tool for measuring CS in cases where there are no qualified personnel for more refined surveys. The high standardization of the method makes it particularly interesting for its clarity (of questions and answers provided) and the ability to relatively easily modify one or more of the individual items of the base version. However, due to its rigidity, it is advisable to use it only after a preliminary analysis of the service's characteristics.

The collected data shows that dissatisfaction is not necessarily attributable to excessive and difficult-to-meet expectations. For example, in the "empathy dimension", low satisfaction has been found even in the presence of low expectations.

The method is characterized by the following strengths:

- The questions are already defined.
- It is easy to change some of the 22 questions to adapt them to any specific reality.
- The use of a questionnaire with only numerical closed answers makes data processing very simple.
- Possibility of using non-professional interviewers and ease of completion.
- Possibility of measuring customers' expectations and perceptions.
- Possibility of easily digitizing the entire procedure by transferring it to a research platform.

Instead, weaknesses of the ServQual model are:

- Closed questions make the questionnaire rigid and unsuitable for highlighting customer aspects that do not fall into the categories used;
- the method should be used after a preliminary analysis of the characteristics of the service.

Negotiation Techniques and Problem Solving

Negotiation is an interpersonal decision-making process that becomes necessary when it is not possible to achieve one's goals unilaterally.

Stages of Negotiation

The premise is that every negotiation, no matter how different, has an identical process. Knowing the techniques to best manage each stage of the process allows for optimizing the result of the negotiation itself.

There are four stages of negotiation:

1. **Preparation**: Among the many negotiation techniques that concern this stage, there is a fundamental one: clarity. You must clarify your needs and set goals to satisfy them. Without

this prerequisite, every subsequent behaviour will be dangerously instinctive and random, wasting time and having a negative impact on results.

2. **Dialogue**: If in the preparation stage, you try to understand your needs, during the dialogue stage, you ensure what the needs of others are, which you have only been able to assume until now and need to verify. People close an agreement only if it satisfies their interests, which they are not available to negotiate on, while they are obviously negotiable on resources or methods necessary to satisfy them. The main technique of dialogue, therefore, is to carefully allocate the available time, using it mostly to ask questions and listen, rather than speak.
3. **Proposal**: In this stage, you try to respond to your needs and those of your interlocutor. In this stage, you can make your proposal for agreements: making the proposal is the queen of negotiation techniques, whose value is essential, by no means taken for granted. You should not expect others to formulate it. See it as the fuel that ignites and nourishes the negotiation. If you have prepared well and dialogued, you will know what you want and what others want. The proposal allows you to start talking in a concrete and constructive way, leaving behind useless controversies and recriminations, which often characterize negotiation dialogues.
4. **Closing**: The closing of the negotiation is a very delicate phase because it is decisive for the entire negotiation. The main rule for this phase is to avoid haste: do not close agreements just to reach a conclusion. Often people discover that they have wasted a lot of time, and unnecessarily. In an attempt to make sense of the poorly invested time, they rush to sign incomplete agreements, on which they will later find themselves complaining. Your goal is not to end the negotiation but to obtain satisfaction with a good agreement. This is possible only by acting methodically and respecting the previous stages. These four techniques – preparing on needs and objectives, keeping quiet and asking questions, making proposals, and not rushing to close agreements – if used with awareness, can already give an important contribution to a successful outcome of negotiations.

The Four Negotiation Strategies

Reaching an agreement through negotiation is a complex activity, which is why shortcuts are often sought, but they lead to counterproductive results. What makes the difference is what you focus on during the negotiation. Depending on whether you are more or less focused on your needs or those of your interlocutor, four main negotiation strategies can be identified:

- accommodating negotiation;
- avoidant negotiation;
- competitive negotiation;
- win-win negotiation (or collaborative).

Accommodative Negotiation

It is the type of negotiation where "I lose, he wins". This type of negotiation occurs when you focus heavily on the needs of your interlocutor, at the expense of your own needs. It is a characteristic of all situations where you want to avoid conflicts and heated confrontations. It often happens when your interlocutor is an authoritative person, perhaps with an aggressive

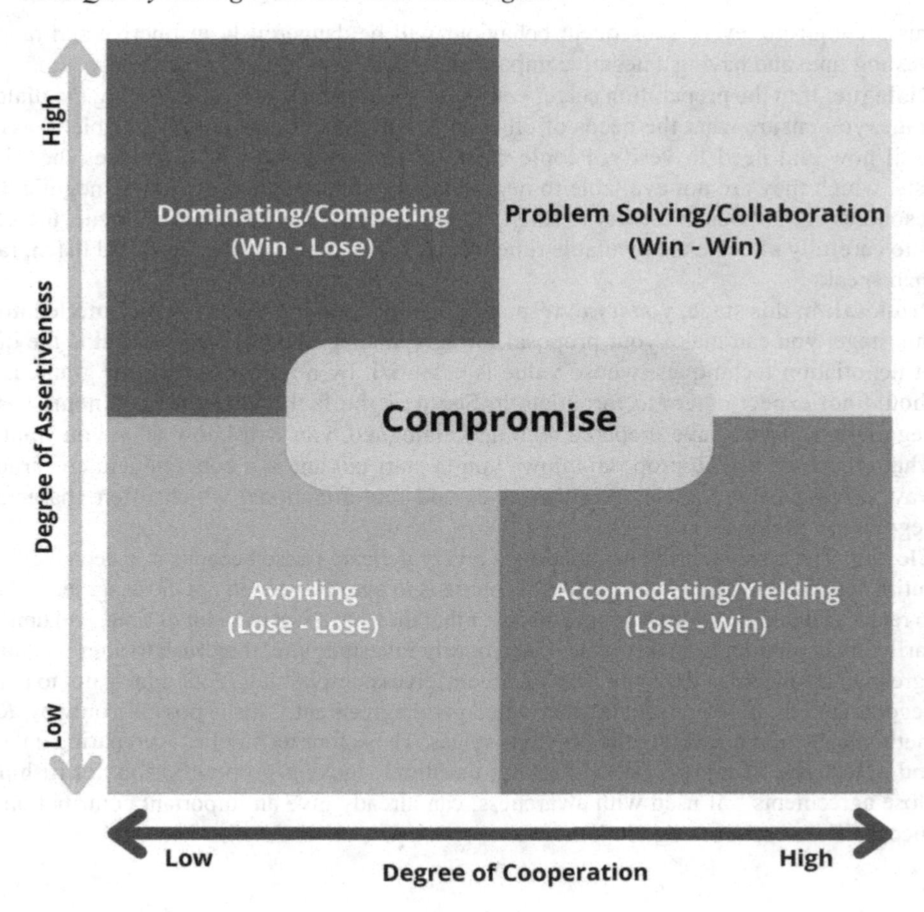

Figure 8.2 The segmented negotiating behaviour matrix.

Source: authors' own elaboration

personality, so to avoid heated confrontations, you decide to comply with their requests and put your own needs on the back burner. It is also a characteristic of people who seek harmony with their interlocutor and therefore focus mainly on their needs. This type of negotiation can help to retain clients and build relationships, but often leads to sacrificing your own needs and, consequently, your own results.

Elusive Negotiation

It is the type of negotiation where "I lose, you lose". This type of negotiation occurs when you have limited focus on both the needs of your interlocutor and your own needs. It is characteristic of situations where you have little interest in that deal, so you commit minimal energy to the negotiation and adopt a submissive and elusive style. By putting little energy into achieving your objectives and having limited attention to the needs of your interlocutor, you only get what the other party is willing to concede. This type of negotiation allows you to achieve limited and difficult-to-plan results (the helm is in the hands of the other party) and hardly allows you to retain the client.

Competitive Negotiation

It is the type of negotiation where "I win, you lose". It is a zero-sum game, meaning that what you gain is a loss for your interlocutor and vice versa. For example, you get an advance payment, and your interlocutor has to pay passive interests. Or your interlocutor gets a discount from you, and in this way, reduces your margin and therefore your profit. In this type of negotiation (still widely used today), each party defends their position by making concessions to reach a compromise.

Negotiation is seen as a battle and the other party as an "enemy". In a competition-oriented negotiation, since each party seeks to maximize their gain at the expense of the other, each party tries to exploit any potential advantage and often settles for "losing as little as possible". Furthermore, since each party is interested in their own interests and pays little attention to the interests of the other party, competitive negotiation often achieves these results:

- it is frustrating for you and your interlocutor;
- it only leads to temporary victories;
- it limits or prevents future relationships.

Generally, competitive negotiation is structured through four phases. The analysis phase involves:

- clarifying your own objectives;
- identifying the minimum acceptable result;
- preparing a strategy to implement during the negotiation.

The beginning of the negotiation involves:

- asking questions of the other party;
- identifying the other party's interests and needs.

During the development of the negotiation, you:

- make concessions to the other party;
- obtain concessions from the other party;
- get closer to an agreement while trying to give up as little as possible and gain as much as possible.

Reaching an agreement:

- defines a meeting point between the parties;
- often satisfies only one of the parties and for this reason, it may limit or prevent future relationships.

Win-Win Negotiation

It is a negotiation where "I win, you win". This approach allows you to win without fighting. The negotiation is seen as a collaboration to find benefits for both parties (in fact, it's often called collaborative negotiation). In this context, the negotiator is seen as an "ally". The logic of compromise is surpassed. Instead, a solution is sought through which the gain that each party obtains from the other's concessions is greater than the cost of their own concessions. The goal is to create the maximum value for the group formed by you and your negotiator and obtain an important part of this value. This is achieved first by identifying the interests of both parties, collaborating to obtain the maximum benefits, and sharing the obtained value. It's then necessary to enrich the negotiating table, so the goal is not to reduce the distance between the parties (divide the cake) but to expand the range of options (increase the cake). Through win-win negotiation, you can achieve these results:

- a pleasant negotiation for both you and your negotiator;
- obtain lasting victories;
- lay the foundation for future relationships.

The Six Pillars of Win-Win Negotiation

Win-win negotiation is an effective negotiation because both parties end up in a better position than before. It's important to follow some strategic steps.

Know your emotions:

- avoid stereotypes (think personally);
- be assertive (express your emotions and opinions clearly and effectively without offending or attacking your negotiator);
- send positive signals to your negotiator;
- reason and be open-minded;
- separate people from the problem.

Know your negotiator (their culture, way of thinking, and values);

- apply active listening;
- ask open-ended questions (open-ended questions help create collaboration and lead to seeking an answer);
- limit statements (as they often block dialogue);
- put yourself in your negotiator's shoes (be empathetic);
- allow your negotiator to express all their reactions without reacting, recognize your negotiator's positive intentions;
- analyse the reasons and starting point of your negotiator;
- explore the interests (explicit and implicit) and needs of your negotiator.

Identify your goal:

- Clarify what you want to achieve;
- Ask yourself what you are not willing to give up.

Build a relationship with your negotiator:

- make your negotiator comfortable;
- show that you're attentive to listening and enriching the range of options;
- avoid insurmountable limits;
- offer alternatives;
- build bridges between thinking styles.

Develop more options with your negotiator (you will make the actual choice later):

- make proposals;
- focus on objective criteria;
- observe from a distance and broaden the scenario;
- find satisfactory solutions for both you and your negotiator;
- if behaviours seem opposed, seek a different solution;
- if there are heated discussions, accept criticism;
- focus on common interests, not positions;
- solve the problem together with your negotiator.

The approach of win-win negotiation revolutionizes the traditional concept of negotiation that saw opposing parties. It's no longer a customer and a supplier facing each other, but two partners who, side by side, seek advantageous solutions for both. Precisely for all these reasons, despite having a high focus on its goals, win-win negotiation is the one that best manages to create a relationship with the negotiator and lay the foundation for customer loyalty.

Prospect Theory

In 1978, Herbert Simon, an American psychologist, won the Nobel Prize in economics for his research on decision-making processes in economic organizations. In the same years, Daniel Kahneman, an Israeli psychologist, together with his colleague Amos Tversky, questioned how decision-making processes change when people are in situations of risk, leading to the development of Prospect Theory.

Until then, the commonly shared theory was that of expected utility, a model of rational choice used to describe the economic behaviour of individuals, who would choose according to the actual probability of gaining profit from their decision.

However, the two psychologists noticed that this model does not work in cases where the individual is placed in conditions of risk, providing some examples. Participants in the research were presented with different dilemmas, and systematic violations of the expected utility principle were observed.

For example, they were asked to choose between the following possibilities: (A) 50% chance of winning 1000, 50% chance of winning nothing; and (B) a certain gain of 450.

The subjects did not consider the actual possible gain derived from the probabilistic calculation. The researchers observed a regularity that they defined as the "certainty effect": participants, when comparing certain utility with probable utility, overestimate the gain when it is certain, choosing option B in the majority of cases.

In addition to verifying the subjects' preference in case of gains, the researchers tested the decision-making processes involved in case of possible loss. They noticed a different behaviour by the participants, so much so that they called this phenomenon the "reflection effect": subjects prefer to risk a large loss that is only probable, rather than accept the certainty of a smaller loss. The same principle – the overestimation of the certain data – favours risk aversion when it comes to gains and the pursuit of risk when losses are at stake.

This discovery is analysed by the authors who list the various repercussions of Prospect Theory in every activity of daily life. People do not reflect rationally on the actual probabilities of an event, but select information based on subjective individual patterns, resulting in different choices. The researchers define this mode as the "isolation effect". For example, a person may invest their money in an activity with a probability of losing the entire capital. On the other hand, there is the possibility of earning a fixed amount or having a percentage of the gains. The certainty of a profit increases the attractiveness of this option, while equally probable alternatives are not considered.

Two Phases of Decision-Making Processes in Prospect Theory

The prospect theory, proposed by the authors, describes decision-making processes consisting of two phases: (1) framing, which involves gathering information and analysing different perspectives, and (2) evaluating different possible scenarios and choosing the one that represents the subject's best alternative.

Subjective choices therefore arise from simplification, cancellation, and consideration of the influence of the context: the same person can make different choices when faced with the same problem precisely because of the presence of an underlying process that is unscientific and difficult to repeat.

The researchers then create a new equation that considers the different emerged components, obtaining a formula that is more adherent to real decision-making processes. This equation shows remarkable applicability in different fields, so much so that Daniel Kahneman won the Nobel Prize in economics in 2002, along with American economist Vernon Smith, for demonstrating how human decision-making processes are guided by heuristics and biases.

He was the second psychologist to receive such recognition for his contribution to science, after Herbert Simon, who had demonstrated the inefficiency of the human brain in reasoning processes and the consequent tendency to make satisfactory but not optimal choices.

Thus, in this theory, subjective choices result from simplification, cancellation, and consideration of the influence of the context. The same person can make different choices when faced with the same problem due to the presence of an underlying process that is unscientific and difficult to repeat. Researchers thus create a new equation that takes into account the different components that have emerged, resulting in a formula that is more closely aligned with real decision-making processes.

Mediation Strategies

Negotiating allows us to reduce the differences between our position and that of the other party, so it is necessary for the negotiation dynamics to be oriented towards an exchange of information that strengthens the relationship between the parties and facilitates the achievement of an agreement. However, negotiations can become deadlocked, and it is in these situations that the intervention of a third party capable of redesigning the negotiation situation may be required. The third party can be:

- An agent who has interests closer to those of the seller/buyer, for example, a real estate agent who takes a percentage of the sale.
- A representative, a similar figure to that of the agent. It is carried out by a subject belonging to the organization being represented. Their power is derived from whoever delegated the function, and as a result, the representative will be motivated to please whoever they represent.
- A mediator is a neutral intermediary who facilitates the achievement of an agreement, which must be freely agreed upon by the parties.
- An arbitrator is a neutral intermediary who will make decisions acting as a higher authority. He/she can adopt two lines of conduct: conventional, which involves consulting the parties and asking for their final position and imposing an agreement on an intermediate point, or unconventional, in which the arbitrator asks the parties for their proposed resolution and chooses the one that seems most fair.

Communication with a third party is very important because through the exchange of information, it is possible to influence the negotiation. The mediator's task will only be truly useful when there is a relationship of trust, established in stages, which will transfer into the negotiation process and from there into the relationship between the parties. According to Professor Carnevale, the four fundamental strategies used by the mediator to reach an agreement between the parties are:

- Pressure: that is, trying to reduce the levels of aspiration of the conflicting parties.
- Compensation: that is, trying to induce the parties to an agreement through additional rewards.
- Integration: that is, trying to reconcile the divergent interests of the parties, guaranteeing them high joint benefits.
- Inaction, that is, allowing the parties to manage their disputes directly.

The choice of strategies depends on the ambition and likelihood of agreement of the parties and the time available (the mediator will tend to pressure the parties if there is little time, while if there is time available, they will tend towards inaction). In addition to strategies, mediators also have a whole range of specific tactics for specific contexts, grouped into four categories:

- Tactics aimed at understanding the issues of the negotiation (what it consists of, what the positions of the parties are, what the climate is).
- Tactics for establishing the priorities of the parties (what is important and what is not?).
- Tactics aimed at acquiring information on the intentions and underlying interests of the priorities (desires, intentions, needs).
- Movement tactics to bring negotiating parties closer together (based on communication).

Problem Solving

Problem solving concerns the set of techniques and methodologies necessary for analysing a problematic situation in order to identify and implement the best solution.

1. Define the problem: Understand the problem starting from the apparent problem, which is often not the real problem, but only a symptom. Analyse a situation well, go deep and identify the original critical situation to reach an effective solution. Go to the root of the problem with the 5 Whys, i.e. progressively proceed from the apparent problem to subsequent problems by asking "why" five times.
2. Generate alternatives: This is the creative phase, the one of designing solutions to the questions posed by the problem. It is also about organizing information and identifying resources to implement an action plan. It can be very useful to use design thinking methodologies.
3. Evaluate and select alternatives: Consider different alternative solutions and then select the one that seems most in line with expectations of success and failure tolerance. In this phase, decision-making comes into play, i.e. all that cognitive and emotional process that allows reaching a final choice.
4. Implement solutions: Once the solution has been chosen and an action plan has been created, it must be implemented, i.e. carried out. The entire problem-solving process finds complete expression. The quality of the relationship is determined by trust between the parties. Trust can come from various factors (belonging to the same group, trust in the other, common interests, the perception of community of intentions in the other . . .).

The quality of communication influences the quality of agreement and is characterized by:

- empathy;
- asking questions to guide mental processes;
- setting an example;
- multiple proposals.

TRIZ: The Theory of Inventive Problem Solving

Making innovation in companies is essential to remain competitive today. For some time, companies have introduced methods to efficiently manage some phases of the product development cycle and related industrial processes (e.g. 6 Sigma, Lean Production, FMEA-FMECA, etc.).

Today, faced with concept generation processes, often carried out in an intuitive and occasional way, many companies realize the need to introduce tools capable of making the early stages of a new product development cycle efficient and providing methodological support to ideation and problem-solving processes in a technical context.

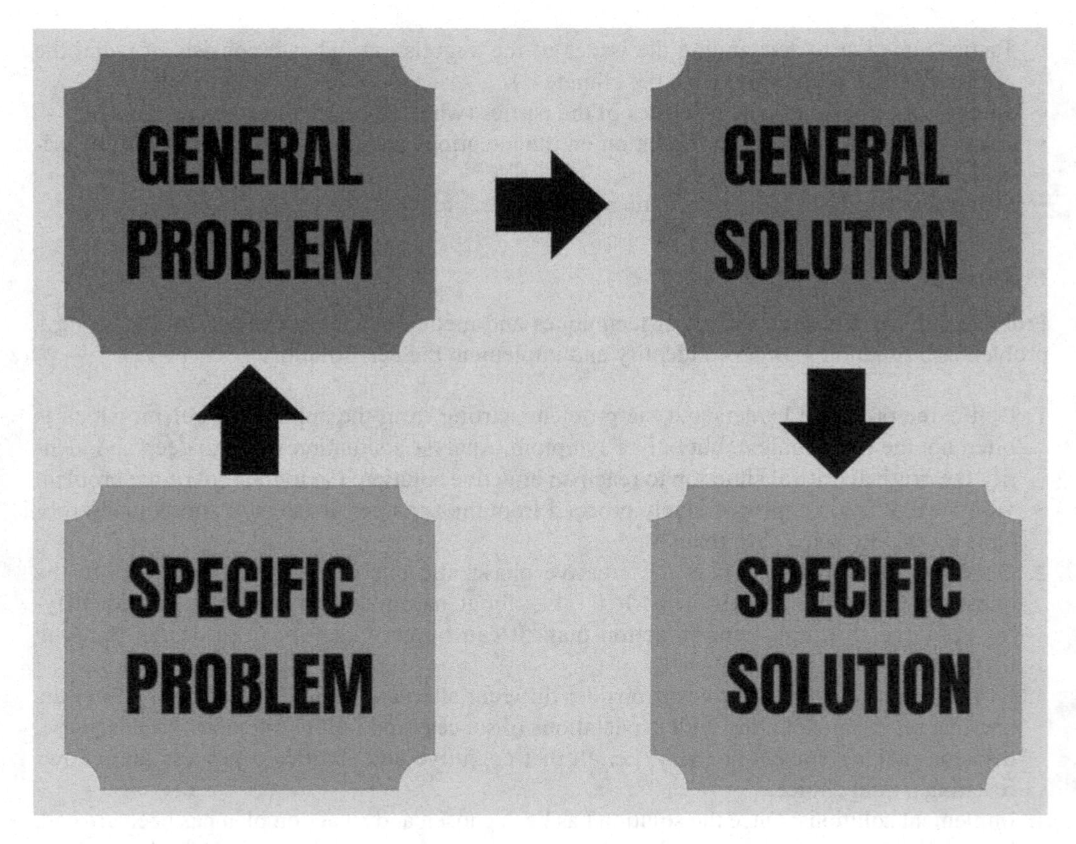

Figure 8.3 Description of the TRIZ process. The arrows represent transformation from one formulation of the problem or solution to another.

Source: authors' own elaboration

The techniques grouped under the scope of Systematic Innovation are characterized by a scientific approach to problem-solving: the original problem is traced back to a problem model, through a path of abstraction, and then general solution models are applied to formulate a detailed solution.

The TRIZ method consists of a series of tools that allow for systematic and scientific problem-solving for technical and technological issues.

The objective is to help companies establish a long-term technological strategy in order to maintain a constant competitive advantage supported by systematic innovation of products and processes.

Compromises and contradictions The TRIZ method was developed by Genrikh Altshuller, a brilliant Russian patent expert. Analysing hundreds of thousands of patents, Altshuller discovered how to systematically solve the engineering contradictions inherent in the innovation process.

He noticed that the most ambitious problems already include contradictory requests (contradictions) from the outset. Therefore, he came to believe that many solutions resort to compromise, where the improvement of one parameter is achieved at the expense of others. Such a solution means that at least one of the system's solutions is not satisfied.

When the contradiction at the basis of the compromise can be formulated, it becomes possible to overcome the problem in an innovative way.

With the TRIZ method, it is possible to enhance the performance of technological products and processes by reducing costs and waste through the development of a specific problem identification and analysis path, its abstraction as a general problem of a principle (engineering contradiction), identification of TRIZ's problem-solving patterns, and application of the most effective solution to solve the initial problem.

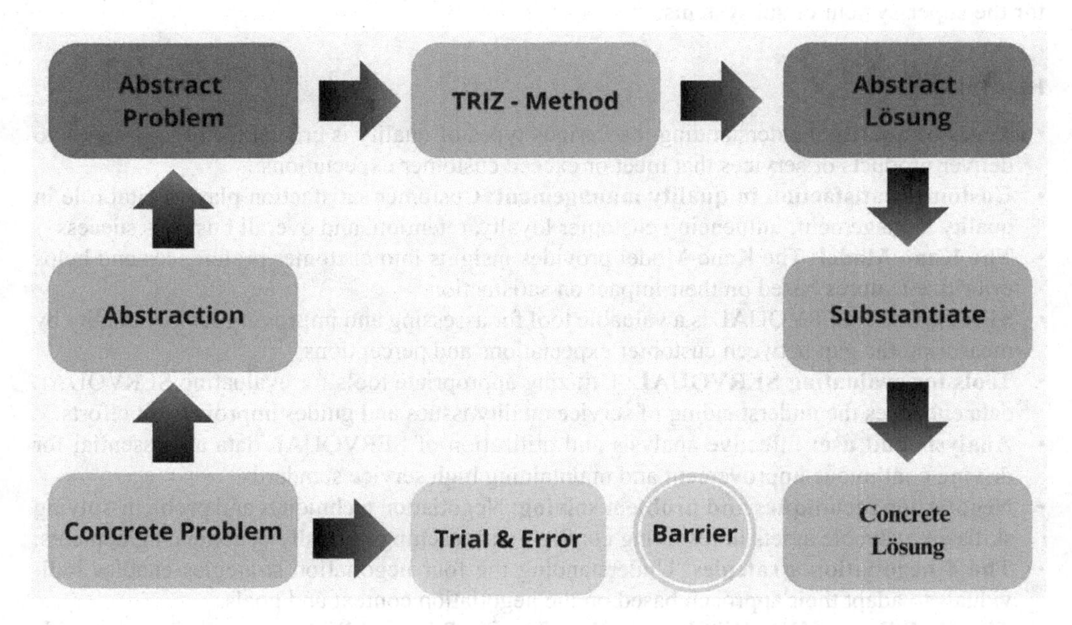

Figure 8.4 TRIZ Method.
Source: authors' own elaboration

To innovate effectively, it is essential to identify the right problem to work on and apply TRIZ tools, including parametric technical contradictions, identification and analysis, 40 inventive principles, and Altshuller's Matrix of Contradictions.

Altshuller proposed a scheme, the Contradiction Matrix. The rows of the matrix report the parameters that may need improvement, while the columns indicate the undesired effects that may arise following an improvement of the previous parameters. The intersection between rows and columns shows the most statistically effective inventive principles for solving the problem at hand. These principles, 40 in total, emerged from the initial study of two hundred thousand patents. This study has been carried out even after Altshuller, and today more than 3 million patents have been examined, confirming, refining, and finding new developments for the approach proposed by the founder.

- Root Cause Analysis;
- Simplification ("trimming");
- ARIZ.

ARIZ, which stands for Algorithm of Inventive Problem Solving, is a component of TRIZ that is not as widely utilized as other TRIZ methods due to its demanding nature and requirement for significant intellectual effort. This technique is reserved for resolving intricate problems that cannot be addressed using other TRIZ methods. It is a tool for exploring alternative approaches to solve complex and non-standard problems. Alternatively, ARIZ is a problem-solving approach that entails redefining and restructuring the problem. Although problems can be tackled using other methods, it is advisable to assess the proposed solution using ARIZ. ARIZ guides you back and forth through the functional domains of the super-system, system, and subsystem to ensure that the derived solution does not create issues for the super-system or subsystems.

Key Takeaways

- **Types of quality**: Understanding the various types of quality is crucial for organizations to deliver products or services that meet or exceed customer expectations.
- **Customer satisfaction in quality management**: Customer satisfaction plays a vital role in quality management, influencing customer loyalty, retention, and overall business success.
- **The Kano Model**: The Kano Model provides insights into customer preferences and helps prioritize features based on their impact on satisfaction.
- **SERVQUAL**: SERVQUAL is a valuable tool for assessing and improving service quality by measuring the gap between customer expectations and perceptions.
- **Tools for evaluating SERVQUAL**: Utilizing appropriate tools for evaluating SERVQUAL data enhances the understanding of service quality issues and guides improvement efforts.
- **Analysis and use**: Effective analysis and utilization of SERVQUAL data are essential for driving continuous improvement and maintaining high service standards.
- **Negotiation techniques and problem solving**: Negotiation techniques and problem-solving skills are valuable assets in resolving conflicts and reaching mutually beneficial agreements.
- **The 4 negotiation strategies**: Understanding the four negotiation strategies enables individuals to adapt their approach based on the negotiation context and goals.
- **The Six Pillars of Win-Win Negotiation**: The Six Pillars of Win-Win Negotiation provide a framework for achieving outcomes that satisfy the interests of all parties involved.

- **Prospect theory**: Prospect Theory highlights how individuals perceive and evaluate risks and rewards, influencing decision-making processes.
- **Two phases of decision-making processes in prospect theory**: Recognizing the two phases of decision-making processes in Prospect Theory helps in understanding and predicting human behaviour in uncertain situations.
- **Mediation strategies**: Mediation strategies offer effective approaches for resolving disputes and facilitating agreements between conflicting parties.
- **Problem solving**: Leveraging problem-solving techniques like TRIZ empowers individuals to generate innovative solutions to complex challenges.

Review Questions

1. What are the different types of quality, and how do they impact product or service delivery?
2. How can organizations ensure consistency in delivering high-quality products or services across different types of quality?
3. What strategies can organizations employ to measure and improve customer satisfaction?
4. How does the Kano Model classify customer preferences?
5. What role does the Kano Model play in product development and customer satisfaction management?
6. What is the SERVQUAL model, and how does it measure service quality?
7. What are some tools commonly used for evaluating SERVQUAL data?
8. What are the key steps involved in using SERVQUAL data for improving service quality?
9. What are negotiation techniques, and how do they contribute to problem-solving?
10. What are the four negotiation strategies, and when is each strategy most appropriate?
11. What are the six pillars of win-win negotiation?
12. What is Prospect Theory, and how does it influence decision-making processes?
13. What are the two phases of decision-making processes in Prospect Theory?
14. What are mediation strategies, and when are they used?
15. What is problem-solving, and why is it important?
16. What is TRIZ, and how does it differ from conventional problem-solving approaches?

9 Case Studies

TQM, Lean Thinking Practice and Six Sigma

1. Driving Force Analysis of Participants in Lean Management Implementation

---

Sector: Manufacturing

Description: This case study focuses on the implementation of lean management in a large engine manufacturing company. The company, founded in 1946, has become a leading manufacturer in the industry. The study examines the driving force of participants in the management innovation process during the implementation of lean management. The implementation process is divided into four stages: point, line, plane, and cube. The study collects interview data from consulting teams and promotion groups within the company to analyse participant characteristics at each implementation stage.

Benefits: The analysis of participant characteristics in the management innovation process provides insights into the roles and contributions of participants at different stages of lean management implementation. This understanding can help identify areas of improvement and optimize the effectiveness of lean management practices. The case study also highlights the importance of participant engagement and driving force in successful management innovation.

Tool as best practice: The study utilizes a driving force analysis system that combines terms mining techniques (TMT) and fuzzy proximity (FP). TMT is used to analyse interview texts and identify participant character patterns, while FP is applied to evaluate the driving force of participants in the management innovation process. This integrated approach provides a comprehensive understanding of participant dynamics and their impact on lean management implementation.

This case study focuses on the implementation of lean management in a large engine manufacturing company. The company, which was founded in 1946, has become a leading manufacturer in the industry with annual sales income of $17 billion. This case study aims to analyse the driving force of participants in the management innovation process during the implementation of lean management.

The implementation process is divided into four stages: point, line, plane, and cube. The study collects interview data from consulting teams and promotion groups within the company to analyse participant characteristics at each implementation stage. The interviews are focused on understanding the roles and contributions of participants in the management innovation process.

The analysis of participant characteristics in the management innovation process provides insights into the roles and contributions of participants at different stages of lean management implementation. This understanding can help identify areas of improvement and optimize the

DOI: 10.4324/9781032726748-9

effectiveness of lean management practices. The case study also highlights the importance of participant engagement and driving force in successful management innovation.

The study utilizes a driving force analysis system that combines term mining techniques (TMT) and fuzzy proximity (FP). TMT is used to analyse interview texts and identify participant character patterns, while FP is applied to evaluate the driving force of participants in the management innovation process. This integrated approach provides a comprehensive understanding of participant dynamics and their impact on lean management implementation.

The case study provides valuable insights into the implementation of lean management in a large engine manufacturing company. By analysing participant characteristics and driving forces, the study offers practical recommendations for improving the effectiveness of lean management practices. The findings can be used as a best practice tool for other companies in the manufacturing sector looking to implement lean management strategies.

This case study contributes to the existing body of knowledge on lean management implementation and highlights the importance of participant engagement and driving force in successful management innovation. The integrated approach of term mining techniques and fuzzy proximity analysis provides a comprehensive understanding of participant dynamics and their impact on lean management implementation.

The TMT-FP analysis system combines TMT and FP to evaluate the driving force of participants in management innovation.

The TMT component of the analysis system is used to analyse interview texts and identify participant character patterns. It applies non-linear integer programming (NIP) to minimize differences between pair-words for each category and extracts key words from the texts. The extracted key words are then filtered and their frequency is calculated.

The FP component of the analysis system is based on fuzzy set theory and is used to assess the driving force of participants. It utilizes unsymmetrical proximity to evaluate the driving force of participants in the processes of management innovation.

The TMT-FP analysis system works in two stages. In the first stage, interview cases are collected from enterprises that have implemented lean management. TMT is used to analyse the interview texts and identify participant character patterns. In the second stage, FP is applied to build a driving force analysis model and analyse the driving force of participants in the management innovation process.

By combining TMT and FP, the TMT-FP analysis system provides a comprehensive understanding of participant dynamics and their impact on management innovation. It allows for the evaluation of the driving force of participants and the design of a scientific implementation mechanism for lean management.

Overall, the TMT-FP analysis system utilizes text analysis and fuzzy proximity to assess the driving force of participants in management innovation, providing valuable insights for optimizing lean management practices.

According to the case study, the driving force of different participants in lean management implementation varied at different stages and processes of management innovation. The results showed that:

1. At the "point" stage of lean management implementation, the driving forces of lean experts on the management innovation processes of searching for innovation motivation and designing innovation plan were the highest. The driving forces of basic production personnel and senior managers on the implementation of innovation process were also significant.

2. At the "line" stage of lean management implementation, the driving forces of basic production personnel, lean experts, and junior managers on searching for innovation motivation and designing of innovation plan were higher compared to senior managers and middle managers.
3. At the "plane" stage of lean management implementation, the driving forces of basic production personnel, lean experts, and junior managers on searching for innovation motivation and designing of innovation plan were higher compared to senior managers and middle managers.
4. At the "cube" stage of lean management implementation, the driving forces of basic production personnel, lean experts, and junior managers on searching for innovation motivation and designing of innovation plan were higher compared to senior managers and middle managers.

Overall, the results indicated that the driving forces of participants varied depending on their roles and stages of lean management implementation. Lean experts and basic production personnel consistently showed higher driving forces in the management innovation processes, while senior managers and middle managers had lower driving forces.

Understanding participant characteristics and designing implementation mechanisms accordingly can enhance the effectiveness of lean management implementation and improve management innovation in enterprises in several ways:

1. **Tailored approach**: By understanding the characteristics of different participants, such as senior managers, middle managers, junior managers, basic production personnel, and lean experts, companies can design lean management strategies that are tailored to the specific needs and capabilities of each group. This ensures that the implementation process is aligned with the skills and motivations of the participants, increasing the likelihood of successful adoption and implementation.
2. **Increased engagement**: When participants feel that their roles and contributions are recognized and valued, they are more likely to be engaged and committed to the implementation process. By designing implementation mechanisms that take into account participant characteristics, companies can create an environment that fosters active participation and collaboration, leading to higher levels of engagement and motivation.
3. **Effective communication**: Understanding participant characteristics can help companies develop effective communication strategies during the implementation process. Different participants may have different communication preferences and styles, and by tailoring communication approaches to their needs, companies can ensure that information is effectively conveyed and understood. This promotes clarity, alignment, and a shared understanding of the goals and objectives of lean management implementation.
4. **Overcoming resistance**: Participant characteristics can also provide insights into potential sources of resistance to change. By identifying potential barriers and understanding the underlying reasons for resistance, companies can proactively address concerns and develop strategies to overcome resistance. This may involve providing additional training, addressing misconceptions, or involving participants in decision-making processes to increase their buy-in and support for the implementation.

Overall, understanding participant characteristics and designing implementation mechanisms accordingly can create a supportive and conducive environment for lean management implementation. This enhances participant engagement, communication, and collaboration, leading to improved effectiveness and successful management innovation in enterprises.

2. Integrating Resilience Engineering and Lean Management to Improve the Performance of Clinical Departments

---

Sector: Healthcare

Description: This case study aimed to evaluate and enhance the performance of clinical departments in a healthcare setting by integrating principles of Resilience Engineering (RE) and Lean Management (LM). The study utilized a comprehensive framework that included the identification of key indicators, data collection through a validated questionnaire, and the application of the Data Envelopment Analysis (DEA) model to assess performance. The impact of each indicator on the clinical department was analysed using robustness analysis and statistical tests. The study also proposed appropriate strategies for improvement based on a strengths-weaknesses-opportunities-threats (SWOT) analysis.

Benefits: The study provided a quantitative evaluation of clinical department performance in terms of RE and LM indicators. By identifying the strengths, weaknesses, opportunities, and threats of the clinical department, the study offered insights for managers and decision-makers to enhance performance and make the department more resilient and cost-beneficial. The framework presented in the study can be implemented in various healthcare units to improve their RE- and LM-related performance.

Tool as best practice: The integration of RE and LM principles in evaluating and improving the performance of clinical departments can be considered a best practice in the healthcare sector. The study utilized the Best-Worst Method (BWM) for determining the relative importance of indicators, the DEA model for performance evaluation, and SWOT analysis for strategy development. These tools can serve as a guide for healthcare organizations seeking to enhance their clinical department performance.

In this case study, the authors propose a comprehensive framework to evaluate and improve the performance of clinical departments in healthcare. The framework integrates principles of Resilience Engineering (RE) and Lean Management (LM) to assess the performance of clinical departments and identify areas for improvement.

The first step in evaluating the performance of clinical departments is to determine the relative importance of each indicator. The authors utilize the Best-Worst Method (BWM) to establish the significance of each indicator in relation to RE and LM principles. This helps prioritize the indicators and focus on the most critical aspects of performance.

Next, a validated questionnaire is administered to clinical department specialists to collect data on various indicators. This data collection process allows for a comprehensive assessment of the clinical department's performance. The authors ensure the reliability of the data by using Cronbach's alpha to validate the questionnaire.

To evaluate the performance of the clinical department, the authors employ the Data Envelopment Analysis (DEA) model. This model assesses the efficiency and effectiveness of the clinical department based on the collected data. It provides a quantitative evaluation of performance and allows for comparisons between different clinical departments.

The impact of each indicator on the clinical department's performance is analysed through robustness analysis and statistical tests. This analysis helps identify the strengths and weaknesses of the clinical department and provides insights for improvement. By understanding the effects of each indicator, managers and decision-makers can prioritize areas for enhancement.

Additionally, a strengths-weaknesses-opportunities-threats (SWOT) analysis is conducted to propose appropriate strategies for improving the performance of the clinical department. This analysis considers the internal strengths and weaknesses of the department, as well as external opportunities and threats. The SWOT analysis helps identify potential strategies that can enhance performance and make the department more resilient and cost-beneficial.

Overall, this case study proposes a comprehensive approach to evaluate and improve the performance of clinical departments in healthcare. By integrating principles of RE and LM, the study provides a robust framework for assessing performance, identifying areas for improvement, and implementing strategies to enhance the resilience and efficiency of clinical departments.

In the framework proposed in the case study, several methods are used to assess the performance of clinical departments. These methods include:

1. Best-Worst Method (BWM): The BWM is used to determine the relative importance of each indicator in relation to RE and LM principles.
2. Validated Questionnaire: A standardized questionnaire is developed and administered to clinical department specialists to collect data on various indicators.
3. Data Envelopment Analysis (DEA) Model: The DEA model is employed to evaluate the efficiency and effectiveness of the clinical department based on the collected data. It provides a quantitative evaluation of performance and allows for comparisons between different clinical departments.
4. Robustness Analysis: Robustness analysis is performed to determine the impact of each indicator on the clinical department's performance. Efficiency scores are obtained by removing each indicator one at a time and running the DEA model for the remaining indicators.
5. Statistical Tests: Statistical tests, such as the Wilcoxon rank-sum test, are conducted to assess the significance of the indicators on the clinical department's performance.

Overall, these methods are used in combination to comprehensively assess the performance of clinical departments and identify areas for improvement.

The study identifies the best and worst indicators for each clinical department using the BWM. The BWM is a multi-criteria decision-making method that allows decision-makers to determine the relative importance of different criteria or indicators.

In the case study, a group of experts, including managers of the clinics, chief residents, and head nurses, participated in the BWM process. They were asked to compare and rank the indicators based on their perceived importance in relation to RE and LM principles.

The experts assigned a number ranging from 1 to 9 to each pairwise comparison, indicating the preference of one indicator over another. The weighted geometric mean of these numbers was then calculated to generate the best-to-others vector for each indicator.

By analysing the best-to-others vectors, the study determined the best and worst indicators for each clinical department. The indicators with higher weights in the BWM analysis were considered the best indicators, while those with lower weights were considered the worst indicators.

This process allowed the study to identify the indicators that had the most significant impact on the performance of each clinical department, providing insights into areas of strength and areas that required improvement.

3. Value Stream Mapping in Precast Component Manufacturing

Sector: Precast component manufacturing

Description: The case study focused on the application of Value Stream Mapping (VSM) in precast component manufacturing. It aimed to recognize and eradicate waste in the manufacturing process to improve efficiency and productivity. The study used empirical analysis to gather data and analyse the current state of the manufacturing process, and then implemented VSM to identify areas of improvement and develop a future state map.

Benefits: The application of VSM in precast component manufacturing resulted in several benefits. It helped to find and remove non-value-added activities, lessen lead time, improve resource utilization, and enhance overall process efficiency. The case study demonstrated the effectiveness of VSM in improving the manufacturing process and achieving lean objectives.

Tool as best practice: VSM was used as a best practice tool in this case study to analyse and improve the precast component manufacturing process.

This case study focused on the application of Value Stream Mapping (VSM) in the precast component manufacturing sector. The study aimed to recognize and remove waste in the manufacturing process to improve efficiency and productivity. It utilized empirical analysis to gather data and analyse the current state of the manufacturing process, and then implemented VSM to identify areas of improvement and develop a future state map.

The benefits of applying VSM in precast component manufacturing were significant and, resulted in increased productivity and cost savings for the company.

The case study highlighted the use of VSM as a best practice tool in precast component manufacturing. VSM is a lean manufacturing technique that allows companies to visualize and analyse their production processes, identify bottlenecks and waste, and develop strategies for improvement. By implementing VSM, companies can streamline their operations, reduce waste, and improve overall efficiency.

In terms of key research dimensions and gaps identified in the application of lean principles in prefabricated construction, several studies have been conducted in this field. One research dimension is the data-driven intelligent decision-making in prefabricated construction. This involves using data analytics and advanced technologies to optimize decision-making processes and improve project outcomes.

Another research dimension is construction sustainability in prefabricated construction. This involves integrating lean principles with sustainable practices to minimize environmental impact, reduce waste, and promote sustainable development in the construction industry.

Activity process optimization is another research dimension identified in the application of lean principles in prefabricated construction. This involves analysing and optimizing the various activities involved in the prefabrication process to improve efficiency and productivity.

The adoption of lean construction principles in prefabricated construction is also a research dimension. This involves studying the implementation of lean techniques and tools in prefabricated construction projects and evaluating their impact on project performance.

Other research dimensions include exploring lean adoption strategies such as circular economy, flexible field factory, and on-site industrialization in prefabricated construction.

Despite the progress made in the application of lean principles in prefabricated construction, there are still some research gaps that need to be addressed. One gap is the need for a systematic framework that can be used to promote the application of lean manufacturing principles in the construction.

The main tools adopted in this case study VSM and Process Flow Mapping (PFM).

VSM is a lean manufacturing tool used to visualize and analyse the flow of materials and information in a production process. It helps identify waste, bottlenecks, and areas for improvement. In the case study, VSM was used to analyse the current state of the precast component manufacturing process and develop a future state map to improve efficiency and productivity.

PFM is another lean tool used to map out the sequence of activities in a process. It helps identify the steps involved, their sequence, and the interactions between different activities. In the case study, PFM was used to map out the current state of the precast component manufacturing process and identify areas of improvement.

These tools were used in combination to analyse the current state of the manufacturing process, identify waste and inefficiencies, and develop a future state map to improve the process. By utilizing VSM and PFM, the case study aimed to optimize the flow of materials and information, reduce lead time, and improve overall process efficiency.

Lean construction contributes to energy savings, low carbon emissions, and economic benefits in prefabricated construction projects through several mechanisms:

1. Waste reduction: Lean construction principles aim to eliminate waste in all forms, including material waste, time waste, and energy waste. By optimizing processes and reducing unnecessary activities, lean construction minimizes the consumption of energy and resources, leading to energy savings and reduced carbon emissions.
2. Streamlined production: Lean construction emphasizes the efficient flow of materials and information throughout the construction process. In prefabricated construction, this translates to streamlined production in controlled factory environments. By optimizing the production flow and reducing transportation and handling, lean construction reduces energy consumption and carbon emissions associated with on-site construction activities.
3. Standardization and modularization: Lean construction promotes standardization and modularization in design and construction processes. Standardized components and modules in prefabricated construction enable efficient production, reduced material waste, and improved energy efficiency. Additionally, modular construction allows for better energy performance through enhanced insulation and reduced thermal bridging.
4. Continuous improvement: Lean construction encourages a culture of continuous improvement, where teams actively seek opportunities to enhance efficiency and reduce waste. This includes identifying energy-saving measures, optimizing equipment and machinery, and implementing sustainable practices. Continuous improvement efforts in lean construction contribute to long-term energy savings and economic benefits.
5. Life cycle perspective: Lean construction considers the entire life cycle of a building, including its operation and maintenance. By optimizing design and construction processes, lean construction can improve the energy performance of prefabricated buildings, leading to reduced energy consumption and lower carbon emissions throughout the building's life cycle.

Lean construction in prefabricated construction projects contributes to energy savings, low carbon emissions, and economic benefits by reducing waste, streamlining production, promoting standardization and modularization, fostering continuous improvement, and adopting a life cycle perspective.

Potential research directions and opportunities for future studies in lean prefabricated construction include:

- **Digital innovation**: Future research can explore the integration of digital technologies, such as Building Information Modelling (BIM), Internet of Things (IoT), and Artificial Intelligence

(AI), in lean prefabricated construction. This can involve developing digital platforms for information integration, optimizing supply chain management through real-time monitoring and data analytics, and utilizing digital tools for lean decision-making.

- **Sustainability goals**: Research can focus on aligning lean principles with sustainability goals in prefabricated construction. This can involve investigating the environmental impacts of prefabrication processes, exploring strategies to reduce carbon emissions and waste, and evaluating the life cycle sustainability performance of prefabricated buildings.
- **Lean-oriented simulation modules**: Future studies can develop simulation modules specifically designed for lean prefabricated construction. These modules can simulate and optimize various aspects of the prefabrication process, such as production flow, resource allocation, and logistics planning. The use of simulation can help identify bottlenecks, improve efficiency, and support decision-making in lean prefabricated construction projects.
- **Supply chain management**: Research can focus on lean supply chain management in prefabricated construction. This can involve studying strategies to optimize the coordination and collaboration among different stakeholders in the supply chain, exploring innovative procurement models, and investigating the impact of lean principles on supply chain performance.
- **Lean performance measurement**: Future studies can develop comprehensive and standardized performance measurement frameworks for lean prefabricated construction. This can involve identifying key performance indicators (KPIs) that capture the efficiency, productivity, quality, and sustainability aspects of lean prefabricated construction projects. The development of robust performance measurement frameworks can facilitate benchmarking, continuous improvement, and knowledge sharing in the industry.

Overall, future research in lean prefabricated construction can explore the potential of digital innovation, align with sustainability goals, develop lean-oriented simulation modules, investigate supply chain management strategies, and establish comprehensive performance measurement frameworks.

4. Implementation of Lean Construction Practices in an Industrial Building Construction Project

---

Sector: Construction industry

Description: The case study was conducted on an industrial building construction site in Suzhou, China. The project involved the construction of a six-storey car park with two spiral ramps. The study focused on the implementation of Lean Construction Management (LCM) methods and Lean Construction (LC) techniques throughout the project process.

Benefits: The implementation of lean practices in the case study project brought significant benefits in terms of schedule, workflow, quality, and safety. The use of prefabricated elements led to more precise engineering, reduced on-site waste generation, improved mechanical and durability performance, avoidance of hazards, and accelerated construction progress. The adoption of digital technology and IoT allowed for better control of the project, including 3D visualization, construction process simulation, clash detection, and in-time inspection. The reduction of waiting time and defects were identified as the most distinguishable benefits of implementing lean practices in the project. Other benefits included improved construction workflow and schedule, project quality and productivity, on-site health and safety, and communication and collaboration among stakeholders.

Tool as best practice: The case study employed various lean construction techniques and methods, including Last Planner System (LPS), Kanban system, Just-in-Time (JIT), concurrent engineering, constraint analysis, quality and safety management, 5S method, and target value delivery. These practices were identified as best practices in lean construction and were found to contribute to the project's success.

This case study discusses the implementation of lean construction techniques and management methods in the Chinese construction industry, using a case study in Suzhou, China. The study found that the adoption of lean practices has led to improved project performance, including maximizing project value, shortening project schedules, improving project quality, and reducing waste. The study also identified challenges such as lack of trust and stakeholder abilities in implementing lean practices in China.

In the case study conducted in Suzhou, China, several specific lean practices were implemented in the industrial building construction project. These practices included:

1. Last Planner System (LPS): LPS was widely recognized and utilized in the case study project. It is a lean construction technique that focuses on collaborative planning and scheduling, involving all project stakeholders. LPS helps to improve workflow, coordination, and communication among team members.
2. Kanban system: The Kanban system was identified as a core method that supported construction progress in the case study project. Kanban is a visual management tool that helps to control and monitor the flow of materials and tasks in construction projects. It improves efficiency, reduces waste, and enhances communication between stakeholders.
3. Just-in-Time (JIT): JIT was implemented in the case study project to minimize waste and reduce inventory. JIT involves delivering materials and resources to the construction site exactly when they are needed, eliminating the need for excess inventory and reducing storage space requirements.
4. Prefabrication: The case study project utilized prefabricated elements, which is a lean construction technique that involves manufacturing components off-site and then assembling them on-site. Prefabrication improves construction quality, reduces waste, and accelerates construction progress.
5. Internet of Things (IoT): The implementation of IoT in the case study project allowed for better control and monitoring of the construction process. IoT technologies, such as sensors and data analytics, provided real-time information on construction progress, quality, and safety. This enabled proactive decision-making and improved project management.
6. Quality and safety management: The case study project emphasized the importance of quality and safety management as part of lean practices. By implementing robust quality control measures and safety protocols, the project aimed to minimize defects, accidents, and rework, leading to improved project outcomes.

These specific lean practices were identified as best practices in lean construction and were found to contribute to the success of the project in terms of improved workflow, schedule, quality, and safety.

The main tools adopted in the case study project in Suzhou, China include:

1. Building Information Modelling (BIM): BIM was used as a tool to present a clear visual representation of the design scheme, facilitate clash detection, and improve communication among stakeholders.
2. Aconex and Aconex Field: Aconex is an information collaboration platform that provides project information and process management services. Aconex Field, a Plan-Do-Check-Act (PDCA) inspection tool, was used for site inspections and issue tracking.
3. I-site and QR code: I-site is a content management platform connected to intelligent safety helmets worn by site personnel. It provides a flow heat map, manpower management, and attendance tracking. QR codes were used to record production and inspection procedures for prefabricated components.

These tools were utilized to support the implementation of lean practices and improve project management, quality control, and safety monitoring.

The main benefits of implementing lean practices in the Chinese construction industry include:

1. Improved construction workflow and schedule: Lean practices such as Last Planner System (LPS) and Kanban system help to streamline the construction process, reduce waiting time, and improve overall project schedule.
2. Enhanced project quality and productivity: Lean practices focus on eliminating waste and improving efficiency, leading to higher quality construction and increased productivity.
3. Reduced on-site waste generation: Lean practices, including prefabrication and just-in-time delivery, minimize waste and optimize the use of resources, resulting in cost savings and environmental benefits.
4. Improved on-site health and safety: Lean practices emphasize safety management and the identification and mitigation of hazards, leading to a safer working environment for construction workers.
5. Better communication and collaboration among stakeholders: Lean practices promote collaboration and communication among project stakeholders, leading to improved coordination and decision-making.

These benefits were identified in the case study conducted in Suzhou, China, where lean practices were implemented in an industrial building construction project.

According to the case study, the challenges identified in implementing lean practices in China include:

1. Difficulty of convincing users to align their requirements with lean thinking: Users may be resistant to change and may not fully understand the benefits of lean practices, making it challenging to align their requirements with lean principles.
2. Stakeholders' lack of time and willingness to learn lean thinking, techniques, and management: Stakeholders may not prioritize learning about lean practices or may not have the time to invest in understanding and implementing them.
3. Educational level and customs: Traditional customs and knowledge of conventional construction methods can be difficult to change, hindering the adoption of lean techniques and management.
4. Manpower management: Managing the workload and worktime of construction personnel can be challenging, as lean practices prioritize respect for humanity and may require adjustments to traditional work practices.
5. Profit distribution: Implementing lean practices may initially cut down on contractors' profits, which can be a barrier to their adoption.
6. Insufficient design and construction time: High demand and rapid expansion of the architecture, engineering, and construction (AEC) industry in China can lead to insufficient time for design and construction, making it challenging to implement lean practices effectively.
7. Policy adjustments: Changes in government policies can delay project schedules and increase costs, potentially leading to project abandonment.

These challenges highlight the need for addressing cultural, organizational, and systemic barriers to the successful implementation of lean practices in the Chinese construction industry.

5. Unleashing Innovation: The Power of Lean Innovation Capability in Resource-Constrained Environments

---

Sector: Start-up enterprises operating within vibrant sectors in the United States

Description: The case study centred on innovation under resource constraints within organizations, particularly focusing on start-up firms. It encompassed 10 start-ups from diverse industries in the US, with a specific focus on the innovation journey of the participants. The study sought to comprehend the essence and workings of lean innovation in environments with limited resources, where firms commonly confront shortages in financial, human, and material resources.

Benefits: The study provided insights into how start-up firms in resource-constrained environments can leverage lean innovation capability to drive innovation and develop new competences or skills. It emphasized the significance of a targeted, validity-oriented approach in the innovation process, along with the beneficial outcomes of embracing a "can-do" mindset through the utilization of available resources in innovative combinations. This involves rapid and iterative learning from market feedback to address new challenges and opportunities.

Tool as best practice: The project adopted a multi-case, discovery-oriented, theory-in-use, grounded theory methodology to explore lean innovation capability within resource-constrained environments. It employed theoretical sampling to determine case selection and utilized semi-structured interviews, industry documents, archival data, and company documents as primary and supplementary sources of data.

This case study discusses the concept of Lean Innovation Capability (LIC) and its importance in enabling firms to innovate with limited resources. The work identifies four dimensions of LIC: focus on product-market fit, experimentation culture, mission-oriented leadership, and network learning capability. The research fills a gap in understanding how firms can successfully innovate under resource constraints and has implications for innovation management practices.

The case study focuses on the concept of Lean Innovation Capability (LIC) and its impact on innovation outcomes in resource-constrained environments, particularly within start-up firms. The study aims to fill a gap in understanding how firms can successfully innovate under resource constraints and has implications for innovation management practices.

The case study identified four dimensions of LIC: focus on product-market fit, experimentation culture, mission-oriented leadership, and network learning capability. These dimensions highlight the importance of aligning products with market needs, fostering a culture of experimentation, having leadership that is driven by a clear mission, and leveraging network learning for successful innovation.

The research involved the development and validation of the LIC construct through item generation, expert review, and pilot testing. The findings suggest that resource-constrained companies thrive in an experimentation culture, have mission-oriented leadership, and leverage network learning for successful innovation. This indicates that LIC is crucial for firms to overcome resource limitations and achieve innovation success, especially in a munificent environment.

The study also conducted a large sample survey to measure LIC and assess its nomological validity. The results showed that LIC has a significant moderating effect on the relationship between relative resource constraints and innovation performance. This suggests that LIC can mitigate the negative impact of resource constraints on innovation performance.

Additionally, the research revealed that Lean Innovation Capability (LIC) moderates the correlation between resource limitations and innovation performance. Moreover, it identified a three-way interaction effect among resource constraints, LIC, and environmental abundance on innovation performance.

This indicates that LIC plays a crucial role in enhancing innovation performance, particularly in resource-constrained environments.

The case study recognizes its limitations and proposes directions for future research. It explores theoretical, methodological, and managerial implications of the findings, offering practical recommendations for firms to assess and enhance Lean Innovation Capability (LIC). In summary, the case study offers valuable insights into the role of LIC in facilitating successful innovation under resource constraints. It emphasizes LIC's dimensions and its moderating influence on the connection between resource constraints and innovation performance, providing actionable insights for firms and innovation strategies.

The study also acknowledges the potential of LIC to drive successful innovation in resource-constrained environments and suggests future research directions in this area.

Indeed, the four dimensions of Lean Innovation Capability (LIC) identified in the study are:

1. Focus on Product-Market Fit: This dimension emphasizes the organization's ability to align its products with the needs and demands of the market. It involves continuously iterating and refining offerings based on market feedback to ensure that the products or services meet core customer needs. This dimension reflects the organization's agility and responsiveness to market dynamics, enabling it to achieve sustainable innovation performance.
2. Experimentation Culture: This dimension refers to the organization's culture of experimentation and continuous learning. It involves a willingness to take calculated risks, test new ideas, and learn from both successes and failures. An experimentation culture fosters an environment where employees are encouraged to explore new approaches, challenge assumptions, and adapt quickly based on the insights gained from experimentation.
3. Mission-Oriented Leadership: This dimension highlights the importance of leadership that is driven by a clear mission. Mission-oriented leadership provides a sense of purpose and direction, aligning the organization's efforts towards achieving its innovation goals. It involves leaders who inspire and motivate their teams, communicate a compelling vision, and create a supportive environment for innovation.
4. Network Learning Capability: This dimension focuses on the organization's ability to leverage networks for learning and knowledge exchange. It involves building and nurturing multi-stakeholder networks, collaborating with external partners, and tapping into diverse sources of expertise and insights. Network learning capability enables the organization to access valuable resources, information, and perspectives that contribute to its innovation efforts.

These dimensions collectively represent the core competencies of firms that excel in managing innovation with limited resources. They underscore the importance of aligning products with market needs, fostering a culture of experimentation, having mission-oriented leadership, and leveraging network learning for successful innovation in resource-constrained environments.

Lean Innovation Capability (LIC) alleviates the adverse effects of resource constraints on innovation performance by moderating the correlation between resource constraints and innovation outcomes. The research indicates that LIC diminishes the detrimental impact of relative resource limitations on innovation, especially within resource-constrained environments.

This moderation effect suggests that firms with a strong LIC are better equipped to overcome resource limitations and achieve innovation success, even when facing significant constraints. Furthermore, the research pinpointed a three-way interaction effect involving relative resource constraints, LIC, and environmental abundance on innovation performance. This underscores the significance of LIC in augmenting innovation results, particularly in resource-limited environments characterized by high environmental abundance.

Additionally, the case study demonstrated that LIC firms in munificent environments significantly outperform non-LIC firms, indicating that LIC proves especially beneficial in environments experiencing increasing industry demand. This suggests that LIC not only mitigates the negative impact of resource constraints but also enables firms to take advantage of favourable external conditions and innovate in a competitively advantageous manner, even with limited resources.

In summary, LIC mitigates the negative impact of resource constraints on innovation performance by moderating the relationship between resource constraints and innovation outcomes, enabling firms to perform better in resource-constrained environments and capitalize on favourable external conditions for innovation.

The theoretical implications of the study's findings on Lean Innovation Capability (LIC) and innovation outcomes in resource-constrained environments are significant. The study enriches innovation literature by showcasing Lean Innovation Capability (LIC) and its moderating role in the link between resource limitations and innovation outcomes. It underscores the significance of the "focus" concept inherent in LIC, highlighting its importance as a determinant in managing innovation within resource constraints. Additionally, the research offers insights into the optimal conditions for LIC effectiveness, especially in environments with abundant resources and increasing industry demand. These findings provide theoretical insights valuable for firms innovating under resource constraints.

Overall, the study's findings have theoretical implications for understanding the role of LIC in managing innovation in resource-constrained environments and offer practical insights for firms looking to enhance their lean innovation capabilities and improve innovation outcomes.

The study's theoretical findings confirm the existence of Lean Innovation Capability (LIC) and underscore the significance of the pervasive "focus" concept within LIC, highlighting its role as a determinant for managing innovation in resource-constrained settings.

From a managerial standpoint, the study's insights hold practical significance for firms aiming to enhance their lean innovation endeavours. Recognizing LIC's importance in influencing innovation performance suggests that companies should actively assess and cultivate LIC subdimensions within their organizations. This indicates that firms can improve their capacity to manage lean innovation, potentially resulting in enhanced innovation outcomes despite resource limitations.

Moreover, the study provides recommendations for firm and innovation strategies rooted in established frameworks like Porter's and Miles and Snow's, offering direction for future research and actionable guidance for firms seeking to navigate resource constraints and foster successful innovation.

6. Lean-Green Implementation Practices in Indian SMEs

Sector: Waste Management

Description: The case study investigated the implementation of lean-green practices in Indian SMEs using an Analytical Hierarchy Process (AHP) approach. The study aimed to improve the sustainability of waste management by integrating lean and green practices within the SMEs.

Benefits: The study aimed to recognize the benefits of implementing lean-green practices in Indian SMEs, particularly in the context of waste management. The benefits included improved sustainability, waste reduction, and environmental impact mitigation.

Tool as best practice: The Analytical Hierarchy Process (AHP) approach was utilized as a best practice tool to investigate and evaluate the implementation of lean-green practices in Indian SMEs for waste management for sustainability.

This case study presents a systematic literature review on the relationship between Lean Supply Chain Management (LSCM) and performance. The review identifies two research lines: LSCM performance-based models and LSCM's impact on performance. The analysis classifies the literature into research sublines and highlights gaps in the research. The paper also discusses the importance of defining key concepts, formulating research questions, and using a systematic approach in conducting a literature review.

The main research lines identified in the literature on LSCM and performance include:

1. LSCM performance-based models;
2. LSCM's impact on performance.

These research lines have enabled the identification of gaps in the literature and have determined directions for future research.

The number of publications in the field of LSCM and performance has evolved over time. The two main research lines, namely LSCM's impact on performance and LSCM performance-based models, have shown growth in a similar way, although the difference between the publication of papers studying LSCM's impact on performance and the line of LSCM performance-based models can be observed to widen from 2014 onwards.

The gaps in the research on LSCM and performance include the lack of attention to the interrelationships between LSCM and performance, as well as the absence of studies analysing the social and environmental outcomes of LSCM. Future research directions to address these gaps could involve delving into the social and environmental outcomes of LSCM, as well as exploring and identifying the issues addressed in the interrelationships between LSCM and performance.

Additionally, there is a need for empirical studies to address the role of uncertainty and supply uncertainty in LSCM performance, to help managers and practitioners better understand the new challenges and opportunities provided by LSCM. Future research could also focus on exploring unexplored contextual factors such as the country where the LSC operates, the current implementation stage, and firm size, as well as differences across industries.

The provided case study contains a comprehensive overview of a systematic literature review on the relationship between LSCM and performance. The study aims to evaluate the state-of-the-art of research in this area, identify research lines, and propose future research directions. It provides a structured analysis of the literature and offers insights for further research in the field.

The study is structured and systematic, following a well-defined methodology for conducting the literature review. The authors began by defining key concepts and formulating research questions before conducting the review. This approach is essential for ensuring the reliability and validity of the results. The third stage involved defining criteria for including and excluding papers related to the research question. A search string was designed to identify relevant papers, and filters were applied to narrow down the search. Duplicate studies were eliminated, and the remaining papers were examined for relevance. After applying inclusion and exclusion criteria, a total of 162 papers were selected for the literature review. In the fourth stage, the selected papers were analysed and synthesized, and a structured coding process was used to mine relevant details from each study. In the final stage, the authors presented the results of the literature review and proposed a classification of the literature based on the identified research lines and sublines.

The literature on LSCM and its impact on performance is classified into two main research lines: LSCM performance-based models and LSCM's impact on performance. The number of studies on LSCM's impact on performance is greater than those on performance-based models. The research lines have evolved similarly over time, with an increasing number of publications. The papers have been further classified into specific sublines and groups based on their research questions and focus. This classification aims to provide a comprehensive view of the literature and identify research gaps and future challenges.

The largest number of articles focus on models that improve or optimize LSCM performance, with a significant difference in the evolution of publications of the three research sublines. The studies have also been classified according to the methodology used, with most of the studies using quantitative methods.

The results of the paper include the identification of two main research lines in the literature on LSCM and performance: LSCM performance-based models and LSCM's impact on performance. The study also presents a novel classification of the literature on LSCM and performance relationships, providing a structured analysis of the literature and offering insights for further research in the field.

The conclusions drawn from the paper include the identification of gaps in the literature, such as the lack of attention to the interrelationships between LSCM and performance, as well as the absence of studies analysing the social and environmental outcomes of LSCM. The paper also suggests future research directions to address these gaps, such as exploring the social and environmental outcomes of LSCM and delving into the role of uncertainty and supply uncertainty in LSCM performance.

7. Lean Management Implementation in the Hospitality Industry

---

Sector: Hospitality Industry

Description: The case study focuses on the implementation of Lean Management (LM) in the hospitality industry, specifically in homestay establishments. The study explores the application of LM practices in the service operations of five different homestay establishments using the case study methodology. It aims to offer an overview of LM research in the hospitality industry and detect and propose LM research lines in this sector through a systematic literature review.

Benefits: The study measures lean implementation through waste elimination, transparency, quality management, and workforce empowerment and treatment. It identifies and observes three different practitioners of lean management as methodical, admirer, and sustained believer. The study also highlights the potential benefits of LM in the hospitality industry, such as enhancing staff well-being through cost management in crisis periods.

Tool as best practice: The study emphasizes the use of Value Stream Mapping (VSM) and the 5S applications as standout tools for LM in the hospitality industry. It also suggests that more research is required on specific applications in many hotel activities and some under-reported practices.

The case study focuses on the application of Lean Management (LM) practices in the hospitality industry, specifically in homestay establishments. This study employs the case study approach to investigate how LM is applied in the service operations of five distinct homestay establishments. It evaluates the adoption of lean principles by examining the reduction of waste, transparency, quality control, and the empowerment and treatment of the workforce. The research identifies three types of lean management practitioners: methodical adopters, enthusiastic admirers, and consistent believers. The case study also aims to offer an overview of LM research in the hospitality industry and detect and propose LM research lines in this sector through a systematic literature review. It highlights the potential benefits of LM in the hospitality industry, such as enhancing staff well-being through cost management in crisis periods.

The use of Value Stream Mapping (VSM) and the 5S applications as standout tools for LM in the hospitality industry are adopted.

Overall, the case study provides valuable insights into the implementation and potential benefits of Lean Management in the hospitality industry, particularly in homestay establishments. It also highlights the need for further research and exploration of specific LM applications in various hotel activities.

The current state of LM implementation in the hospitality industry, particularly in hotels, is characterized by limited research and application of LM tools. A systematic literature review found that there is a paucity of research studies in the field of LM in the hospitality industry, with only a few references identified. The use of LM tools in hotels appears to be very limited, with VSM and the 5S applications standing out as the most commonly used tools. However, there is a clear need for more research on specific applications of LM in various hotel activities and some under-reported practices.

Research on specific LM tools and their application in the hospitality industry seems to be limited, despite the potential benefits that LM can bring to this sector. The literature review also highlights that lean practices are more applicable to some particular service industries than others, and there is a need for further research to identify and propose lines of research on LM in hotels.

Overall, the current state of LM implementation in the hospitality industry, particularly in hotels, indicates a lack of comprehensive research and limited application of LM tools. There is a clear need for more studies and practical applications to explore the potential benefits of LM in various hotel activities and to address the existing gaps in knowledge.

The potential benefits of implementing LM in the hotel industry are numerous. With the recent COVID-19 pandemic causing a significant crisis in the tourism industry, including the hospitality sector, there is a need for innovative transformation with competitive costs and profitable operations to adapt to the "new normal". Implementing LM can be one of the most attractive and valuable responses for hotels to tackle the need to reduce costs and improve customer service in the current context. Some potential benefits of LM implementation in the hotel industry include:

1. **High Responsiveness to Customer Demand**: LM aims to be highly responsive to customer demand by reducing waste, which can help hotels in providing safe experiences for their customers and adapting to changing customer needs.
2. **Cost Reduction**: LM can help hotels in reducing costs through the elimination of waste and the optimization of processes, which is particularly important in times of crisis such as the COVID-19 pandemic.
3. **Improved Customer Service**: By streamlining processes and focusing on customer value, LM can help hotels in improving customer service and satisfaction.
4. **Efficiency and Productivity**: LM can increase efficiency and productivity in operational processes within the hospitality industry, leading to improved overall performance.
5. **Competitiveness and Efficiency**: Managing hotel processes and activities under LM principles can improve competitiveness and efficiency, which is crucial for the success of hotels in the current market.
6. **Innovation and Technology Integration**: LM can provide an opportunity to improve competitiveness and efficiency in the hotel industry, where innovation and technology, including Industry 4.0 concepts, can play a fundamental role.

These potential benefits highlight the value of implementing LM in the hotel industry, particularly in the current challenging business environment.

The barriers to implementing LM in hotels are multifaceted and include factors such as lack of knowledge, resistance to change, investment costs, long implementation times, and the fragmented nature of hotel departments. Specific barriers identified in the context of the hotel industry include senior management's lack of involvement, wrong project selection, lack of incentives for application, resistance to change, additional workload, extra expenses, lack of knowledge, and customer and supplier involvement.

In terms of specific LM tools and practices that could be beneficial for hotel managers, VSM and the 5S applications are highlighted as standout tools for LM in the hospitality industry. Additionally, practices such as Kaizen, Cellular layout, Poka-yoke, Set-up time reduction, and Kanban control are mentioned as specific applications that require more research in a wide range of hotel activities. These tools and practices can help in streamlining processes, reducing waste, and improving overall efficiency and customer service in the hotel industry.

Furthermore, integrating Industry 4.0 concepts, such as the internet-of-things (IoT), cyber-physical systems (CPS), and big data, with LM is identified as a promising line of research that can lead to the adoption of new business models and provide competitive advantages for hotels.

8. Impact of Industry 4.0 Technologies on Lean and Agile Supply Chain Strategies in Spanish Firms

---

Sector: Supply Chain Management

Description: The case study explores the impact of Industry 4.0 (I4.0) technologies on Lean and Agile supply chain (SC) strategies in Spanish firms. It also delves into the consequences of these strategies on the operational performance of the firms. The study emphasizes the importance of understanding the processes behind the development of Lean and Agile SC strategies and their role in enhancing firm performance.

Benefits: The study provides valuable insights into the relationship between Lean and Agile SC strategies and their consequences for firm operational performance. It also offers managerial guidance to supply chain managers on the value of I4.0 base technologies for supporting Lean and Agile SC strategies and the expected effects on focal firm operational performance.

Tool as best practice: The study serves as a best practice for understanding the impact of I4.0 technologies on SC strategies and firm performance. It also sheds light on the processes behind the development of Lean and Agile SC strategies, thereby providing valuable insights for supply chain managers.

The case study explores the impact of Industry 4.0 (I4.0) technologies on Lean and Agile supply chain (SC) strategies in Spanish firms and their consequences on operational performance. It focuses on the impact of Lean Supply Chain (LSC), Agile Supply Chain (ASC), and Industry 4.0 base technologies on operational performance. The case study uses quantitative methods and data collected through a questionnaire to analyse the influence of these factors. The findings suggest that LSC, ASC, and Industry 4.0 technologies have a positive influence on operational performance. The case study uses a quantitative approach and data collected via a questionnaire to investigate the interrelationships between Industry 4.0 base technologies, Lean and Agile supply chain strategies, and operational performance. The measures used in the study were taken from validated items from prior research, and the case study used confirmatory factor analysis and structural equation modelling to analyse the data.

The findings suggest that Industry 4.0 base technologies enable Lean supply chain strategies, which in turn positively impact Agile supply chain strategies and operational performance. However, the direct effect of Industry 4.0 technologies on operational performance was found to be non-significant. The study also explores the interaction between Lean and Agile supply chain strategies and their impact on operational performance, indicating a direct relationship between Lean and Agile supply chain strategies and that Agile supply chain generates mediation effects between Lean supply chain and operational performance.

The case study provides theoretical and practical implications for supply chain managers, offering valuable insights into the relationship between Industry 4.0 technologies and Lean and Agile supply chain strategies, as well as their effects on a firm's operational performance. It also provides insights into the integration of Industry 4.0 technologies into supply chain management models and their effects on firm performance.

Additionally, it provides a list of academic articles related to supply chain strategies, information technology, and organizational performance, covering topics such as lean and agile supply chain strategies, the role of information technology, industry 4.0 technologies, and the impact of various strategies on business performance. This demonstrates a comprehensive review of the existing literature and a strong theoretical foundation for the study.

The case study offers valuable insights into the influence of Industry 4.0 technologies on supply chain (SC) strategies and company performance, serving as a benchmark for supply chain managers.

Key academic articles in the field encompass topics such as:

1. The importance of strategic supplier partnerships and postponement strategies in supply chain management.
2. The impact of supply chain information systems strategy on both supply chain and firm performance.
3. The role of IT application orchestration capability in improving agility and performance.
4. Industry 4.0 technologies as enablers of lean and agile supply chain strategies.
5. Modelling the relationship of digital technologies with lean and agile strategies.
6. Big data analytics as a mediator in lean, agile, resilient, and green (LARG) practices effects on sustainable supply chains.
7. Creating a conceptual framework for lean supply chain planning within the context of Industry 4.0.
8. Ambidextrous supply chain strategy and supply chain flexibility.
9. Supply chain flexibility in dynamic environments: the enabling role of operational absorptive capacity and organizational learning.
10. Forecasting the correlations between virtual enterprises and supply chain agility.
11. Industry 4.0 implies lean manufacturing: research activities in Industry 4.0 function as enablers for lean manufacturing.
12. Competence in business intelligence, agile capabilities, and the agile performance within the supply chain.
13. Developing a business model for knowledge management and sharing to address disruptions.
14. Exploring the relationship between lean manufacturing and the Internet of Things: synergy or antagonism?

15. Investigating the significance of big data analytics in the development of supply chain agility.
16. Analysing the factors driving and outcomes resulting from the adoption of innovative technologies in the supply chain: cloud computing and supply chain integration.

These topics cover a wide range of issues related to supply chain strategies, information technology, and organizational performance, providing insights into the integration of technology into supply chain management and its impact on firm performance.

Lean Supply Chain (LSC) and Agile Supply Chain (ASC) strategies, along with Industry 4.0 (I4.0) base technologies, have significant implications for a firm's operational performance. LSC focuses on reducing costs and non-value-added activities, which requires collaboration with customers and suppliers, and aims to improve efficiency in terms of cost reduction and delivery performance. LSC management can improve firm efficiency by decreasing costs, cycle times, and improving inventory turnover. On the other hand, ASC focuses on quickly adapting and responding to customer requirements, and it generates mediation effects between LSC and operational performance.

The case study suggests that SC managers who encourage the development of LSC as an initial step will obtain benefits when it comes to implementing ASC and, ultimately, better operational performance. The results demonstrate that LSC is fundamental to achieving ASC in an I4.0 environment. I4.0 base technologies have the potential to improve forecasting capability and increase supply chain visibility, which can lead to greater compatibility between the two SC strategies. Thus, the leaner an SC is, thanks to the I4.0 base technologies, the more Agile it will be over time, which will boost its performance.

In summary, LSC and ASC, when supported by I4.0 base technologies, can enhance a firm's operational performance by reducing costs, improving efficiency, and increasing agility in responding to customer requirements.

The implications of using Industry 4.0 (I4.0) base technologies for lean and agile supply chains and performance are significant. The case study suggests that the implementation of I4.0 base technologies, such as cloud computing, Internet of Things, and Big Data analytics, can enhance the data collection, storage, and sharing, as well as the analysis processes. The findings indicate that I4.0 base technologies can make supply chains leaner, leading to improvements in lean supply chain (LSC) implementation. However, the study also highlights that I4.0 base technologies do not have a significant direct effect on agile supply chain (ASC) implementation. Furthermore, the study indicates that ASC mediates the relationship between LSC and operational performance, emphasizing the importance of considering the relationship between I4.0 and Lean strategies for maximizing operational performance.

The research also provides examples of ways in which the LSC strategy can be enhanced by the use of I4.0 base technologies. It suggests that automation achieved with the support of technology must occur in processes that genuinely add value to the customer, taking into account Lean principles. Additionally, the study emphasizes the importance of considering the I4.0 – Lean relationship for the maximization of operational performance, highlighting the potential for I4.0 base technologies to enhance LSC and, consequently, operational performance.

In summary, the implications of using I4.0 base technologies for lean and agile supply chains and performance include the potential for making supply chains leaner, enhancing LSC strategy, and ultimately contributing to operational performance improvements. However, the direct impact on ASC implementation may be limited, and the study underscores the importance of understanding the relationship between I4.0 and Lean strategies for optimizing operational performance.

9. Supply Chain Management with Lean Production and RFID Application

---

Sector: Supply Chain Management

Description: The case study delves into a three-tier spare parts supply chain characterized by inefficient transportation, storage, and retrieval processes. To enhance supply chain management, it employs lean production techniques and integrates radio frequency identification (RFID) technologies. Utilizing Value Stream Mapping (VSM), the study depicts current and future state mappings, encompassing material, information, and time flows. Additionally, it assesses efficiency enhancements and conducts return on investment (ROI) analysis.

Benefits: Initial experiments indicate that integrating RFID and lean methodologies can reduce total operation time by 81% from the present to the projected future stage. Furthermore, incorporating cross-docking can elevate the time savings to 89%. By leveraging RFID technology, labour costs can be notably decreased without compromising service capacity among the supply chain members under study. Return-on-investment (ROI) analysis confirms the efficacy and viability of the proposed approach.

Tool as best practice: The study demonstrates the best practice of using RFID technology to improve supply chain operations. It shows that RFID implementation in receiving and shipping processes in distribution centres and local distribution centres can significantly improve the average process time of operations. The adoption of RFID technologies prompts the centres to operate in real-time mode, eliminating manual activities such as counting or data input, and improving overall efficiency.

This case study explores the integration of lean production and RFID technology in a spare parts supply chain, analysing the current state of operations and proposing a future state with lean production and RFID. The study shows significant improvements in efficiency and reduction in operation time with the integration of lean production and RFID technology. Value Stream Mapping (VSM) is used to analyse supply chain operations, and the study suggests further improvements in supply chain and logistics systems through better management philosophy and modern technology.

The benefits of the study include significant time savings in operations, cost reduction in labour, and maintaining service capacity. The study also provides a comprehensive ROI analysis, demonstrating the effectiveness and feasibility of the proposed method.

The case study demonstrates the best practice of using RFID technology to improve supply chain operations. It shows that RFID implementation in receiving and shipping processes in distribution centres and local distribution centres can significantly improve the average process time of operations. The adoption of RFID technologies prompts the centres to operate in real-time mode, eliminating manual activities such as counting or data input, and improving overall efficiency.

The integration of lean production and RFID technology improves efficiency in a spare parts supply chain in several ways. Firstly, the application of lean production principles helps to identify and eliminate waste in the supply chain processes, leading to streamlined operations and reduced lead times. Additionally, RFID technology facilitates real-time tracking and visibility of inventory, enabling better inventory management and reducing the time spent on manual tracking and counting processes. The combination of lean production and RFID technology also allows for improved data accuracy and faster information flow, leading to enhanced decision-making and reduced operational delays.

Furthermore, the integration of lean production and RFID technology enables the implementation of cross-docking, a process in which goods are directly transferred from inbound to outbound without intermediate storage., leading to further reductions in operation time and improved overall efficiency.

In summary, the integration of lean production and RFID technology improves efficiency in a spare parts supply chain by reducing waste, improving inventory management, enhancing data accuracy and visibility, enabling faster information flow, and facilitating the implementation of cross-docking.

VSM plays a crucial role in analysing supply chain operations and implementing lean production and RFID technology. VSM is used to draw current state mapping and future state mapping with material, information, and time flows, allowing for a comprehensive visualization of the entire supply chain process.

In the context of lean production, VSM helps identify and eliminate waste in human effort, inventory, time to market, and manufacturing space, enabling the implementation of lean principles to reduce operational inefficiencies and improve responsiveness to customer demand.

When integrating RFID technology, VSM facilitates the visualization of material and information flows, enabling the identification of areas where RFID can be effectively utilized for real-time tracking and visibility of inventory. This visualization helps in identifying opportunities for RFID implementation to streamline processes and reduce manual tracking and counting activities, ultimately improving overall supply chain efficiency and effectiveness.

The potential benefits of implementing lean production and RFID technology in a supply chain include:

1. Efficiency Improvement: The integration of lean production and RFID technology can lead to significant efficiency improvements in supply chain operations.
2. Labor Cost Reduction: Utilizing RFID technology can significantly reduce the cost of labour leading to substantial cost savings for the organization.
3. Real-time Tracking and Visibility: RFID technology enables real-time tracking and visibility of inventory, leading to improved inventory management and reduced time spent on manual tracking and counting processes.
4. Return on Investment (ROI): Return-on-investment (ROI) analysis shows that the proposed method of integrating lean production and RFID technology is effective and feasible. The ROI analysis results in a value of 2.6, indicating a worthwhile investment in lean and RFID for the spare parts supply chain.

In conclusion, the implementation of lean production and RFID technology in a supply chain offers potential benefits such as efficiency improvement, labour cost reduction, improved inventory management, and a positive return on investment.

10. Integrated Decision-Making System for Lean Six Sigma Success Factors and Critical Business Process Selection in Food Companies with Corporate Identity in Istanbul

Sector: Food industry in Istanbul, Turkey

Description: The case study focused on the selection of the ideal Critical Business Process (CBP) using Lean Six Sigma (LSS) success factors in the food industry in Istanbul, Turkey. The study aimed to develop an integrated decision-making system based on ARAS, CRITIC, and q-ROFSs to assess and rank the CBPs, and then select the most ideal CBP in food companies with corporate identity in Istanbul following the LSS success factors.

(*Continued*)

Benefits: The study provided a comprehensive approach to selecting the most ideal CBP in food companies, contributing to numerous Multiple Criteria Decision Making (MCDM) models in intelligent systems and Decision Experts (DEs) which are able to treat inherent fuzziness using more efficient ways. The approach can handle more complex uncertainty and contribute to various MCDM models, providing logical and reliable results.

Tool as best practice: The study utilized various MCDM methods such as BWM, DEMATEL, UTASTAR, EATWOS, MULTIMOORA, ITARA, MABAC, and CoCoSo, showcasing a best practice approach to decision-making in the food industry with corporate identity in Istanbul. This case study is significant as it addresses the limited number of studies on LSS success factors and the selection of CBPs, making it an important comparison unit for future research works. The study's approach and findings also enhanced the significance of the contribution to the subject matter and the general literature.

This case study aims to develop an integrated decision-making system based on ARAS, CRITIC, and q-ROFSs to assess and rank Critical Business Processes (CBPs) in food companies with corporate identity in Istanbul following Lean Six Sigma success factors. The proposed methodology involves the use of q-rung orthopair fuzzy sets to tackle the decision-making problem in the food business.

The study utilized various Multiple Criteria Decision Making (MCDM) methods such as BWM, DEMATEL, UTASTAR, EATWOS, MULTIMOORA, ITARA, MABAC, and CoCoSo, showcasing a best practice approach to decision-making in the food industry with corporate identity in Istanbul. The approach provided a comprehensive method for selecting the most ideal CBP in food companies, contributing to numerous MCDM models in intelligent systems and Decision Experts (DEs) which are able to treat inherent fuzziness using more efficient ways. The approach can handle more complex uncertainty and contribute to various MCDM models, providing logical and reliable results.

The study's findings and approach are significant as they address the limited number of studies on LSS success factors and the selection of CBPs, making it an important comparison unit for future research works. The study's approach and findings also enhanced the significance of the contribution to the subject matter and the general literature.

The study's methodology involved a comprehensive literature review to identify the gaps in existing research and to establish the novelty and significance of the study.

The study's limitations included the scope being limited to a single province in Turkey and one sector, as well as the choice of LSS success factors and CBPs as the areas of focus, which left out other aspects of LSS and business processes. The experimental findings were provided to accommodate the expectations of the DMs during the interviews with expert groups.

The proposed methodology in the study utilizes q-rung orthopair fuzzy (q-ROF) sets to address the decision-making problem in the food business by employing a q-ROF integrated weighting model to assess the importance levels of Lean Six Sigma (LSS) success factors. This method is based on three techniques of the q-ROF CRITIC method, objective weighting, and the q-ROF subjective weighting. Additionally, the q-ROF ARAS method is applied to evaluate Critical Business Process (CBP) alternatives. The q-ROF integrated weighting model and the q-ROF ARAS method provide a comprehensive approach to decision-making in the food business, allowing for the assessment and ranking of CBPs based on LSS success factors.

11. Lean Management Practices, Dynamic Capabilities, and Sustainable Business Performance in Italian Manufacturing Firms

---

Sector: Manufacturing

Description: The case study concentrated on exploring the connections among lean management practices, dynamic capabilities, and sustainable business performance within Italian manufacturing companies. Its objective was to comprehend the reasons behind the inability of numerous lean management adopters to maintain favourable results over time, especially within the context of a rapidly evolving and competitive environment marked by technological advancements and shifting customer demands.

Benefits: The research offers an understanding of how lean management influences sustainable business performance. It differentiates between two groups: "lean adopters", who achieve lasting results, and "lean duplicators", who only attain short-term gains. Moreover, it underscores the significance of cultivating advanced capabilities, termed "lean-related dynamic capabilities", for companies to transition into "lean adopters" and realize enduring advantages.

Tool as best practice: The research employs a survey questionnaire administered to a sample of 99 Italian manufacturing firms. It utilizes partial least squares structural equation modelling to analyse the data and derive conclusions regarding the interconnections among lean management practices, dynamic capabilities, and sustainable business performance.

The case study explores the relationship between lean management practices, dynamic capabilities, and sustainable business performance in Italian manufacturing firms. It discusses the implementation of lean manufacturing and total productive maintenance practices and their impact on sustainability. The study emphasizes the need for dynamic capabilities to achieve sustainable performance and proposes a set of "lean-related dynamic capabilities". The findings suggest that lean management practices positively affect dynamic capabilities, which in turn positively affect economic, environmental, and social performance.

The study utilizes data from the AIDA Bureau Van Dijk database and employs partial least square structural equation modelling (PLS-SEM) to analyse the relationships between the variables.

The findings of the study provide empirical support for the positive association between lean management practices and dynamic capabilities, as well as the positive association between dynamic capabilities and sustainable business performance. It distinguishes between "lean adopters" achieving sustainable outcomes and "lean duplicators" obtaining only short-term outcomes and quick wins. The study emphasizes the importance of developing higher-order or "lean-related dynamic capabilities" for firms to become "lean adopters" and achieve sustainable benefits.

The study also highlights the need for organizations to fully implement lean management practices and continuously develop lean-related dynamic capabilities to maintain a sustainable market position. It identifies lean routines as "ordinary" capabilities that bring short-term success but fail to contribute to sustainable outcomes, and higher-order dynamic capabilities are required for long-term solutions and sustainable development.

The study's theoretical framework suggests that lean management practices, when coupled with dynamic capabilities, can contribute to sustainable competitive advantage in the form of economic, environmental, and social performance. It also emphasizes the importance of considering lean management as an integrated socio-technical system and identifies systematic problem-solving, agile manufacturing, and continuous improvement as "lean-related dynamic capabilities" that contribute to sustainable development.

The study's findings have implications for both theory and practice in operations and supply chain management. It provides insights for practitioners and contributes to the operations management and supply chain management literature. Additionally, the study provides a list of continuous improvement, agile manufacturing, customer involvement, supplier partnership, human resource management, unique just-in-time practices, and unique total quality management practices as potential "lean-related dynamic capabilities" that contribute to sustainable business performance.

Overall, the study offers valuable insights into the mechanisms through which lean management contributes to sustainable business performance and emphasizes the importance of developing and integrating dynamic capabilities to achieve long-term sustainable outcomes.

Lean management practices contribute to sustainable business performance through their ability to develop higher-order or "lean-related dynamic capabilities" that are essential for achieving sustainable outcomes. Lean management practices, when fully implemented and integrated with dynamic capabilities, can lead to sustainable competitive advantage in terms of economic, environmental, and social performance.

The study emphasizes that lean management practices, when coupled with dynamic capabilities, can contribute to sustainable business performance. It distinguishes between "lean adopters" achieving sustainable outcomes and "lean duplicators" obtaining only short-term outcomes and quick wins. The development of higher-order or "lean-related dynamic capabilities" is crucial for firms to become "lean adopters" and achieve sustainable benefits.

Furthermore, the study highlights the importance of considering lean management as an integrated socio-technical system and identifies systematic problem-solving, agile manufacturing, and continuous improvement as "lean-related dynamic capabilities" that contribute to sustainable development.

In summary, lean management practices contribute to sustainable business performance by enabling firms to develop and integrate dynamic capabilities, thereby achieving long-term sustainable outcomes in economic, environmental, and social aspects.

The study identifies several examples of "lean-related dynamic capabilities" that can enhance sustainable outcomes. These include systematic problem-solving, agile manufacturing, and continuous improvement. These capabilities are considered higher-order or "lean-related dynamic capabilities" that are effective for achieving sustainable benefits.

Systematic problem-solving involves the ability to identify and address issues in a structured and methodical manner, leading to sustainable improvements in processes and products. Agile manufacturing, or change proficiency, refers to the capability to adapt quickly to changing market demands and technological advancements, enabling firms to sustain their competitive position in dynamic environments. Continuous improvement involves the ongoing effort to enhance processes, products, and services, leading to sustained advancements in efficiency, quality, and customer satisfaction.

These "lean-related dynamic capabilities" are essential for firms to become "lean adopters" and achieve sustainable benefits, contributing to economic, environmental, and social sustainability.

The relationship between lean management practices and sustainable business performance differs across economic, environmental, and social dimensions. Existing studies have shown that lean management practices can improve operational performance through the elimination of waste, leading to short-term economic benefits. However, the relationship between lean management and environmental and social sustainability is more complex and has yielded contradictory results.

Some studies have reported positive associations between lean management and environmental performance, indicating that lean practices can reduce environmental impacts such as emissions and waste. On the other hand, there are conflicting findings, with some studies showing negative associations between lean management and environmental performance, suggesting that lean practices may not always lead to positive environmental outcomes.

Similarly, with respect to social performance, while some studies have found positive associations between lean management and social sustainability due to improved workplace conditions, others have reported negative effects such as increased stress and reduced autonomy for employees.

Overall, the relationship between lean management practices and sustainable business performance differs across economic, environmental, and social dimensions. While lean practices may lead to short-term economic benefits, the impact on environmental and social sustainability is more nuanced and may vary depending on specific organizational contexts and practices.

Existing studies focusing on the relationship between lean management and sustainability suffer from two limitations. First, despite the effectiveness of lean management for operational performance, relatively less attention is paid in the literature to how lean management can simultaneously embrace the triple dimensions of sustainable business performance (economic, environmental, and social performance).

One potentially neglected reason of mixed results for the lean-sustainability relationship could be that lean management practices per se are not sufficient for sustainable outcomes and there might be a mechanism by which lean management can appropriately explain sustainable business performance. However, a lean-sustainability link or the mechanism through which lean management simultaneously leads to balancing economic, environmental, and social performance is missing in the literature.

12. Sustainable Impacts of Lean Office in University Administrative Processes

---

Sector: The case study focuses on the application of Lean Office principles in the administrative processes of a Federal Public University in Brazil

Description: The research is a case study conducted at a Federal Public University in Brazil, analysing two service processes aimed at students and employees (teachers and technicians) . The study is supported by the Lean Evaluation and Future Improvement (LEFI) method, specifically designed for administrative services, and involves the use of Value Stream Mapping to identify and eliminate waste in administrative processes.

Benefits: The application of Lean Office principles in the university's administrative processes resulted in significant benefits, including the saving of 444,754 sheets of printed paper, indirectly impacting the economy of 1,380,000 L of water, non-emission of 43.82 kg of CO2, and the preservation of 67 trees.

Tool as best practice: The LEFI method is highlighted as a best practice tool for the application of Lean Office principles in administrative services, as it proved to be valid, viable, and extremely useful in reducing costs, eliminating rework, minimizing communication problems, and increasing the efficiency of administrative functions.

The case study discusses the application of lean principles in various industries, including higher education, to improve administrative processes and promote sustainability in the administrative processes of a Federal Public University in Brazil. It highlights the use of methods such as Lean Evaluation and Future Improvement (LEFI) and Value Stream Mapping (VSM) to identify waste, propose improvements, and achieve environmental, economic, and social impacts. It explores the use of lean office principles in public universities to improve administrative processes and promote sustainability. The work also emphasizes the importance of leadership and stakeholder involvement in the successful implementation of lean practices in higher education.

The study specifically focuses on two service processes for students and employees (teachers and technicians). The application of Lean Office principles in the university's administrative processes resulted in significant benefits, including the saving of 444,754 sheets of printed paper, indirectly impacting the economy of 1,380,000 L of water, non-emission of 43.82 kg of CO_2, and the preservation of 67 trees.

The LEFI method is highlighted as a best practice tool for the application of Lean Office principles in administrative services. The method proved to be valid, viable, and extremely useful in reducing costs, eliminating rework, minimizing communication problems, and increasing the efficiency of administrative functions.

The research limitations include the fact that it is a single case study, and the findings cannot be statistically generalized to other universities. However, the authors believe that the steps developed in the study allow similar procedures to be adopted and conducted by other researchers and professionals working with research involving Lean Office applied to universities.

The case study provides valuable insights into the application of Lean Office principles in the public university administrative processes, demonstrating the potential for significant sustainable impacts and the effectiveness of the LEFI method as a best practice tool for lean administrative processes.

Overall, the case study contributes to the understanding of the benefits and best practices associated with the application of Lean Office principles in the administrative processes of a Federal Public University in Brazil, providing valuable insights for future research and practical implementation in similar settings.

The positive results of the study highlight the potential for lean principles and practices to improve administrative processes and add value to students, faculty, and staff in the university setting.

The application of lean principles in public universities led to significant environmental, economic, and social impacts. From an environmental perspective, the application of Lean Office principles in the administrative processes of a Federal Public University in Brazil resulted in the saving of 444,754 sheets of printed paper, indirectly impacting the economy of 1,380,000 litres of water, non-emission of 43.82 kg of CO_2, and the preservation of 67 trees. This demonstrates the potential for lean practices to contribute to environmental sustainability by reducing resource consumption and minimizing waste. Economically, the application of Lean Office principles led to cost savings, including the reduction of printer ink and paper usage, as well as savings in fuel and the use of plastic boxes for storing processes. These economic benefits contribute to the efficient use of resources within the university's administrative processes.

In the social sphere, the application of lean principles resulted in quicker completion times for processes, bringing benefits to students, faculty, and staff directly involved in the administrative processes. This efficiency improvement positively impacted user satisfaction and the overall experience of stakeholders within the university.

Overall, the application of lean principles in public universities has demonstrated the potential to generate positive impacts across environmental, economic, and social dimensions, contributing to sustainability and efficiency within the university setting.

The key tool discussed in the paper for identifying waste and improving administrative processes in higher education institutions is the LEFI method. This method, developed by Beckers (2015), is designed to reduce waste in administrative processes at universities and is based on lean principles and practices. The LEFI method involves the use of VSM as its main tool to identify the flow of value and waste, allowing for the visualization of activities that add and those that do not add value. Additionally, the LEFI method incorporates continuous improvement through Kaizens and the creation of continuous flow, aligning with the basic premises of the Lean philosophy.

Furthermore, the paper discusses the application of Lean Office principles and practices, which aim to improve operational efficiency and productivity by reviewing processes and identifying and eliminating waste. Value Stream Mapping, a technique commonly associated with Lean principles, is utilized to analyse administrative processes and information flows, with the goal of minimizing or eliminating activities that do not add value.

The LEFI method and Value Stream Mapping are highlighted as key tools and techniques for identifying waste and improving administrative processes in higher education institutions, contributing to the reduction of waste and the promotion of sustainability in the social, environmental, and economic dimensions.

The potential benefits of integrating lean principles in universities, particularly in terms of sustainability and waste reduction, are significant. The application of Lean Office principles in administrative processes at universities can lead to the reduction of waste, improved efficiency, and cost savings, contributing to sustainability in the social, environmental, and economic dimensions.

From an environmental perspective, the integration of lean principles can result in substantial reductions in resource consumption and waste generation. For example, the application of Lean Office principles in a Federal Public University in Brazil led to the saving of 444,754 sheets of printed paper, indirectly impacting the economy of 1,380,000 L of water, non-emission of 43.82 kg of CO2, and the preservation of 67 trees. These outcomes demonstrate the potential for lean practices to contribute to environmental sustainability by minimizing resource usage and waste generation.

Economically, the integration of lean principles can lead to cost savings and improved resource utilization. For instance, the application of Lean Office principles in universities can result in reduced printer ink and paper usage, as well as savings in fuel and other resources, contributing to more efficient and sustainable operations.

Furthermore, the integration of lean principles in universities can lead to social benefits, such as improved service delivery and user satisfaction. For example, the application of Lean Office principles can result in quicker completion times for administrative processes, positively impacting the satisfaction of students, faculty, and staff within the university.

Overall, the integration of lean principles in universities has the potential to generate substantial benefits in terms of sustainability and waste reduction, contributing to environmental conservation, cost savings, and improved service delivery.

13. Integration of Lean, Green, and Smart Practices in Aerospace Manufacturing

---

Sector: Aerospace

Description: The case study focused on the integration of lean, green, and smart practices in the aerospace manufacturing sector. The study aimed to identify and analyse the relationships between these practices and their impact on operational performance within the aerospace industry.

Benefits: The study highlighted the potential utility of a guiding framework to integrate the use of lean, green, and smart practices. It was found that such a framework could support firms in early decision-making processes for undertaking potential improvement actions aimed at fulfilling lean green practices.

Tool as best practice: The study utilized a semi-structured interview method to gather insights from industry experts and practitioners. This method allowed for in-depth exploration of specific themes and responses, providing the freedom for informants to express their views on their own terms.

This research focuses on the relationship between smart, lean, and green practices in the context of operational performance. It presents a methodology for mapping these relationships and defines dimensions for each practice. The case study discusses the impact of these practices on operational performance, the potential synergies between them, and the role of smart technologies in supporting lean and green practices. It also provides a framework for integrating these practices and highlights the need for further validation and development.

The case study focuses on the integration of lean, green, and smart practices in the aerospace manufacturing sector, with the aim of identifying and analysing the relationships between these practices and their impact on operational performance within the industry. The study utilizes a methodology that involves a literature review, expert interviews, and the development of a conceptual framework to map the relationships between these practices and their impact on operational performance.

The study presents a mapping of strong and weak impacts of various practices on operational performance, as well as the residual dimensions involved in the final step of the analysis. It also discusses the impact of advanced analytics supporting value stream mapping and Poka-Yoke within design for environment practices. Additionally, it presents the Smart-Lean-Green triplets emerged from the analysis and the dimensions supporting operational performance.

The framework developed in the study aims to support companies in integrating lean, green, and smart practices to improve operational performance. The framework was validated through interviews with Polish and Swiss SMEs, and it was found that companies are still focused on lean practices and prioritize financial competitiveness and customer satisfaction. The framework was seen as a valuable tool for early decision-making and increasing awareness of the benefits of integrated practices.

The study also discusses the impact of smart technologies on lean and green practices in the context of operational performance. It identifies specific smart technologies that support lean and green practices and highlights the potential synergies between them. The study provides examples of how advanced analytics can support total productive maintenance programs and environmental management systems to achieve operational performance benefits.

Overall, the case study provides valuable insights into the integration of lean, green, and smart practices in the aerospace manufacturing sector and offers a framework that can guide practitioners in selecting a mix of approaches to optimize operational performance. The study's methodology, which includes a literature review, expert interviews, and the development of a conceptual framework, provides a comprehensive and robust analysis of the relationships between these practices and their impact on operational performance within the aerospace industry.

The dimensions of lean, green, and smart practices defined in the paper's reference framework are as follows:

- **Lean practices**:
  - waste reduction;
  - continuous improvement;
  - value stream mapping;
  - total productive maintenance;
  - Poka-Yoke.
- **Green practices**:
  - design for environment;
  - environmental management systems;

- eco-design;
- life cycle assessment;
- closed-loop supply chain.

- **Smart practices**:
  - advanced analytics;
  - Internet of Things (IoT);
  - artificial intelligence;
  - 3D printing;
  - cyber-physical systems.

These dimensions were identified through a comprehensive literature review and expert insights from practitioners with extensive experience in the aerospace manufacturing sector.

Smart technologies support lean and green practices by enabling data collection, analysis, and decision-making processes that can lead to improved operational performance and environmental sustainability. For example, IoT platforms, sensors, and cyber-physical systems enable the collection of data regarding emissions, energy consumption, and waste, which can be used as input for triggering numerous green practices. Additionally, advanced analytics and Industrial Internet of Things (IIoT) technologies have been identified as impactful smart technologies that can strengthen the positive impacts on operational performance, particularly in the context of lean practices.

One potential synergy is the use of advanced analytics to support value stream mapping and Poka-Yoke within design for environment practices. This can lead to improved waste reduction, material and time consumption, supply chain management, and product life cycle optimization, thereby enhancing both lean and green practices. Another example is the use of IoT platforms to collect data that can support the implementation of lean practices such as total productive maintenance and continuous improvement, while also enabling the implementation of green practices related to environmental management systems and eco-design.

Overall, smart technologies provide the means to collect and analyse data in a smart manner, with the aim of pursuing reduced consumption and waste, thereby fostering the integration and synergies between lean and green practices.

The proposed framework for integrating lean, green, and smart practices aims to improve operational performance by providing a structured approach to identify mutually supportive relationships among these practices. The framework helps companies in early decision-making processes for undertaking potential improvement actions aimed at fulfilling lean green practices, thereby maximizing operational performance objectives.

Findings from interviews with manufacturing SMEs revealed that the framework could greatly support companies in early decision-making for evaluating potential improvement actions. The interviews highlighted that most companies are still focused on lean practices as the main lever of improvement, with a priority on improving financial competitiveness and customer satisfaction. However, there was a recognized advantage of the holistic framework, as it could increase awareness of the benefits of integrated practices and support companies in exploiting multiple levers to achieve their goals.

The framework acts not only at the level of practice selection but also increases the awareness of companies about the benefits of integrated practices, supporting them in exploiting multiple levers to achieve their goals. The interviews also indicated that companies had difficulties listing possible combinations of practices, and they would see value in a framework that presents a list from which to identify what could be useful for them.

Overall, the framework provides a valuable tool for companies to integrate lean, green, and smart practices, thereby improving operational performance and increasing awareness of the benefits of integrated practices.

14. Lean-Based Health Information Management Model for the Municipality of Ituverava

Sector: Health Information Management

Description: The case study focuses on proposing a health information management model based on Lean thinking for the Municipality of Ituverava in Brazil. The study integrates health information systems across government, state, and regional levels to improve data reliability, provide greater visibility for managers, and enhance the decision-making process. The research protocol aims to improve health information quality and identify bottlenecks within the hierarchical and regionalized system.

Benefits: The proposed model aims to act as an auditable instrument for the analysis, representation, and improvement of the quality of health information in the municipality. It is expected to provide means for using audit filters to assess the quality of care provided to patients, identify strengths and weaknesses, and allow corrective actions. Additionally, the model is intended to enhance the security and reliability of health information, ultimately leading to improved decision-making processes.

Tool as best practice: The study aims to integrate Lean tools and techniques into the health information management model. It seeks to identify the state of the art of using Lean tools in health information management and apply best practices from the literature to support the proposed methodology. The use of Lean thinking is expected to provide a systematic approach to identify and eliminate waste, improve processes, and enhance the overall quality of health information.

This case study discusses the proposal for a health information management model based on Lean thinking in the municipality of Ituverava. It explores the challenges of assessing quality in the health field and the need for appropriate assessment systems and methods. The article also discusses the application of Lean principles in healthcare and its potential benefits in improving patient care, reducing waste, and optimizing resources. The main goals of the proposal are to review different models and philosophies applied to health information systems, evaluate existing information flows and practices, and identify areas for improvement in the municipality's health information systems.

The case study focuses on proposing a health information management model based on Lean thinking for the Municipality of Ituverava in Brazil. The study integrates health information systems across government, state, and regional levels to improve data reliability, provide greater visibility for managers, and enhance the decision-making process.

The proposed model aims to act as an auditable instrument for the analysis, representation, and improvement of the quality of health information in the municipality. It is expected to provide means for using audit filters to assess the quality of care provided to patients, identify strengths and weaknesses, and allow corrective actions. Additionally, the model is intended to enhance the security and reliability of health information, ultimately leading to improved decision-making processes.

The study integrates Lean tools and techniques into the health information management model. It seeks to identify the state of the art of using Lean tools in health information management and apply best practices from the literature to support the proposed methodology. The use of Lean thinking is expected to provide a systematic approach to identify and eliminate waste, improve processes, and enhance the overall quality of health information.

The research protocol includes a study design aimed at improving the quality of health information and identifying bottlenecks within the hierarchical and regionalized system. The study aims to improve the reliability of information, provide greater visibility for managers through the use of visual tools, and enhance security in the decision-making process. It also aims to provide means for using audit filters to assess the quality of care provided to patients, identify strengths and weaknesses, and allow corrective actions.

The case study presents a comprehensive approach to integrating Lean thinking into health information management in the Municipality of Ituverava. It addresses the challenges of assessing quality in the health field, proposes a model to improve the reliability and security of health information, and outlines the potential benefits of implementing Lean principles in healthcare. The study provides valuable insights into the application of Lean thinking in health information systems to improve reliability, visibility, and security in decision-making processes.

In summary, the case study provides a detailed analysis of the proposed health information management model based on Lean thinking for the Municipality of Ituverava. It outlines the potential benefits of the model, the integration of Lean tools and techniques, and the research protocol designed to improve the quality of health information and identify bottlenecks within the system.

Assessing quality in the health field presents significant challenges, including the need to select appropriate assessment systems and methods to assist in the administration of services and provide decision-making with the least degree of uncertainty possible.

One of the challenges is the highly heterogeneous and sometimes ambiguous nature of the medical language, as well as its constant evolution. This complexity can make it difficult to assess and monitor the activities carried out by health services, as well as to identify the degree of risk of the occurrence of a certain event or health problem. Additionally, the high amount of data generated constantly by the automation of processes and the emergence of new technologies further complicates the assessment of quality in the health field.

It is important to have appropriate assessment systems and methods in place to assist in the administration of services and provide decision-making with the least degree of uncertainty possible. Health indicators are useful for assessing and monitoring the activities carried out by health services, contributing to the identification of the degree of risk of the occurrence of a certain event or health problem, and checking values and acquiring information that allows intervening in the reality to achieve goals and objectives. Therefore, these systems and methods are essential for improving the quality of care provided at all levels of health services.

In summary, the challenges in assessing quality in the health field stem from the complexity and constant evolution of medical language, the high volume of data generated by automation, and the need to ensure the least degree of uncertainty in decision-making. Appropriate assessment systems and methods are crucial for monitoring activities, identifying risks, and intervening to achieve healthcare goals and objectives.

The application of Lean principles in healthcare can improve patient care, reduce waste, and optimize resources in several ways. Firstly, Lean thinking promotes patient-focused assistance, which can lead to improvements in operational management, satisfaction of the technical team and patients, and reduction of waste and costs. By eliminating waste and non-value-added activities, Lean principles can streamline processes, making them more efficient and effective. This can result in faster and more accurate service to patients, ultimately improving patient care.

Additionally, Lean principles emphasize the importance of continuous improvement and respect for the people who develop the work. This can lead to a culture of ongoing enhancement of healthcare processes and services, ultimately benefiting patient care. Furthermore, by standardizing processes with flexibility, Lean principles can help in reducing errors and indirect costs, optimizing the use of resources, and improving the workplace, all of which contribute to better patient care and resource optimization.

In summary, the application of Lean principles in healthcare can improve patient care by promoting patient-focused assistance, streamlining processes, fostering a culture of continuous improvement, reducing errors and indirect costs, and optimizing the use of resources.

The main goals of the proposal for a health information management model based on Lean thinking in the municipality of Ituverava are as follows:

1. Review the different models, thoughts, and philosophies applied to health information systems to support the proposed methodology based on best practices exposed in the literature.
2. Evaluate the different information flows and practices present in the health information systems within the location of the study.
3. Identify the levels of flow bottlenecks within a hierarchical and regionalized system to identify the waste (seedlings) of information present in the municipality.

These goals aim to improve the reliability of health information, provide greater visibility for managers, and enhance the decision-making process in the municipality of Ituverava.

15. Lean Construction Techniques and BIM Technology

Sector: Construction and Building Information Modelling (BIM)

Description: The case study conducted a systematic literature review to analyse the integration of lean construction techniques and BIM technology. The study aimed to systematize the current knowledge, identify hot topics, and indicate potential future applications in the field of construction and BIM technology.

Benefits: The study identified various benefits related to the joint use of lean construction and BIM, such as enhanced project management, improved efficiency, and better coordination during the construction process.

Tool as best practice: The study utilized a systematic literature review as the best practice to analyse and synthesize the existing knowledge and research in the field of lean construction and BIM technology.

This case study aimed to analyse the integration of lean construction techniques and Building Information Modelling (BIM) technology. The study utilized a systematic literature review to systematize the current knowledge, identify hot topics, and indicate potential future applications in the field of construction and BIM technology.

The systematic literature review consisted of four general phases composed of seven stages. The first phase involved identifying the purpose of the study, followed by the identification of the literature sample, analytical activities, and the development of the research report. The study aimed to demonstrate various hot topics, such as the benefits related to the joint use of lean construction and BIM, issues related to simultaneous implementation and utilization of lean construction and BIM, instruments supporting implementation and utilization, and potential future applications.

The research was based on databases such as Scopus, ProQuest, and scholar.google.com, which were used to identify similarities and differences in the approach of authors to the subject of cooperation of lean construction techniques and BIM technology in theoretical and case study foundations. The study identified 58 relevant publications for further analysis, demonstrating a comprehensive approach to gathering and analysing existing literature on the topic.

The review identified various benefits of integrating lean construction and BIM technology, including enhanced project management, improved efficiency, better coordination during the construction process, cost control, production management, project delivery, design management, and sustainability. Additionally, the potential of integrating lean construction and BIM with emerging technologies such as the Internet of Things, augmented reality, virtual reality, and blockchain was discussed.

The study also highlighted the challenges associated with this integration, indicating the need for further exploration and research in this area. The research indicated the significance and support for this topic in the academic and industry sectors, as it was funded by the Polish National Agency for Academic Exchange.

In conclusion, the systematic literature review conducted in this case study provided valuable insights into the integration of lean construction techniques and BIM technology, identifying benefits, challenges, and potential future applications. The study's comprehensive approach to gathering and analysing existing literature on the topic makes it a valuable resource for researchers and practitioners in the construction and BIM industry.

The paper discusses the combination of lean construction and Building Information Modelling (BIM) technology in the construction industry. It presents a systematic literature review, highlighting the benefits, challenges, and future demand of integrating these two concepts. The review also identifies research gaps and the need for further exploration in this area.

The benefits of combining lean construction and BIM technology in the construction industry include enhanced project management, improved efficiency, better coordination during the construction process, cost control, production management, project delivery, design management, and sustainability. This integration offers a comprehensive approach to addressing various aspects of construction projects, leading to improved overall performance and outcomes.

The challenges associated with integrating lean construction and BIM include resistance to change, lack of IT technical skills among project teams using BIM tools for project development, and a significant problem in education among potential users due to a lack of subjects and teachers. Additionally, the implementation of BIM in construction is not occurring as quickly as anticipated, and there is a need for a roadmap for implementing BIM and lean construction in construction project implementation.

The combination of lean construction and BIM technology contributes to cost control, production management, project delivery, design management, and sustainability in construction projects through various means. Lean construction techniques focus on minimizing waste and maximizing value, which directly impacts cost control and production management. BIM technology provides a platform for efficient information flow among internal stakeholders, aiding in project delivery and design management. Additionally, the integration of lean construction and BIM supports sustainability by optimizing management in construction companies and innovative entities, and by enabling energy and environment analysis, constructability analysis, and structural analysis.

16. Lean Maintenance and Repair Implementation in Seven Automotive Service Suppliers

---

Sector: Automotive Industry

Description: The case study aimed to comprehend the current state of lean management implementation in maintenance and repair shops within the automotive sector. The research methodology comprised a systematic literature review (SLR) alongside a multiple-case study approach. The empirical phase rigorously examined seven automotive repair and maintenance shops, with the objective of pinpointing performance opportunities and offering recommendations for enhancing industrial maintenance and repair practices.

Benefits: The study aimed to address the gap in literature regarding lean management applications in service environments, particularly in the automotive industry. By conducting a cross-case analysis, the research provided insights into the extent to which lean principles can be applied in after-sales maintenance and repair services, offering valuable recommendations for operational excellence in the automotive sector.

Tool as best practice: The research employed a multiple case-study approach, ideal for addressing "how" or "why" research inquiries in contexts where the researcher cannot establish a controlled environment for the studied event, especially when it entails significant behavioural elements. This method facilitated a thorough investigation of lean implementation within automotive service providers, enabling diverse data collection techniques to ensure robust triangulation.

The case study discussed in the provided citations focuses on the application of lean management principles in after-sales services, specifically maintenance and repair shops in the automotive industry. The research methodology employed in the study includes a systematic literature review and a multiple-case study approach, which allowed for a comprehensive exploration of the lean implementation in automotive service suppliers.

The study highlights the importance of after-sales services in commercial settings and the potential for lean implementation in the automotive after-sales market to improve operational performance. It notes that the lack of research on applying lean principles to professional services, such as legal services, is partially due to the high cost and lower significance of improving operational performance. However, the automotive sector, particularly after-sales services, has seen a shift towards lean principles due to increased competition and the impact of customer satisfaction on repurchase decisions.

The case study identified defects, waiting, and overproduction as key areas for improvement in the maintenance and repair process in automotive service-shops. Defects were found to be a significant source of waste, leading to job extensions and rework, with the lack of standardized quality controls throughout the repair process identified as a root cause. Waiting was mainly attributed to waiting for vehicles, parts, and technician capacity. These findings highlight the potential for lean implementation in after-sales services to address these issues and improve operational performance.

The study also compares the root causes of waste in services with those in manufacturing, acknowledging that while some lean principles can be applied directly to service environments, others may require adaptation due to the unique nature of services. It emphasizes the challenges and potential benefits of implementing lean practices in service environments and suggests areas for future research.

Additionally, the provided list of academic articles and research papers related to lean service and lean manufacturing, as well as their integration with agile manufacturing, further supports the importance and relevance of lean principles in service environments. The articles cover various aspects of lean service, including its impact on customer loyalty, waste reduction, and performance tracking in repair shops, indicating a growing interest in applying lean principles to after-sales services.

In summary, the case study provides valuable insights into the potential for lean implementation in after-sales services in the automotive industry, highlighting the challenges, benefits, and areas for further research. The systematic literature review and multiple-case study approach contribute to the robustness of the findings and recommendations for operational excellence in the automotive sector.

This study discusses the application of lean management principles in after-sales services, specifically in the automotive industry. It explores the challenges and potential benefits of implementing lean practices in service environments, compares waste in services with manufacturing, and suggests areas for future research. The lack of research on applying lean principles to professional services is also addressed.

The key areas for improvement identified in the case study of maintenance and repair process in automotive service-shops include defects, waiting, and overproduction. Defects were found to be a significant source of waste, leading to job extensions and rework, with the lack of standardized quality controls throughout the repair process identified as a root cause. Waiting was mainly attributed to waiting for vehicles, parts, and technician capacity. These findings highlight the potential for lean implementation in after-sales services to address these issues and improve operational performance.

Additionally, the study emphasizes the importance of addressing operational inefficiencies and waste in the maintenance and repair process, particularly in the context of after-sales services in the automotive industry. The lack of standardized quality controls and the prevalence of job extensions and rework were identified as significant areas for improvement, highlighting the potential for lean implementation to enhance performance in automotive service-shops.

The automotive sector, particularly after-sales services, has seen a shift towards lean principles due to the increasing importance of after-sales services for the bottom-line of manufacturing firms. The shift towards services, especially after-sales, is influenced by market overcapacity in regions like Western Europe, prompting manufacturers to diversify. Authorized dealerships play a vital role in this shift. Customer satisfaction with after-sales significantly impacts repeat purchases, urging manufacturers to focus on operational performance in automotive after-sales. While professional services like legal services often prioritize premiums over operational improvement, this is not the case for after-sales in the automotive sector. Due to limited growth opportunities in vehicle sales, after-sales services have become essential revenue sources, fostering customer satisfaction and loyalty. This has further emphasized the need for operational excellence in after-sales services, leading to the adoption of lean principles.

17. Eco Lean Management Implementation in a Consortium

Sector: Manufacturing

Description: The case study involved a consortium cooperation between four companies in southwestern Germany, including electrical engineering, machine tool manufacturing, conveying & handling technology, and mechanical engineering. Over the course of two years, the companies continuously exchanged information about the introduction of ELM in regular consortium workshops. A total of 96 problem areas were identified and improved, including issues in assembly, incoming and outgoing goods, order picking, factory layout, material, and energy efficiency.

(Continued)

Benefits: The implementation of ELM led to significant improvements in the internal material flow, factory layout optimization, reduction of throughput times, minimization of material waste, and minimized distances. Additionally, the introduction of ELM resulted in reduced search time for responsible employees, further optimizations of the factory layout and the storage system, and extended usage of cooling lubricant. These improvements led to cost savings and environmentally relevant savings in emissions.

Tool as best practice: The case study utilized Eco Lean Management (ELM) as a best practice for simultaneously achieving economic and ecological improvements in production systems. The implementation of ELM allowed for the identification and improvement of various problem areas, leading to tangible benefits in terms of cost savings and environmental impact.

The case study focuses on the implementation of ELM in a consortium cooperation between four manufacturing companies in southwestern Germany. The companies involved in the consortium represent various sectors, including electrical engineering, machine tool manufacturing, conveying & handling technology, and mechanical engineering. The implementation process took place over a two-year period and involved continuous information exchange and collaboration in regular consortium workshops.

The case study identified a total of 96 problem areas within the companies, spanning across different aspects of their operations such as assembly, incoming and outgoing goods, order picking, factory layout, material, and energy efficiency. Through the implementation of ELM, the companies were able to address and improve these problem areas, leading to significant benefits in their operations.

The benefits of implementing ELM included improvements in internal material flow, optimization of factory layout, reduction of throughput times, minimization of material waste, and minimized distances. Additionally, the introduction of ELM resulted in reduced search time for responsible employees, further optimizations of the factory layout and the storage system, and extended usage of cooling lubricant. These improvements led to cost savings and environmentally relevant savings in emissions.

The case study demonstrates the successful application of ELM in a real-world manufacturing setting, showcasing the potential for economic and ecological improvements through the integration of lean management principles with ecological optimization. The implementation of ELM not only resulted in tangible benefits for the companies involved but also contributed to environmental goals by reducing emissions and waste.

As a best practice, the case study highlights the effectiveness of ELM in identifying and addressing specific problem areas within manufacturing operations, leading to improvements in both economic and ecological performance. The successful implementation of ELM in the consortium serves as a practical example of how companies can achieve cost savings and environmental impact through the adoption of ELM principles.

Overall, the case study provides valuable insights into the potential of ELM to support environmental goals and drive sustainable practices within the manufacturing sector. It also underscores the importance of continuous collaboration and information exchange among companies to successfully implement ELM and realize its benefits.

This case study provides an overview of regulations, research, and strategies related to environmental and sustainable practices in manufacturing, including lean and green operations, energy efficiency, and supply chain management. It also discusses the integration of lean and green principles, the potential impact on greenhouse gas emissions and climate change, and the concept of bio intelligent production.

The implementation of ELM in companies is associated with several challenges and potential improvements. Some of these are:

1. **Low-cost improvements**: ELM aims to continuously improve the economic and ecological performance of a company through cost-effective organizational improvements. However, the implementation of individual solutions with manageable organizational effort may lead to the possibility of greenwashing, where companies implement superficial changes to appear environmentally friendly without making substantial improvements.
2. **Holistic added value**: The implementation of ELM often leads to the implementation of single solutions that solve specific problem areas holistically but may not provide holistic added value at the micro and macro levels. This can limit the overall impact of ELM on the economic and ecological sustainability of production systems.
3. **Macro-level diffusion**: The overall economic dissemination of the ELM concept has only been partially successful. To increase the potential and resulting expansion of the concept, there is a need for further incentives that promote not only economic improvements but also the pursuit of ecological improvements. This may require political decisions at the macro level to create additional incentives for companies to adopt ELM.
4. **Environmental factors and policy guidance**: In addition to policy guidance, initiatives such as the Science Based Targets Initiative need to be created to help set voluntary greenhouse gas reduction commitments for industry. ELM should be placed to support the achievement of environmental goals with the incentive of economic savings. This would help in addressing the challenges associated with the implementation of ELM and promote its adoption in companies.
5. **Technological and systemic innovations**: To adequately solve the inherent challenges of future production systems towards sustainable production, technological and systemic innovations are necessary. These innovations should cause production to be completely rethought, and may include novel challenges for management, such as bio intelligent production, which is considered one of the most innovative ways for a comprehensive reorientation of existing industrial patterns.

In summary, the challenges associated with implementing ELM in companies include the need for holistic added value, macro-level diffusion, and addressing environmental factors. Potential improvements include addressing the issue of low-cost improvements, creating policy guidance and initiatives, and embracing technological and systemic innovations to promote sustainable production.

Eco Lean Management (ELM) combines lean management principles with ecological optimization to improve the economic and ecological performance of a company through a holistic approach. ELM aims to continuously improve the economic and ecological performance of a company in small steps (kaizen) through cost-effective organizational improvements. It integrates the objectives of classic lean management with aspects of ecological optimization, assuming environmental preservation as an implicit customer requirement. ELM introduces environmentally relevant aspects as a clear extension of lean management, ensuring that improvement measures are implemented only if the resource efficiency requirement is met. This approach sensitizes managers and employees for sustainable development of the manufacturing environment by addressing issues at the economic-ecological interface on a daily basis.

ELM is characterized by defined principles, methods, and a tailored indicator system as a comprehensive approach for the holistic economic and ecological optimization of production systems at the factory level. It relies on a broad methodological basis and a tailored indicator system with 18 aggregated indicators and their correlation to implement and continuously optimize ecological criteria synergistically with economic aspects as part of a holistic day-to-day operation. By addressing issues at the economic-ecological interface on a day-to-day basis, managers and employees are sensitized to sustainable development of the production environment.

In summary, ELM combines lean management principles with ecological optimization by integrating environmentally relevant aspects into lean management, ensuring that improvement measures meet resource efficiency requirements, and sensitizing managers and employees to sustainable development through the systematic introduction of ecological criteria alongside economic aspects.

18. Development of the Toolbox Lean 4.0 for Industrial Engineering in Producing Companies

Sector: Industrial Engineering and Manufacturing

Description: The case study focuses on creating the Toolbox Lean 4.0, a holistic solution crafted to assist industrial engineering in manufacturing firms. It offers a systematic compilation and explanation of various methods and tools for designing production systems, precisely customized to meet the requirements of industrial engineering.

Benefits: The toolbox seeks to furnish users with a range of viable solution options for tackling current production process challenges. It presents a coherent framework, a database model, and a prototype implementation of the database approach, with the goal of laying the groundwork for a software solution ready for the market. The toolbox is designed to be customizable, adaptable to company-specifics, and independent from IT systems, making it suitable for training purposes and comprehensible for users.

Tool as best practice: The Toolbox Lean 4.0 stands out as a top-tier tool for industrial engineering support in manufacturing firms. It boasts an array of features, such as a rating system for best practice identification, auditing capabilities across multiple production sites, and performance measurement functionalities. Moreover, it enables seamless navigation and filtering through various design principles, methods, tools, technologies, use cases, problem types, and solution patterns, offering users a comprehensive and customizable toolbox.

The case study focuses on the development and implementation of the Toolbox Lean 4.0, a digital catalogue of methods and tools for industrial engineering in Lean Production Systems 4.0 (LPS 4.0).

The Toolbox Lean 4.0 allows users to select problem types and process steps, suggests use cases, and provides feasible solution patterns. The prototype is filled with reliable input sources and is designed for further expansion. The initial implementation consists of over 50 methods, over 50 tools and technologies, and over 80 use cases.

The case study discusses the implementation of a database approach for the management of digital methods and tools in Lean Production Systems 4.0 (LPS 4.0). It outlines the requirements, concept, and development roadmap of the Toolbox Lean 4.0, as well as the prototypical implementation of the database approach.

Overall, the case study highlights the importance of a structured and comprehensive solution to support industrial engineering in producing companies. The Toolbox Lean 4.0 is designed to provide users with a set of suitable solution alternatives to address existing problems in the production process. It offers a logical structure, a database model, and a prototypical implementation of the database approach, making it a valuable tool for industrial engineering purposes.

The development and implementation of the Toolbox Lean 4.0 demonstrate a best practice approach in providing a comprehensive yet individually adjustable toolbox for industrial engineering. It offers a wide range of features, including a rating feature for identification of best practices, auditing multiple production sites, and realizing performance measurements. Additionally, it allows for free navigation and filtering providing users with a comprehensive yet individually adjustable toolbox.

In conclusion, the Toolbox Lean 4.0 serves as a valuable tool for industrial engineering in producing companies, providing a structured and comprehensive model.

The Toolbox Lean 4.0 is a systematic solution designed to support industrial engineering in managing the complexity inherent in Lean Production Systems 4.0 (LPS 4.0). It offers a structured collection of methods, tools, technologies, and use cases crucial for designing and optimizing corporate processes, integrating Lean Production and Industry 4.0 approaches. Developed with input from industrial practitioners, it undergoes iterative development and evaluation to meet specific enterprise needs and advance applied research on production system design.

With over 50 methods, tools, and technologies, along with over 80 use cases in its initial prototype, the Toolbox Lean 4.0 is continually evolving to address industrial engineering complexity and reduce operational challenges. However, some limitations persist, including restricted knowledge base and customization features, which are being addressed through expanded content creation and improved database management.

Future development plans include enhancing user roles and relations, usability, and customization features, transitioning from a click dummy solution to a mature software platform. The roadmap also involves testing various application processes and mechanisms for analysing problem classes and solution patterns. Continuous testing with industrial engineering professionals and real production systems will ensure its effectiveness in improving production processes tailored to individual business objectives.

19. Integrating Behavioural Health in Primary Care Using Lean Workflow Analysis

---

Sector: Healthcare

Description: The case study focuses on integrating behavioural health in primary care using lean workflow analysis. It outlines the process of implementing Lean Management (LM) in a healthcare setting to improve the level of coordination and integration of care in the healthcare system, particularly focusing on chronic pathways.

Benefits: The case study demonstrates the benefits of integrating behavioural health in primary care using LM, such as improved care coordination, enhanced patient engagement, and better health behaviour change support.

Tool as best practice: The case study highlights the use of lean workflow analysis as a best practice for integrating behavioural health in primary care. It emphasizes the importance of a system of values, principles, activities, and managerial instruments combined with lean tools and practices to achieve successful projects.

The case study discusses the implementation of Lean principles in healthcare settings, particularly in primary care and healthcare networks. It highlights the outcomes, benefits, enablers, and challenges of Lean healthcare projects. The text also emphasizes the importance of care integration strategies and the need for standardized metrics and tools in LM implementation.

The case study focuses on the implementation of Lean Management (LM) in healthcare networks (HNs) to improve the integration and continuity of care, particularly for chronic diseases. It addresses gaps in the current literature and proposes a framework for LM implementation in healthcare networks. The study emphasizes the importance of care integration strategies and the need for standardized metrics and tools. It also highlights the need for extending LM projects beyond hospital boundaries and emphasizes the importance of integrating care delivery systems across the HN. The lack of standardized approaches for LM implementation in HNs is also addressed, and the case study aims to fill the gap in the existing literature by providing insights and guidance for future research and practical applications.

The research aims to address the gaps in the literature about LM implementations in healthcare networks (HNs) by answering four research questions. A systematic literature review was conducted, resulting in 70 relevant papers. The content analysis of the literature revealed the inner and outer settings, project characteristics, process of implementation, outcomes, and enablers of LM projects in HNs. The majority of LM projects were set in primary care, with the main motivation being accessibility and coordination barriers. The goals of the projects were mainly focused on improving effectiveness, timeliness, and efficiency of care processes. Multidisciplinary teams were found to be essential for LM projects in HNs. The study also identified the need for more comprehensive and holistic LM implementations across different levels and sites of care within HNs.

The study provides valuable insights into the implementation of LM in HNs and emphasizes the need for clear objectives, team training, and the use of LM tools and practices. It also discusses the overlap between LM and care integration principles and the potential synergies between the two approaches. The research highlights the importance of leadership support, multidisciplinary team involvement, and a quality improvement culture in achieving positive outcomes. It also identifies enablers for successful implementation and emphasizes the need for extending LM projects beyond hospital boundaries.

The implementation of Lean Management in healthcare networks (HNs) involves the engagement of social workers, health coaches, and patient representatives to promote self-care and empower patients and families. However, there is significant heterogeneity in the steps conducted to implement Lean Management in HNs, with many fundamental activities still missing in the process. The lack of shared knowledge about the procedure to be followed in these projects results in fragmented and challenging implementations. Organizational practices that foster teamwork and strengthen communication channels have been implemented by project teams, including continuous reviews, regular team meetings, group presentations, emails and communication forums and full-time engagement of the project management.

The main outcomes and benefits of implementing Lean principles in healthcare settings, particularly in primary care and healthcare networks, include improved quality of care, efficiency, timeliness, effectiveness, integration, patient-centeredness, safety, and equitability. The implementation of Lean principles has been associated with successful outcomes, including enhanced patient satisfaction, improved care coordination, and increased efficiency in performing common clinical tasks. Additionally, the use of Lean tools and practices has led to positive results such as reduced emergency department use, decreased hospital admissions, and improved wait times for individuals with complex conditions. Furthermore, the implementation of Lean principles has been linked to the achievement of secondary objectives, including increased patient and employee satisfaction, enhanced communication, and the sustainability of the Lean journey. Overall, the application of Lean principles in healthcare settings has demonstrated significant improvements in various aspects of care delivery and organizational performance.

The implementation of Lean healthcare projects involves various enablers and challenges. Enablers for successful Lean healthcare projects include the engagement and empowerment of frontline staff in multidisciplinary teams, supportive leadership, a team-based approach, alignment of system values and resources, and the use of specific tools and practices to foster engagement and continuous improvement. Additionally, the creation of an implementation climate that enables improvement, the involvement and engagement of team members in every project step and decision, and the establishment of ownership of improvements made are crucial enablers for Lean healthcare projects.

Challenges encountered in the implementation of Lean healthcare projects include the lack of leadership engagement, insufficient time and resources dedicated to the project, fragmented and challenging implementations due to the lack of shared knowledge about the procedure to be followed, and the need for standardized approaches for Lean Management implementation in healthcare networks. Furthermore, the difficulties in defining multidisciplinary teams, applied practices, and project enablers, as well as the lack of care integration as the main motivation for Lean Management projects, represent significant challenges in the implementation of Lean healthcare projects.

The successful implementation of Lean healthcare projects requires addressing these challenges and leveraging enablers to create an environment conducive to improvement and change.

Care integration strategies and standardized metrics and tools can contribute to the successful implementation of Lean principles in healthcare networks in several ways. Firstly, care integration strategies can help remove inefficiencies and improve communication and coordination within the healthcare network. By addressing fragmentation of care and communication barriers, care integration practices can enhance the outcomes of LM projects, particularly in healthcare networks where these issues are prevalent.

Standardized metrics and tools play a crucial role in providing a clear framework for measuring the success of Lean principles implementation. They enable healthcare networks to define specific objectives and target metrics, such as cycle or turnaround times, readmission rates, screening rates, and activities lead times, among others. These standardized metrics provide a quantifiable way to assess the impact of Lean principles on care integration, patient-centeredness, equitability, and safety, which are essential aspects of healthcare delivery.

Furthermore, the use of standardized tools and practices in combination with care integration strategies can create an integrated methodology that fosters the achievement of common objectives, engages and empowers patients and staff, and promotes standardization and a team-based approach. This integrated approach aligns with the principles of Lean Management and can lead to improved quality of care, accessibility, and equitability within healthcare networks.

In summary, care integration strategies and standardized metrics and tools contribute to the successful implementation of Lean principles in healthcare networks by addressing fragmentation of care, improving communication and coordination, providing a clear framework for measuring success, and fostering an integrated methodology that aligns with the principles of Lean Management.

20. Implementation of Lean-Six Sigma Philosophy in Market Penetration Model for Mobile Surveillance Devices

---

Sector: Surveillance and Security

Description: The case study investigates the implementation of a new product development strategy for mobile surveillance devices in the Indian market using the Define-Measure-Analyse-Design-Validate (DMADV) methodology. The study emphasizes the integration of lean-six sigma philosophy and the use of various tools such as project charter, brainstorming, voice of customer, market analysis, benchmarking, quality function deployment, and system diagram to stabilize startups effectively.

Benefits: The research uncovers valuable insights into the market penetration model for lean startups, emphasizing its importance. Moreover, it identifies a cost difference of $94 in product manufacturing between the lowest available model price, showcasing potential cost-saving advantages for entrepreneurs and businesses.

Tool as best practice: The DMADV methodology is highlighted as a best practice for the new product development process, ensuring that the product evolves with customer-needed features by following lean philosophy.

The case study concerns the development of mobile surveillance devices in the Indian market using the DMADV methodology for market penetration. It explores the implementation strategy using lean-six sigma philosophy and provides insights into the market conditions, customer preferences, and product design considerations. Findings emphasize the significance of the market penetration model for lean startups.

The case study investigates the implementation of a new product development strategy for mobile surveillance devices in the Indian market using the Define-Measure-Analyse-Design-Validate (DMADV) methodology. The study emphasizes the integration of lean-six sigma philosophy and the use of various tools such as project charter, brainstorming, voice of customer, market analysis, benchmarking, quality function deployment, and system diagram to stabilize startups effectively.

The study focuses on the development of mobile surveillance devices in the Indian market. It discusses the use of lean tools and institutional entrepreneurship in the development process. The research methodology includes the DMADV methodology and a case study approach, providing a detailed analysis of the market conditions, customer preferences, and product design considerations for the mobile surveillance devices.

It discusses the cost analysis, testing, and comparison with existing models in the market. The cost analysis reveals a significant cost gap compared to existing models in the market, highlighting potential cost-saving benefits for entrepreneurs and businesses. The product is tested in lab conditions and real-world situations, with good performance. The finished product is compared with different models in the market, showing sustainable value in most factors. The study concludes that the iterative DMADV methodology ensures the evolution of the product with customer-needed features.

Overall, the case study provides a comprehensive analysis of the implementation of lean-six sigma philosophy in the market penetration model for mobile surveillance devices, highlighting its significance and potential benefits for businesses and entrepreneurs.

The study proposes implementing the lean-six sigma philosophy in the development of mobile surveillance devices through the use of the Define-Measure-Analyse-Design-Validate (DMADV) methodology. This methodology integrates lean tools such as project charter, brainstorming, voice of customer, market analysis, benchmarking, quality function deployment, and system diagram to stabilize startups effectively. The DMADV methodology ensures that the product evolves with customer-needed features by following lean philosophy. Additionally, the study emphasizes the significance of lean startups in customizing the product value as per the customer and improving the startup company's product reach.

The critical implementation insights revealed by the study regarding the market penetration model for lean startups include the following:

1. The study underscores the significance of the market penetration model for lean startups, emphasizing its potential benefits for entrepreneurs and businesses.
2. The study reports a cost gap of 94 US$ in the product manufacturing cost with the lowest available model price, highlighting potential cost-saving benefits for entrepreneurs and businesses.

These insights provide valuable information for entrepreneurs and management teams seeking to stabilize their startups effectively using lean philosophy and the market penetration model.

The DMADV methodology ensures the evolution of the product with customer-needed features by following lean philosophy. This methodology integrates lean tools such as project charter, brainstorming, voice of customer, market analysis, benchmarking, quality function deployment, and system diagram to stabilize startups effectively. By following the DMADV methodology, the product development process is structured to define, measure, analyse, design, and validate, ensuring that the product evolves with features that are aligned with customer needs and preferences. This approach allows for a systematic and customer-centric development process, ultimately leading to the creation of a product that meets the specific requirements and expectations of the target mark.

21. Application of Building Information Modelling (BIM) in Residential Construction Projects

---

Sector: Construction and Architecture

Description: The case study focuses on the application of Building Information Modelling (BIM) in the design and construction of a two-story home in Mumbai, India. The study utilized Autodesk Revit to create a comprehensive three-dimensional (3D) model of the bungalow, including all structural elements, interior areas, and details such as wall finishes and light fixtures. The study also incorporated Lean methods and tools, as well as principles from the Leadership in Energy and Environmental Design (LEED) framework to promote sustainable and efficient construction processes.

Benefits: The utilization of BIM, Lean methods, and LEED principles had a positive impact on the cost, speed, and quality of the residential construction project. It enabled precise estimates of material quantities and building costs, improved communication and collaboration, and increased efficiency and accuracy during the design, construction, and operation phases of the project.

Tool as best practice: The study highlights the use of Autodesk Revit as a best practice for creating detailed 3D models of the building, identifying potential problems and conflicts before construction, and producing realistic representations of the home for marketing purposes.

The case study regards the integration of Building Information Modelling (BIM), Lean methods, and LEED principles in sustainable construction. It proposes a methodological framework that combines these elements to optimize resource utilization, monitor construction progress, and improve sustainability. It also includes specific recommendations based on a literature review. The case study highlights the benefits of using BIM, Lean, and AR/VR tools in residential construction projects.

The case study focuses on the integration of Building Information Modelling (BIM), Lean methods, and Leadership in Energy and Environmental Design (LEED) principles in residential construction projects, with a specific emphasis on the use of Augmented Reality (AR) and Virtual Reality (VR) technologies. The study proposes a methodological framework that combines these elements to optimize resource utilization, monitor construction progress, and improve sustainability in construction projects.

The methodology outlined in the study involves several key steps. First, it involves creating a comprehensive 3D model using BIM, which allows for detailed visualization and planning of the construction project. This 3D model serves as a foundation for integrating Lean methods, which are aimed at streamlining processes, reducing waste, and improving efficiency. Additionally, the study emphasizes the incorporation of LEED principles, which focus on promoting sustainable and energy-efficient construction practices.

The study also highlights the use of AR and VR technologies in conjunction with BIM, Lean methods, and LEED principles. These technologies are shown to offer benefits such as improved collaboration, reduced waste, and increased efficiency. The case study specifically demonstrates the positive impact of using BIM with AR and VR software, Lean methods, and tools with LEED principles on cost, time, and quality in residential construction projects.

Furthermore, the study provides specific recommendations for quality control and energy-efficient materials, as well as a comparison of operational costs between traditional methods and VR technology. The results of the study indicate a decrease in cost and time, as well as an increase in quality when using BIM, Lean, and LEED principles together. This suggests that the integration of these elements can lead to tangible improvements in construction project outcomes.

Overall, the case study presents a comprehensive and integrated approach to sustainable construction, leveraging BIM, Lean methods, LEED principles, and AR/VR technologies. It underscores the potential benefits of this integrated approach, including cost savings, time efficiency, and improved quality in residential construction projects. The study also provides valuable insights for further research and application of these methodologies in the construction industry.

The integration of Building Information Modelling (BIM), Lean methods, and Leadership in Energy and Environmental Design (LEED) principles contributes to resource utilization optimization in construction projects in several ways. Firstly, BIM allows for the creation of comprehensive 3D models, enabling architects, engineers, and construction teams to visualize and plan the project in detail before construction begins. This visualization facilitates more precise and cost-effective design processes, leading to optimized resource utilization.

Additionally, Lean principles contribute to reducing resource waste and improving efficiency in construction projects.

Furthermore, the incorporation of LEED principles emphasizes sustainable and energy-efficient construction practices. This includes the use of recycled materials, resource-efficient engineered materials, and renewable resources, all of which contribute to resource conservation and optimization in construction projects.

The integration of these three elements allows for the efficient and sustainable use of resources throughout the construction project lifecycle. By leveraging BIM for visualization and planning, Lean methods for waste reduction and process optimization, and LEED principles for sustainable material and resource utilization, construction projects can achieve significant resource utilization optimization.

The potential benefits of using Building Information Modelling (BIM), Lean methods, and Augmented Reality/Virtual Reality (AR/VR) tools in residential construction projects are numerous. These include improved visualization and interaction with designs, enhanced communication and collaboration, increased efficiency and accuracy, reduced waste, and improved sustainability.

BIM, as a digital representation of a facility's physical and functional characteristics, improves cooperation, coordination, and efficiency during the design, construction, and operation phases of a project. Lean methods, which focus on eliminating waste and optimizing the building process, contribute to reducing resource waste and improving efficiency in construction projects. Additionally, the application of Leadership in Energy and Environmental Design (LEED) principles, which emphasize sustainable and energy-efficient construction practices, can lead to improved sustainability and environmental friendliness in residential construction projects.

The integration of AR and VR technologies in conjunction with BIM and Lean methods provides new ways to visualize and interact with designs, improving communication and collaboration, and increasing efficiency and accuracy. AR and VR can also be used to create virtual models of buildings and structures, allowing professionals to walk through and explore the design, identify potential issues, and make changes before the physical structure is built, ultimately saving time and money.

Overall, the combination of BIM, Lean methods, and AR/VR tools in residential construction projects can lead to cost savings, time efficiency, improved quality, and enhanced sustainability.

The integration of Building Information Modelling (BIM), Lean methods, and Augmented Reality/Virtual Reality (AR/VR) technologies has resulted in cost and time savings, as well as improved quality in construction projects. For example, in a residential bungalow project, BIM technology was used to create detailed 3D models of the building during the planning and design phase. This allowed for the identification of potential problems and conflicts before construction began, leading to improved efficiency and accuracy.

Additionally, the application of Lean methods, such as Six Sigma, in the construction process has been shown to reduce waste and improve quality. By streamlining and eliminating non-value-added activities, Lean methods contribute to cost savings and improved project outcomes.

Furthermore, the use of AR and VR technologies has revolutionized the way professionals design, plan, and visualize projects. These technologies have been used to create virtual models of buildings and structures, allowing professionals to walk through and explore the design, identify potential issues, and make changes before the physical structure is built. This early identification and correction of problems have led to significant time and cost savings in construction projects.

Moreover, the integration of LEED principles, which focus on sustainability and energy efficiency, has contributed to improved quality and reduced operational costs in construction projects. By applying LEED principles, construction teams can ensure that projects meet the highest standards in terms of sustainability and environmental friendliness, leading to long-term cost savings and improved quality.

In summary, the integration of BIM, Lean methods, and AR/VR technologies has resulted in cost and time savings, as well as improved quality in construction projects by improving efficiency, accuracy, and sustainability.

22. Improving Warehouse Processes Using Lean Six Sigma Approach in a Third-Party Logistics Company

---

Sector: Logistics and Supply Chain Management

Description: The case study focuses on evaluating and improving the existing warehouse processes of a third-party logistics (3PL) operated company in Nigeria. The study uses a lean six sigma (DMAIC) approach to identify and address issues related to productivity and process cycle efficiency. Data collection was based on warehouse operational areas of suppliers, customers, and internal processes, and the optimization of the warehouse processes was based on established lean tools. The study aimed to minimize non-value-added activities and waste in the warehouse processes, ultimately improving productivity and customer satisfaction.

Benefits: The implementation of the lean six sigma approach led to a significant improvement in process cycle efficiency, increasing it from 40% to 70%. Additionally, an improvement framework was established for productivity across the warehouse processes to minimize waste. Unlike previous studies that focus on specific processes, this research contributes to lean warehousing literature by examining all warehouse processes through the six sigma DMAIC approach.

Tool as best practice: The study utilized a combination of lean tools, including six sigma (DMAIC) approach, Qi Macros embedded in Microsoft Excel software, and statistical process control tools to make improvements in the warehouse. These tools were used to diagnose and resolve warehousing problems, minimize non-value-added activities, and optimize process cycle efficiency.

The case study discusses the implementation of Lean Six Sigma in warehouse management, specifically focusing on its impact on supply chain resilience, service performance, and logistics operations. The study presents a framework for improving warehouse processes using the DMAIC approach and highlights the benefits and implications of implementing Lean Six Sigma in warehouse environments.

The case study focuses on the development of an improvement framework for warehouse processes using a lean six sigma (DMAIC) approach, specifically in the context of third-party logistics (3PL) services. The study aims to enhance process cycle efficiency and productivity within the warehouse. Through the use of a case study method, the existing processes were evaluated to identify non-value-added activities. The implementation of the lean six sigma approach resulted in a significant improvement in process cycle efficiency, with a decrease in non-value-added time and an increase in productivity.

The study presents an improved value stream mapping of a warehouse, showing a decrease in non-value-added time and an increase in process cycle efficiency. The improvement framework for warehouse productivity is summarized in a table, providing a clear overview of the proposed improvements. The study also discusses the control phase and the implications of the results, offering a comprehensive analysis of the impact of the lean six sigma approach on the warehouse processes.

Furthermore, the study discusses the limitations and implications of the research, providing insights into the potential replication of the methodology for other process parameters and the suitability of Lean Six Sigma for improving process and organization productivity. This demonstrates a thorough consideration of the broader implications and applicability of the study's findings.

The case study also provides a comprehensive review of lean practices in warehouse management, including their impact on supply chain resilience, service performance, and logistics operations. It discusses the application of Lean Six Sigma in various industries, offering insights into the design and optimization of lean storage and handling systems, as well as the evaluation of lean practices in logistics distribution centres. This comprehensive review adds depth to the understanding of the impact and potential applications of lean practices in warehouse management.

The study also identifies root causes for low process cycle efficiency, such as accumulated non-value-added times leading to prolonged lead times. Improvement actions were proposed, including better storage and labelling of stocks, regular maintenance of equipment, and training and motivation for employees. Control charts were used to monitor process performance, and lean warehouse tools were implemented for improvement, demonstrating a practical application of the lean six sigma approach in addressing identified issues within the warehouse processes.

In summary, the case study provides a detailed and comprehensive analysis of the application of lean six sigma in improving warehouse processes within the context of third-party logistics services. It offers valuable insights into the potential benefits, limitations, and broader implications of the lean six sigma approach, as well as practical recommendations for addressing identified issues within warehouse operations.

The study also identified root causes for low process cycle efficiency, such as accumulated non-value-added times leading to prolonged lead times. Improvement actions were proposed, including better storage and labelling of stocks, regular maintenance of equipment, and training and motivation for employees. Control charts were used to monitor process performance, and lean warehouse tools were implemented for improvement.

The implementation of Lean Six Sigma in warehouse management to a more agile and responsive supply chain, better equipped to handle disruptions and changes in demand. Additionally, the reduction in waste and improved efficiency can lead to cost savings, which can be reinvested into enhancing the resilience of the supply chain.

Furthermore, the implementation of Lean Six Sigma in warehouse management can lead to improved service performance. By optimizing processes and reducing variability, warehouses can enhance their ability to meet customer demands effectively and consistently.

In summary, the implementation of Lean Six Sigma in warehouse management positively impacts supply chain resilience by creating a more efficient and agile supply chain, and it enhances service performance by improving operational effectiveness and customer satisfaction.

The DMAIC (Define, Measure, Analyse, Improve, Control) approach for improving warehouse processes involves several key steps:

1. Define: This step involves defining the problem or opportunity for improvement, setting project goals, and establishing the scope of the improvement effort. It is essential to clearly articulate the objectives and boundaries of the project to ensure a focused and effective improvement process.
2. Measure: In this step, the current state of the warehouse processes is measured using relevant metrics and data collection methods. This involves identifying key performance indicators (KPIs) and gathering data to assess the baseline performance of the processes.
3. Analyse: The analysis phase involves identifying root causes of inefficiencies or non-value-added activities within the warehouse processes. This may involve using tools such as value stream mapping, Ishikawa diagrams, and waste analysis to understand the factors contributing to process inefficiencies.
4. Improve: During this phase, improvement actions are implemented based on the findings from the analysis phase. This may include implementing lean tools, process redesign, training, and other measures aimed at reducing waste, improving efficiency, and optimizing the warehouse processes.

5. Control: The final step involves establishing controls to sustain the improvements and prevent the reoccurrence of previous issues. This may involve developing standard operating procedures, implementing monitoring systems, and providing ongoing training and support to ensure the sustained effectiveness of the improved processes.

These steps form the structured approach of DMAIC, providing a systematic framework for identifying, analysing, and improving warehouse processes to achieve enhanced efficiency and productivity.

23. Lean RPA Implementation in a Portuguese Bank

Sector: Banking

Description: The case study focuses on the implementation of Lean Robotic Process Automation (RPA) in a private Portuguese bank to streamline back-office tasks, particularly in the ABC team responsible for new accounts and account alterations. The study utilized Design Science Research Methodology to identify and solve organizational problems by designing and evaluating IT artifacts. The process involved problem identification, defining objectives for a solution, process selection, process study, and communication. The study involved mapping the current state of processes, identifying waste, and analysing RPA's potential to optimize the processes.

Benefits: The implementation of Lean RPA aimed to reduce the number of full-time employees (FTEs) in the ABC team by automating tasks related to account opening and validation. The study aimed to understand the effectiveness and efficiency of Lean RPA compared to standard RPA projects. The Lean RPA implementation was expected to reduce average processing time, spare FTEs, and streamline the process flow, leading to cost savings and improved operational efficiency.

Tool as best practice: The study utilized Business Process Model and Notation (BPMN) for process modelling due to its convenience in understandability and simulations. Additionally, the study evaluated RPA tools such as Blue Prism and UiPath, concluding that UiPath was the correct choice for RPA starters due to its functional appropriateness and price. The implementation followed a framework with incremental activities to ensure everyone's participation from the project's beginning, aligning with best practices for Lean RPA projects.

The case study discusses the implementation of Lean RPA (Robotic Process Automation) in a bank to improve efficiency and effectiveness compared to traditional RPA. It proposes a lean approach to RPA and evaluates the outcomes of applying the Lean RPA framework in a bank's processes. The study finds mostly positive results, including time savings and process improvements. However, it also identifies limitations and areas for future research. The article provides a theoretical background on RPA, lean management, and RPA tools, and includes a demonstration of the proposed approach in a Portuguese private bank.

The case study focuses on the implementation of Lean Robotic Process Automation (RPA) in a private Portuguese bank to improve process efficiency and effectiveness compared to traditional RPA. The study evaluates the use of RPA tools and the development of a Lean RPA framework, aiming to reduce time and improve utility in the bank's processes.

The study outlines a research methodology and proposes a model for Lean RPA, including RPA tool assessment, Lean RPA framework, and collection of RPA/Lean RPA results. The framework takes advantage of a continuous improvement approach and is successful in reducing time and improving utility. The study also discusses the RPA life cycle, lean principles, wastes, Kaizen, SMART goals, and value stream management, providing a theoretical background on RPA, lean management, and the RPA tools available in the market.

The implementation of Lean RPA aimed to reduce the number of full-time employees (FTEs) in the bank's ABC team by automating tasks related to account opening and validation. The study utilized Business Process Model and Notation (BPMN) for process modelling due to its convenience in understandability and simulations. Additionally, the study evaluated RPA tools such as Blue Prism and UiPath, concluding that UiPath was the correct choice for RPA starters due to its functional appropriateness and price.

The study found that Lean RPA was more efficient and effective than traditional RPA, leading to significant time savings and process improvements. The implementation followed a framework with incremental activities to ensure everyone's participation from the project's beginning, aligning with best practices for Lean RPA projects. However, the study also identified limitations and areas for future research, such as evaluating the framework's understandability by users and merging Lean with Six Sigma.

Overall, the case study provides a comprehensive analysis of the implementation of Lean RPA in a banking environment, showcasing the potential for significant time savings, process improvements, and operational efficiency. The study was supported by the Portuguese funding agency and the authors declare no competing interests. It includes references to related research and a table showing the average time difference for each process with and without RPA and Lean RPA. The article highlights the importance of considering Lean RPA as a valuable technology for organizations and provides a demonstration of the proposed approach in a Portuguese private bank, showcasing the assessment of RPA tools and the application of Lean RPA to a specific process.

The main advantages of implementing Lean RPA in a bank's processes compared to traditional RPA include significant time reduction and improved utility. Furthermore, Lean RPA was found to be more efficient because it reduces processing time and effective in reducing runtime mistakes, leading to a reduction in the number of full-time employees (FTEs) needed to carry out business processes.

Overall, the implementation of Lean RPA in a bank's processes offers advantages in terms of time reduction, utility improvement, and cost-effectiveness compared to traditional RPA. These benefits make Lean RPA a valuable technology for organizations, particularly in the banking sector, where efficiency and operational effectiveness are crucial.

The Lean RPA framework contributes to reducing time and improving utility in banking processes. This approach allows for easy and secure development and implementation, leading to significant time reduction compared to traditional RPA. Additionally, the Lean RPA framework leverages internal teams' help, which already knows the processes and workflows, without significantly increasing the overall cost. The framework also aims to change how processes flow inside an entire organization with the right improvements and considering the processes modelling and execution, leading to improved utility and operational efficiency.

Furthermore, the Lean RPA framework was found to be more efficient and effective than traditional RPA, resulting in a reduction in the number of full-time employees (FTEs) needed to carry out business processes. This reduction in FTEs and improved process flow contributes to the overall time reduction and improved utility in banking processes.

Overall, the Lean RPA framework's continuous improvement approach, easy development and implementation, and focus on process flow improvements contribute to reducing time and improving utility in banking processes compared to traditional RPA approaches.

One limitation is the need for testing the framework on processes in different organizations and business realities, as the study was conducted in one bank for three processes. Additionally, the study highlighted the need for future efforts to evaluate the Lean RPA framework's understandability by users and to merge Lean with Six Sigma for an improved framework. The study also mentioned the importance of describing suitable processes for automation and the need for further research on the integration of Lean with Six Sigma to leverage improvement to its full potential.

Furthermore, the study emphasized the need for future research to evaluate the Lean RPA framework's understandability by users and to merge Lean with Six Sigma for an improved framework. Additionally, the study suggested that future research should focus on testing the framework on processes in different organizations and business realities to validate its effectiveness in various contexts.

In summary, the limitations and areas for future research identified in the study on Lean RPA implementation in a bank include the need for testing the framework in different organizational contexts, evaluating its understandability by users, and exploring the integration of Lean with Six Sigma for improved effectiveness. Additionally, the study highlighted the importance of describing suitable processes for automation and the need for further research on the integration of Lean with Six Sigma to leverage improvement to its full potential.

24. Application of Lean Management Tools with Industry 4.0 Technology Solutions in SAREL Schneider Electric

---

Sector: Manufacturing

Description: The case study focuses on the successful implementation of Lean 4.0 within SAREL Schneider Electric, a manufacturing company. The company has integrated Industry 4.0 technologies with Lean management tools to improve quality, productivity, and operational efficiency. The study provides an in-depth examination of the relationship between Lean management tools and Industry 4.0 technologies, aiming to identify synergies for process improvement.

Benefits: The implementation of Lean 4.0 has led to significant improvements in the company's supply chain transformation journey, resulting in breakthroughs towards customer-centricity, cash flow efficiency, and improved productivity performance. The company aims to eliminate waste, become a factory showcase, improve safety, quality, service level, and productivity, and optimize capital employed. The staff's commitment to operating excellence has produced fruitful results, and the company has seen improvements in various key performance indicators.

Tool as best practice: The company has followed the six principles recognized for the implementation of Lean, including KPI for competence management, problem-solving (G&D) method, SIX Sigma for advanced problem-solving, and a customer satisfaction approach.

The case study explores the relationship between Lean Management tools and Industry 4.0 technologies in the manufacturing industry. It discusses the potential benefits of combining these concepts, presents a case study on their application in a company, and highlights the need for further research in this area.

The case study discussed in the provided citations focuses on the integration of Lean Management tools with Industry 4.0 technology solutions in the manufacturing industry, with a specific focus on the successful implementation of Lean 4.0 within SAREL Schneider Electric. The study aims to demonstrate the correlation between Industry 4.0 technologies and Lean management tools, highlighting the potential benefits of combining these methodologies to improve productivity, eliminate waste, and enhance operational efficiency.

The case study presents SAREL Schneider Electric as a company that has effectively integrated Industry 4.0 technologies with Lean management tools to drive improvements in quality, productivity, and operational efficiency. The company's successful implementation of Lean 4.0 has led to significant advancements in its supply chain transformation journey, resulting in breakthroughs towards customer-centricity, cash flow efficiency, and improved productivity performance. The staff's commitment to operating excellence has produced fruitful results, with improvements seen in various key performance indicators.

The study also emphasizes the need for further research in this area and proposes a holistic integration framework of Lean Management and Industry 4.0 within organizations. It provides an overview of Lean and Industry 4.0, discussing the correlation between Lean tools and Industry 4.0 technologies, and highlights the potential synergy between the two approaches.

The study also covers topics such as smart product development, prioritization of lean and sustainable manufacturing tools and the integration of technology in designing lean value streams. It includes references to specific research studies and publications on the subject, providing a comprehensive overview of the current state of knowledge in this area.

Overall, the case study serves as a valuable example of how companies can achieve operational improvements and competitive advantages in the manufacturing industry. It provides insights into the successful implementation of Lean 4.0 at SAREL Schneider Electric and highlights the potential benefits of this integration, offering valuable perspectives for further research and practical application in organizations.

This integration allows for the optimization of processes, reduction of operational complexity, and enhancement of productivity. By leveraging Lean Management tools such as Six Sigma for advanced problem-solving, KPI for competence management, and customer satisfaction approaches, in conjunction with Industry 4.0 technologies, companies can achieve improved quality, operational efficiency, and waste elimination. The combination of Lean Management tools and Industry 4.0 technologies enables companies to streamline operations, enhance decision-making processes, increase productivity and reduced waste in manufacturing and operational processes.

The potential benefits of combining Lean Management and Industry 4.0 in the manufacturing industry include significant improvements in supply chain transformation, breakthroughs towards customer-centricity, cash flow efficiency, improved productivity performance, and enhancements in quality, service level, and safety. The combination of Lean Management and Industry 4.0 technologies can lead to the elimination of waste, improved operational efficiency, and optimized capital employed. This integration can also result in the development of a conceptual framework for implementation, providing a basis for further research and contributing to the development of a holistic framework for integrating Lean Management and Industry 4.0 within organizations with diverse organizational cultures and processes, while promoting the elimination of waste. The synergies between Lean Management tools and Industry 4.0 technologies have been demonstrated to bring significant benefits to companies.

In the case study of SAREL Schneider Electric, specific Lean Management tools and Industry 4.0 technologies used include:

- **Lean Management tools**:
  - 5S;
  - KPI for competence management;
  - problem-solving (G8D);
  - SIX Sigma;
  - customer satisfaction approach.
- **Industry 4.0 technologies**:
  - CPS AIC (SIM);
  - digital work instruction;

– KL2;
– E-PFMEA;
– AGV & Smart AGV.

These tools and technologies were implemented in SAREL Schneider Electric as part of the Lean 4.0 application, aiming to improve operational excellence and drive significant improvements in supply chain transformation, customer-centricity, cash flow efficiency, and productivity performance.

25. Impact Footwear Customizable Flip Flop

Sector: Consumer product

Description: The case study focuses on Impact Footwear, a startup company that has developed an innovative digital business model for customizable slippers. The company provides a cloud-based platform for customers to select preset slippers or customize them. The cloud-based platform enables the software and hardware to remotely respond to orders, minimizing human interaction, reducing operational costs, and improving design efficiency. The company utilized nTopology software to swiftly respond to requests, iterate product design, and deploy for manufacturing. The software performs iterations of input models to create complex and customized lattice (foam-like) structures for the soles. Manufacturing is done using a plastic-based EOS PBF machine system, printing the slippers from dyed thermoplastic elastomer (TPE) and vapor smoothing them using Dye Mansion vaporfuse. Post-processing is also conducted to enhance functional properties, surface quality, and reduce bacteria growth on the footwear.

Benefits: The use of this innovative digital business model and advanced manufacturing techniques has led to improved design efficiency, reduced operational costs, and enhanced customer satisfaction. The company's approach has also minimized human interaction, which is particularly relevant in the context of the ongoing COVID-19 pandemic.

Tool as best practice: The nTopology software used by Impact Footwear can be considered a best practice tool for its ability to swiftly respond to requests, iterate product design, and deploy for manufacturing. This tool has enabled the company to create complex and customized lattice structures for the soles, contributing to the overall success of their innovative digital business model.

This case study explores various topics related to additive manufacturing, lean production, sustainable manufacturing, and business excellence. The study discusses the integration of lean management and additive manufacturing, the potential benefits of this convergence for sustainability and efficiency, and the challenges and opportunities in additive manufacturing. It also highlights the importance of waste reduction, operational efficiency, and the implementation of lean thinking in optimizing production processes.

The case study explores the innovative approach of Impact Footwear, a startup company that has leveraged additive manufacturing (AM) and digital business models to create customizable flip flops. The company's use of AM, particularly in the form of 3D printing, has allowed for the production of customized lattice structures for the soles of the flip flops, showcasing the design flexibility and material efficiency that AM offers. Additionally, the company has implemented a cloud-based platform for on-demand manufacturing and customization, demonstrating the integration of digital tools to enhance production efficiency and customer satisfaction.

The study emphasizes the benefits of AM, such as reduced material consumption, lead time, and lifecycle costs, as well as the potential to decrease energy consumption and carbon dioxide emissions. Impact Footwear's case study serves as a practical example of how AM can optimize product design, adding value for customers in terms of ecological, economic, and experiential performance. The company's approach aligns with the broader industry trends, as the study references the increasing application of AM in various industrial sectors and its evolution from a prototyping method to product development.

Furthermore, the case study highlights the convergence of Lean Management (LM) and AM, emphasizing the potential for waste reduction, time efficiency, and cost savings. By integrating LM and AM, Impact Footwear has been able to enhance space utilization, reduce waste and emissions, and improve overall cost efficiency. This integration aligns with the study's exploration of the relationship between LM and AM, as well as the potential for achieving competitive advantage through the integration of these concepts.

The study also acknowledges the limitations and challenges associated with AM, such as limited expertise, lack of standardization for new materials, and high machinery costs. However, it emphasizes the potential for further research on sustainable business models, the role of AI, and commercialization processes for an integrative LM and AM strategy.

In summary, the case study of Impact Footwear's customizable flip flop serves as a compelling example of the benefits and potential of AM in the consumer product sector. It showcases the company's innovative use of AM and digital business models to create customized, sustainable footwear, while also highlighting the broader industry trends and the potential for integrating LM and AM to enhance production efficiency and sustainability.

The integration of LM and AM can contribute to environmental, social, and economic sustainability in several ways.

From an environmental perspective, the convergence of LM and AM can lead to resource efficiency and waste reduction. AM promotes resource efficiency and waste minimization through light weighting, reduced scrap rates, and shorter lead times. The digital inventory and energy-efficient parts produced through AM also contribute to environmental sustainability by reducing material waste and energy consumption. Additionally, the integration of LM and AM can lead to reduced emissions and minimized material usage, further enhancing environmental sustainability.

In terms of social sustainability, the integration of LM and AM can contribute to improved workplace safety and ergonomic working conditions. By reducing human fatigue and errors, the convergence of LM and AM can create a safer and more ergonomic working environment, thereby contributing to social sustainability.

Economically, the integration of LM and AM can lead to reduced costs, time, and space usage. This can result in economic sustainability through reduced total cost of ownership, improved supply chain efficiencies, and customized batch production. The on-demand manufacturing capabilities of AM can also contribute to economic sustainability by reducing overproduction and process steps, leading to cost savings and improved operational efficiency.

These combined benefits demonstrate how the integration of LM and AM can contribute to environmental, social, and economic sustainability, making it a valuable strategy for organizations seeking to enhance their overall sustainability performance.

In the context of AM, several tools and principles of LM can be applied to improve organizational performance. These include:

1. **Value Stream Mapping (VSM)**: This tool visualizes material and information flow in Lean Management, aiding waste reduction and efficiency improvement in Additive Manufacturing (AM).
2. **Kaizen**: Continuous improvement underpins Lean Management, driving ongoing refinement of AM processes for better quality, shorter lead times, and cost savings.

3. **Just-In-Time (JIT)**: JIT minimizes inventory and lead times in Lean Management, aligning with AM's on-demand manufacturing capabilities for swift customer response.
4. **5S Methodology**: 5S ensures workplace organization and process standardization in Lean Management, optimizing the AM environment for efficiency.
5. **Total Productive Maintenance (TPM)**: TPM maximizes equipment efficiency in Lean Management, ensuring reliable and available additive manufacturing machines for heightened productivity.

These LM tools enhance organizational performance, optimize production processes, and drive continuous improvement.

The challenges and limitations of lean production practices in the context of additive manufacturing (AM) include:

1. **Semantic misinterpretations and misunderstandings**: There are semantic misinterpretations and misunderstandings between different fields of scientific research in technology and business, which can lead to challenges in implementing lean production practices in the context of AM.
2. **Lack of common definitions**: There is a lack of common definitions between different industries or professionals regarding lean production practices, which can lead to misconceptions and limitations in the implementation of lean principles in the context of AM.
3. **Performance expectations**: Some studies have suggested that organizations that have implemented lean management did not perform as expected, especially in areas such as inventory management. This can pose challenges to the successful implementation of lean production practices in the context of AM.
4. **Inventory management**: Reduction of inventory is essential for successful lean management, but the need for inventory management may increase with growing demand, leading to potential challenges and limitations in the context of AM.
5. **Data collection and research limitations:** One of the limitations of research in this area lies in the data collection, where only specific keywords were used for the search. This may have led to the exclusion of relevant publications that cover the full essence of the topic of the research.

26. Application of Drones in Precision Agriculture

Sector: Agriculture

Description: The case study focuses on the application of precision agriculture in the management of agricultural fields. Precision agriculture involves the use of modern technologies to carry out specific agronomic interventions, taking into account the real needs of crops, soil characteristics, and fertility. The study utilized drones equipped with various sensors, including infrared, thermal, and multispectral, to gather data on the field. The drones were used to capture 96 images of the entire field, which were then processed to create NDVI (Normalized Difference Vegetation Index) maps. These maps provided crucial information on the health and biomass of the plants, allowing for the identification of areas with different productivity limitations. The study demonstrated the versatility and power of drones in supporting agronomic field management, enabling the identification of areas with production limitations, their causes, and the percentage of their impact on the analysed surface.

(*Continued*)

Benefits: The use of drones in precision agriculture offers several benefits, including the optimization of spatial-temporal management of agricultural activities, precise and controlled seeding, efficient distribution of fertilizers and phytosanitary products, and the ability to work at night. Additionally, the drone-collected data enables the creation of prescription maps for differentiated field management. This optimization reduces input costs and groundwater pollution, providing both economic and environmental benefits.

Tool as best practice: Drones equipped with various sensors, including infrared, thermal, and multispectral, are highlighted as a best practice tool in precision agriculture. The use of drones for remote sensing offers opportunities to collect field data easily, quickly, and efficiently, providing crucial information for the application of precision agriculture. The data collected by drones, particularly the NDVI maps, serve as valuable tools for identifying areas with different productivity limitations and for creating prescription maps for differentiated field management.

Precision agriculture utilizes modern technologies to manage specific agronomic interventions, taking into account the needs of crops and soil characteristics. The use of drones and sensors allows for the collection of data to optimize the spatial and temporal management of agricultural activities, identify variability in fields, and identify areas with production limitations. This multidisciplinary approach requires professionals who can support the management decisions of farmers.

The case study focuses on the application of precision agriculture in the management of agricultural fields, specifically utilizing drones and sensors to gather data for optimizing spatial-temporal management of agricultural activities, identifying variability in fields, and pinpointing areas with production limitations. Precision agriculture, as described in the case study, involves the use of modern technologies to carry out specific agronomic interventions, taking into account the real needs of crops and soil characteristics. This approach emphasizes the importance of utilizing drones and sensors to collect data that supports the decision-making process for farmers.

The study highlights the benefits of using drones equipped with various sensors, including infrared, thermal, and multispectral, to capture data on the field. The data collected by drones allows for the creation of NDVI (Normalized Difference Vegetation Index) maps, which provide crucial information on the health and biomass of the plants. These maps enable the identification of areas with different productivity limitations, their causes, and the percentage of their impact on the analysed surface. This information is invaluable for creating prescription maps for differentiated field management to optimize input use, reduce technical input costs, and minimize groundwater pollution.

The benefits of utilizing drones in precision agriculture are extensive. They include the optimization of spatial-temporal management of agricultural activities, precise and controlled seeding, efficient distribution of fertilizers and phytosanitary products, and the ability to work at night. Additionally, the drone-collected data enables the creation of prescription maps for differentiated field management. This optimization reduces input costs and groundwater pollution, providing both economic and environmental benefits.

The case study emphasizes the use of drones equipped with various sensors as a best practice tool in precision agriculture. The use of drones for remote sensing offers opportunities to collect field data easily, quickly, and efficiently, providing crucial information for the application of precision agriculture. The data collected by drones, particularly the NDVI maps, serve as valuable tools for identifying areas with different productivity limitations and for creating prescription maps for differentiated field management.

In conclusion, the case study demonstrates the versatility and power of drones in supporting agronomic field management, enabling the identification of areas with production limitations, their causes, and the percentage of their impact on the analysed surface. The use of drones in precision agriculture offers numerous benefits, and the data collected by drones serves as a valuable tool for creating prescription maps and optimizing field management.

Precision agriculture utilizes drones and sensors to collect data for optimizing agricultural activities by employing modern technologies such as geographic information systems, remote sensing, GPS, and drones equipped with various sensors, including infrared, thermal, and multispectral. These technologies enable the identification of variability in fields and the gathering of crucial information for the application of precision agriculture. Drones equipped with sensors are used to capture data on the field, including creating NDVI (Normalized Difference Vegetation Index) maps, which provide information on the health and biomass of the plants. This data allows for the identification of areas with different productivity limitations, their causes, and the percentage of their impact on the analysed surface. The information collected by drones and sensors supports the creation of prescription maps for differentiated field management to optimize input use, reduce technical input costs, and minimize groundwater pollution.

Additionally, the use of drones for remote sensing offers opportunities to collect field data easily, quickly, and efficiently. The data collected by drones, particularly the NDVI maps, serves as valuable tools for identifying areas with different productivity limitations and for creating prescription maps for differentiated field management to optimize input use, reduce technical input costs, and minimize groundwater pollution, resulting in economic and environmental advantages.

Furthermore, the data collected by drones and sensors can be used by technical consultants for various needs, but it requires professional figures to manage flights, interpret the collected images, and process the data according to agronomic criteria to provide useful suggestions for differentiated field management.

In summary, precision agriculture utilizes drones and sensors to collect data for optimizing agricultural activities by leveraging modern technologies to gather crucial information for field management, identify variability, and create prescription maps for differentiated field management to optimize input use and minimize environmental impact.

Identifying variability in fields through precision agriculture techniques offers several benefits. Firstly, it allows for the optimization of spatial-temporal management of agricultural activities, enabling precise and controlled seeding, efficient distribution of fertilizers and phytosanitary products, and the ability to work at night. Additionally, the identification of variability in fields supports the creation of prescription maps for differentiated field management, optimizing input use, reducing technical input costs, and minimizing groundwater pollution, resulting in economic and environmental advantages.

Furthermore, the data collected through precision agriculture techniques, such as identifying variability in fields, provides crucial information for the application of precision agriculture. This information allows for the creation of prescription maps for differentiated field management, optimizing input use, reducing technical input costs, and minimizing groundwater pollution, resulting in economic and environmental advantages.

In summary, the benefits of identifying variability in fields through precision agriculture techniques include the optimization of spatial-temporal management of agricultural activities, the creation of prescription maps for differentiated field management, and the resulting economic and environmental advantages.

Professionals in precision agriculture play a crucial role in supporting farmers' management decisions. They are essential in managing flights, interpreting the collected images, and processing the data according to agronomic criteria to provide useful suggestions for differentiated field management. These professionals are required to have the expertise to handle the data collected through precision agriculture techniques and provide valuable insights to farmers for optimizing agricultural activities.

Furthermore, the multidisciplinary approach required for precision agriculture necessitates various professional figures capable of supporting farmers' management decisions. These professionals are instrumental in providing the necessary expertise and guidance to ensure that the data collected through precision agriculture techniques is effectively utilized to optimize field management and maximize agricultural productivity.

In summary, professionals in precision agriculture play a vital role in supporting farmers' management decisions by managing data collection, interpreting images, and providing valuable insights for optimized field management.

27. Implementation of Blockchain for Chicken Supply Chain Traceability at Carrefour

---

Sector: Retail/Food Industry

Description: Carrefour, a French supermarket chain, implemented blockchain technology in 2018 to ensure complete traceability of the chicken supply chain. The blockchain system records detailed information about the chicken production process, including birth date, feeding, veterinary care, transportation, and retail sale. This was achieved through collaboration with a French chicken producer and IBM, utilizing the IBM Food Trust platform. The system involves farmers recording data using a smartphone app, which is then shared with the producer for transportation to Carrefour's warehouses. Consumers can access the supply chain information by scanning a QR code on the product label.

Benefits: The implementation of blockchain technology at Carrefour resulted in improved transparency and traceability of the chicken supply chain. This allowed consumers to make more informed choices and have greater confidence in the quality and origin of the product. Additionally, the technology reduced the verification time of the supply chain from several days to just a few seconds.

Tool as best practice: The use of blockchain technology for supply chain traceability in the food industry, as demonstrated by Carrefour's implementation, serves as a best practice for enhancing transparency, consumer trust, and efficiency in the production and distribution process.

Carrefour implemented blockchain technology to improve transparency and traceability in their chicken supply chain. This technology reduced verification times, enhanced collaboration among supply chain actors, and allowed consumers to access detailed information about the product's origin and production process. The economic impact of this implementation is not specified, but the initial cost may have been offset by improved efficiency and reputation. The success of blockchain implementation depends on factors such as effectiveness, consumer adoption, and added value to the company and its customers.

The case study of Carrefour's implementation of blockchain technology for chicken supply chain traceability demonstrates the potential benefits and challenges of utilizing this innovative tool in the retail and food industry sector.

The implementation of blockchain technology at Carrefour resulted in several notable benefits. Firstly, it significantly improved transparency within the chicken supply chain. By recording detailed information about the chicken production process, including birth date, feeding, veterinary care, transportation, and retail sale, the blockchain system provided consumers with access to comprehensive and accurate information about the product's origin and production. This transparency is crucial in building consumer trust and confidence in the quality and safety of the products they purchase.

Secondly, the blockchain technology reduced verification times within the supply chain. What previously took several days for verification could now be accomplished in just a few seconds. This efficiency not only benefits the company by streamlining its operations but also enhances the overall supply chain management process.

Thirdly, the implementation of blockchain technology at Carrefour enhanced collaboration among supply chain actors. By utilizing a shared, immutable ledger, all parties involved in the supply chain could access and contribute to the same set of data, fostering greater cooperation and trust among stakeholders.

The initial cost of implementing blockchain technology can be significant, involving expenses related to technology acquisition, integration, and training. However, these costs may be offset by the improved efficiency and reputation resulting from the implementation. Additionally, the enhanced transparency and consumer trust facilitated by blockchain technology contribute to the economic viability of the implementation.

The success of blockchain implementation, as demonstrated by Carrefour, depends on various factors. Effectiveness in providing accurate and comprehensive traceability information is crucial for consumer adoption and trust. The added value to the company and its customers, in terms of improved efficiency, transparency, and collaboration, is also essential for the success of the implementation. Furthermore, the ability to effectively integrate blockchain technology into existing supply chain processes and infrastructure is a critical factor for successful implementation.

The case study of Carrefour's implementation of blockchain technology for supply chain traceability serves as a best practice for the retail and food industry.

28. Lean Six Sigma: The Textron Case

Sector: The case study focuses on the conglomerate Textron, which operates in various sectors including aviation, industrial, and finance

Description: Textron, a US multinational conglomerate, implemented the Lean Six Sigma methodology to improve its operations. The company's history dates back to 1923 when it was founded as a synthetic yarn corporation. Over the years, Textron diversified its operations, acquiring companies in various sectors such as automotive and aviation, ultimately becoming a conglomerate.

Benefits: The implementation of Lean Six Sigma resulted in significant improvements across various business units. For instance, Avco Lycoming experienced a decrease in absenteeism and employee turnover, along with an increase in Return on Invested Capital (ROIC) and operating margins. Similarly, E-Z-GO saw improvements in first-time quality, employee absenteeism, and productivity. The program also led to managerial benefits, with internal analysis showing a score improvement from 4.6/6 to 5.6/6.

Tool as best practice: The case study highlights the use of the SMED (Single-Minute Exchange of Die) method as a best practice. This method allowed Textron to reduce the required time for evaluation by 30%, from an average of 5 days to 3.5 days. The application of SMED principles to evaluation activities led to significant improvements in various areas of the organization, not limited to production areas.

Textron, a US multinational conglomerate, implemented Lean Six Sigma (LSS) principles to address operational inefficiencies and financial challenges. The company made changes in its evaluation systems, set-up activities, and execution methods to improve operational efficiency and performance. The implementation of LSS resulted in significant improvements in various areas such as absenteeism, employee turnover, return on invested capital (ROIC), operating margins, and first-time quality. The success of Textron's implementation of the Single-Minute Exchange of Die (SMED) method was recognized through the Shingo Prize awarded to E-Z-GO and Avco Lycoming.

The case study focuses on Textron, a US multinational conglomerate that has undergone significant evolution from its origins as a textile company to its current status as a diversified conglomerate with a focus on aerospace, automotive, industrial, and financial sectors. Under the leadership of CEO Lewis B. Campbell, the company adopted Lean Six Sigma (LSS) principles to address its operational inefficiencies and financial challenges. The implementation of LSS involved significant changes in the company's evaluation systems, focusing on interfunctional assessment, empirical evidence-based execution, simplified scoring systems, and reducing set-up times. These changes were aimed at improving operational efficiency and performance.

One of the key methodologies implemented by Textron was the Single-Minute Exchange of Die (SMED) method, which was used to optimize set-up activities and reduce machine downtime. The company saw significant improvements in various areas, including absenteeism, employee turnover, Return on Invested Capital (ROIC), operating margins, and first-time quality. The success of the SMED method was further demonstrated by the recognition received by E-Z-GO and Avco Lycoming, as they were awarded the Shingo Prize, indicating their alignment with the Shingo Model's guiding principles. This recognition underscores the economic results achieved by Textron and its business units, showcasing the success of the SMED method in achieving sustainable and lasting successes for organizations.

Overall, the case study highlights the transformative impact of Lean Six Sigma principles and the SMED method on Textron's operations. The company's adoption of these methodologies led to tangible improvements in various aspects of its business, including financial performance, quality, and employee engagement.

Textron addressed its operational inefficiencies and financial challenges through the implementation of Lean Six Sigma (LSS) principles. Under the leadership of CEO Lewis B. Campbell, the company launched a massive training program aimed at its leaders, with the goal of training 100 Black Belts in one year. However, Textron significantly exceeded this target, managing to train 225 Black Belts in less than a year and setting a goal to train at least another 500. This initiative aimed to equip the organization with the necessary expertise to drive operational improvements and address inefficiencies.

Furthermore, Textron focused on restructuring its evaluation systems, which were deemed antiquated and inefficient, with little attention to secondary activities and a lengthy and complex evaluation process. The company's evaluation systems were reformed to emphasize focus, execution mode, scoring, and time required. This overhaul aimed to modernize the evaluation process and align it with contemporary managerial and economic principles.

From a financial perspective, Textron faced challenges due to the dot-com crisis and internal inefficiencies. Despite the economic crisis and a contraction in revenue, the company continued to support its reprogramming process based on Six Sigma, embracing a medium to long-term vision and not giving priority to short-term results. This strategic approach aimed to address financial challenges and drive sustainable improvements in the company's performance.

In summary, Textron addressed its operational inefficiencies and financial challenges through a comprehensive approach that included training leaders in Lean Six Sigma principles, restructuring evaluation systems, and adopting a strategic vision focused on long-term sustainable improvements. These initiatives were aimed at driving operational efficiency and addressing financial challenges within the organization.

During the implementation of Lean Six Sigma (LSS), Textron made significant changes in its evaluation systems and execution methods. The company's original evaluation systems were restructured to align with LSS principles, focusing on four key activities: Focus, Execution mode, Scoring, and Time required.

1. **Focus**: Textron's original evaluation systems were exclusively focused on production, neglecting the importance of other business functions. The company conducted a survey that revealed the willingness of resources not part of the production area to be involved in evaluation processes. As a result, Textron changed the focus of its evaluation systems to create an interventional assessment, considering the organization as an integrated set of processes aimed at maximizing effectiveness and efficiency.
2. **Execution modes**: Textron's original assessment did not rely on emerging results from analyses performed but was characterized by a check-box mentality. The company shifted its approach to align with LSS principles, which are entirely based on empirical evidence, the results of analyses, and execution flexibility. This change aimed to make the evaluation process more proactive and results-oriented.
3. **Scoring**: Textron simplified its scoring system by removing dimensions such as Depth and Width and replacing them with four levels of maturity. The company also applied Measurement System Analysis (MSA) derived from the Six Sigma methodology to identify the goodness of the methods used and the results emerging from the measurements. This change aimed to make the scoring system less complex and more results-driven.
4. **Time required**: Textron identified the excessive time required to carry out activities as a significant issue in its original evaluation systems. The company aimed to streamline and reduce the time required for evaluation activities to align with the efficiency goals of LSS.

These changes in evaluation systems and execution methods were crucial in aligning Textron's operations with Lean Six Sigma principles, driving improvements in operational efficiency and performance.

The implementation of the Single-Minute Exchange of Die (SMED) method at Textron resulted in significant improvements across various business units. Specifically, Avco Lycoming experienced a decrease in absenteeism from 5% to 0.87%, a reduction in employee turnover from 17.6% to 6.7%, an increase in Return on Invested Capital (ROIC) by 138%, and an increase in operating margins by 24%.

Similarly, the E-Z-GO business unit saw improvements, including an increase in first-time quality from 20% to 98%, a reduction in employee absenteeism by 36%, and a 30% increase in productivity.

These improvements demonstrate the tangible benefits achieved by Textron through the implementation of the SMED method, showcasing the method's effectiveness in driving positive changes in various areas of the organization.

29. Eataly: A High-Quality Food and Wine Supermarket

Sector: The case study pertains to the food and wine retail sector, specifically focusing on high-quality gastronomic products and the large-scale distribution of these items.

Description: Eataly is the largest high-quality food and wine supermarket, with a mission to provide gastronomic excellence on a large scale. The company's business model is characterized by a unique relationship between sales, catering, and education, with a selection of producers based on criteria such as product quality, sustainability of production techniques, and the ability to ensure a high quantity and continuity of supplies. Eataly's partnership with Slow Food, as well as its collaboration with Coop and TIP, has been instrumental in its growth and success. The company is guided by principles of sustainability, responsibility, and sharing of values, positioning itself as a provider of high-quality food products while also promoting responsible consumption.

Benefits: Eataly's approach has several benefits, including the availability of high-quality products to a large number of people, the promotion of sustainable production techniques, and the differentiation from other large-scale distribution companies. The partnerships with Slow Food, Coop, and TIP have provided Eataly with expertise, funding, and visibility, contributing to its success in the industry.

Tool as best practice: Eataly's emphasis on quality, sustainability, and responsible consumption can be considered a best practice in the food and wine retail sector. The company's unique business model, selection criteria for producers, and strategic partnerships serve as a model for other businesses seeking to provide high-quality products on a large scale while promoting sustainable and responsible practices.

Eataly's mission is to offer high-quality gastronomic excellence at a large scale, in line with Slow Food's principles. They prioritize product quality, sustainability, and supply continuity when selecting producers. Eataly has partnered with Coop and TIP for funding and expertise in the food distribution industry. They emphasize sustainability, responsibility, and shared values. Eataly positions itself as a premium food and wine supermarket, with a goal of making high-quality products accessible to a wide audience. Their business model encompasses sales, catering, and education.

The case study of Eataly provides a comprehensive overview of the business model, and strategic partnerships, highlighting its unique approach to providing high-quality gastronomic products on a large scale. Eataly's mission to offer high-quality gastronomic excellence aligns with the values of Slow Food, a global grassroots organization that promotes good, clean, and fair food for all. This alignment underscores Eataly's commitment to sourcing and offering products that meet high standards of quality, sustainability, and ethical production practices.

Eataly's business model is characterized by its meticulous selection of producers based on specific criteria, including product quality, sustainability of production techniques, and the ability to ensure a high quantity and continuity of supplies. This approach not only ensures that Eataly's offerings meet the highest standards but also supports producers who adhere to sustainable and responsible practices. By emphasizing sustainability, responsibility, and the sharing of values, Eataly positions itself as a company that goes beyond traditional retail to promote ethical and responsible consumption.

The strategic partnerships with Coop and TIP further illustrate Eataly's commitment to leveraging expertise and funding in the food distribution industry. These partnerships have likely provided Eataly with valuable resources, knowledge, and visibility, contributing to the company's success in providing high-quality products to a broad consumer base.

Eataly's business model, which encompasses sales, catering, and education, reflects a holistic approach to engaging with consumers. By offering not only products but also culinary experiences and educational opportunities, Eataly creates a multifaceted environment that promotes a deeper understanding and appreciation of high-quality food and wine.

Overall, the case study of Eataly serves as a compelling example of a company that has successfully integrated high-quality gastronomic products, sustainability, and responsible consumption into its business model. Eataly's emphasis on quality, sustainability, and responsible consumption can be considered a best practice in the food and wine retail sector, providing a model for other businesses seeking to offer high-quality products while promoting ethical and sustainable practices.

Eataly selects producers for their products based on three primary criteria. The selection process is based on the following principles:

1. Quality of the products: Eataly evaluates the quality of the products through tastings, ensuring that the selected items meet high standards of taste and overall quality.
2. Sustainability of production techniques: The company prioritizes producers who employ sustainable production techniques, emphasizing environmentally friendly and ethical practices in the creation of their products.
3. Ability to ensure a high quantity and continuity of supplies: Eataly seeks producers who can consistently provide a high quantity of products, ensuring a reliable and continuous supply to meet consumer demand.

These criteria reflect Eataly's commitment to offering high-quality, sustainable products while ensuring a consistent supply to meet the needs of their customers. By prioritizing these factors, Eataly aims to provide a selection of products that align with their mission of high-quality gastronomic excellence on a large scale.

Eataly has formed key partnerships to support its business model, including collaborations with Slow Food, Coop, and TIP. The partnership with Slow Food has been instrumental in providing expertise and aligning with the values of promoting high-quality, sustainable food products. Additionally, Eataly's collaboration with Coop, a company already present in the Organized Large-Scale Distribution (GDO) industry, has provided the advantage of handling perishable goods and avoiding competition between the two entities, while also offering greater visibility and prestige for both parties. Furthermore, the agreement with TIP, Tamburi Investments Partners, a participation finance company, has provided Eataly with investment and support, contributing to the company's growth and success in the food distribution industry.

These partnerships have played a crucial role in providing Eataly with expertise, funding, and visibility, supporting the company's mission to provide high-quality gastronomic excellence on a large scale while promoting sustainability and responsible consumption.

Eataly is the largest high-quality food and wine supermarket. Its mission is to provide high-quality gastronomic excellence on a large scale, demonstrating that high-quality products can be made available to a large number of people. The closely aligned with the values of Slow Food, which gave rise to an important partnership. The business model is characterized by a relationship between sales, catering, and education in a single selection of producers is based on three criteria: value proposition.

Eataly differentiates itself from other supermarkets by focusing on providing high-quality food products and promoting responsible consumption. The company's approach to making high-quality products more accessible is based on the principles of sustainability, responsibility, and sharing of values. Eataly's mission is to provide high-quality gastronomic excellence on a large scale, demonstrating that even high-quality products can be available to a large number of people. This approach positions Eataly as a provider of high-quality food products while also promoting responsible consumption, setting it apart from traditional supermarkets that may not prioritize these values to the same extent.

Eataly's differentiation strategy also involves limiting its product range to food items, which are the central element of the food world. This decision is aimed at focusing consumers' attention on the excellence of the food offer and differentiating Eataly from other large-scale distribution companies in Italy.

Overall, Eataly's differentiation lies in its commitment to sustainability, responsibility, and the sharing of values, as well as its focus on providing high-quality food products while making them accessible to a broad consumer base.

30. Interport: A Complex of Services for Logistics and Intermodal Transportation

---

Sector: Logistics and Transportation

Description: The case study discusses the essential elements and services provided by Interports, which are logistics platforms aimed at exchanging goods using different transportation modes. It highlights the various activities carried out by Interports, such as storage, transit, intermodal transport systems, and logistics services. The study emphasizes the importance of location, integrated offers, and connections with other logistics platforms in the area for an Interport to be effective.

Benefits: Interports provide added value for operator services, effective management of financial structure and inventory movement, and opportunities for sustainable and environmentally friendly transportation modes. They also offer value-added services that can become significant sources of competitive advantage for logistics platforms.

Tool as best practice: The case study presents Interports as a best practice for integrating logistics and intermodal transportation services, emphasizing the importance of seizing the opportunity of intermodality to guarantee value-added services and competitive advantage.

The case study discusses the importance of Interports in facilitating intermodal transportation services and the exchange of goods. It highlights the added value they provide for operator services and the need for them to be located in areas with strong freight traffic. The activities carried out by Interports include storage, transit, intermodal transport systems, and logistics services. The paper also emphasizes the opportunities for value-added services and competitive advantage that intermodality offers for logistics platforms. The challenge lies in creating favourable conditions for companies to settle and benefit from these services.

The case study focuses on Interports, which are logistics platforms that offer intermodal transportation services and play a crucial role in the exchange of goods. Interports are strategically located in areas with strong freight traffic, allowing for efficient movement of goods between different transportation modes. The study emphasizes the essential activities carried out by Interports, including storage, transit, intermodal transport systems, and logistics services.

One of the key benefits highlighted in the case study is the added value for operator services provided by Interports. This includes effective management of financial structure and inventory movement, as well as opportunities for sustainable and environmentally friendly transportation modes. Additionally, Interports offer value-added services that can become significant sources of competitive advantage for logistics platforms.

Overall, the case study presents Interports as a best practice for integrating logistics and intermodal transportation services. It underscores the importance of seizing the opportunity of intermodality to guarantee value-added services and competitive advantage for logistics platforms. The study provides valuable insights into the essential elements and benefits of Interports, positioning them as crucial components in the logistics and transportation sector.

The key activities performed by Interports include storage and transit, characterized by intermodal transport systems and logistics services. Additionally, they offer total logistics services, such as inventory control, packaging, and various equipment, as well as support services, including support for vehicles, repair and maintenance support, and support for people and businesses.

Interports need to be located in areas with strong freight traffic for several reasons. Firstly, the presence of a strong freight traffic basin ensures a high volume of goods moving through the area, providing ample opportunities for the Interport to facilitate the exchange of goods using different transportation modes (). Additionally, being located in an area with strong freight traffic allows Interports to capitalize on the diverse transportation modes available, such as rail, road, and sea, enabling efficient movement of goods from origin to destination. This strategic location also facilitates integrated offers with services useful for businesses, people, and goods, as well as connections with other logistics platforms in the area, creating a network that enhances the overall efficiency and effectiveness of the Interport.

Furthermore, the strong freight traffic basin provides a larger market to serve, allowing the Interport to address the spatial extension of the market and meet the needs expressed by it within the time interval required, thus ensuring timely and efficient service delivery.

In summary, the location of Interports in areas with strong freight traffic is essential for maximizing the potential for efficient and effective exchange of goods, leveraging diverse transportation modes, and creating a network of integrated logistics platforms to serve a larger market within the required time interval.

Intermodality can provide a competitive advantage for logistics platforms by offering value-added services that go beyond traditional intermodal transportation. This can include improved and enhanced sustainable modes of transportation, which are increasingly important in addressing environmental issues related to freight traffic. Specific logistics platforms, such as Interports, provide considerable added value for operator services, allowing them to become significant sources of competitive advantage.

Intermodality also presents opportunities for efficiency and sustainability, which can be leveraged to differentiate logistics platforms from their competitors. By offering a diverse range of transportation modes and value-added services, logistics platforms can attract and retain customers seeking environmentally friendly and efficient transportation solutions. This can contribute to the overall competitiveness and success of the logistics platform in the market.

In summary, intermodality provides a competitive advantage for logistics platforms by offering value-added services, addressing environmental concerns, and providing efficient and sustainable transportation solutions, which can differentiate them from competitors and attract customers seeking these benefits.

31. The Gazzetta del Sud Case

Sector: Manufacturing

Description: The Gazzetta del Sud case study discusses the challenges and inefficiencies associated with excessive material stocks in the manufacturing sector. It highlights the negative impacts of excess stocks on capital immobilization, increased costs, energy consumption, obsolescence, and rigidity in adapting to changes. The case emphasizes the importance of adopting the Just-In-Time (JIT) system to optimize material and information management activities within the production context.

Benefits: The case study emphasizes the benefits of implementing the JIT system, such as reducing capital immobilization, minimizing costs and energy consumption, preventing material obsolescence, and increasing flexibility in responding to changes. It also highlights how JIT helps in revealing plant problems, making their analysis and resolution less complicated.

Tool as best practice: The JIT system is presented as the best practice for inventory management, ensuring that the required material is made available at the exact moment when the process requires it, thereby meeting customer demand efficiently and minimizing the negative impacts of excessive stocks.

The study discusses the concept of Just-In-Time (JIT) inventory management and its benefits in reducing inefficiency and costs associated with excess stocks. It highlights how Gazzetta del Sud utilizes JIT to order materials based on demand, thereby minimizing the negative impacts of excess stocks.

The Gazzetta del Sud case study highlights the challenges and negative impacts associated with excessive material stocks in the manufacturing sector. Excess stocks can lead to inefficiency and increased costs, as they tie up capital, require more personnel and space, and can lead to obsolescence and damage. The case emphasizes the importance of adopting the Just-In-Time (JIT) inventory management system to optimize material and information management activities within the production context.

The benefits of implementing the JIT system are extensively discussed in the case study. JIT aims to provide materials exactly when needed, reducing the negative impacts of excess stocks. By ordering only what is necessary to satisfy demand, Gazzetta del Sud has been able to reduce capital immobilization, minimize costs and energy consumption, prevent material obsolescence, and increase flexibility in responding to changes. Additionally, JIT helps in revealing plant problems, making their analysis and resolution less complicated.

The case study presents the JIT system as the best practice for inventory management in the manufacturing sector. By ensuring that the required material is made available at the exact moment when the process requires it, JIT meets customer demand efficiently and minimizes the negative impacts of excessive stocks. This approach not only optimizes material and information management but also contributes to overall operational efficiency and cost-effectiveness.

In conclusion, the Gazzetta del Sud case study serves as a compelling example of the benefits of implementing the JIT system in the manufacturing sector. It underscores the importance of efficient inventory management in reducing costs, preventing obsolescence, and increasing flexibility, ultimately contributing to improved operational performance and customer satisfaction.

The JIT inventory management system helps in reducing inefficiency and costs by ensuring that the required material is made available at the exact moment when the process requires it, thereby meeting customer demand efficiently and minimizing the negative impacts of excessive stocks. This approach reduces capital immobilization, minimizes costs and energy consumption, prevents material obsolescence, and increases flexibility in responding to changes. By ordering only what is necessary to satisfy demand, JIT optimizes material and information management activities within the production context, contributing to improved operational performance and cost-effectiveness.

Maintaining excess stocks can lead to several potential drawbacks. These include:

1. Capital immobilization: Excess stocks tie up capital necessary to finance production, leading to an increase in the financial cycle and a higher cost of capital.
2. Increased costs and energy consumption: More personnel are required to manage excess stocks, and additional space is occupied, leading to increased costs and energy consumption.
3. Material obsolescence: Excess stocks may lead to materials becoming obsolete, resulting in potential waste and financial losses.
4. Rigidity in adapting to changes: Excess stocks can lead to rigidity when it becomes necessary to introduce modifications to the original project, making it difficult to adapt to changes in demand or production requirements.
5. Increased risk of input damage: The presence of excess stocks increases the risk of damage to materials, potentially leading to further financial losses.

In addition, excess stocks can be used as a tool to conceal plant problems, making their analysis and resolution more complicated.

Overall, maintaining excess stocks can lead to inefficiency, increased costs, and reduced flexibility in responding to changes in the production environment. Therefore, it is essential to carefully manage inventory levels to avoid these potential drawbacks.

Several companies and industries have successfully implemented JIT inventory management. One notable example is Toyota, which is often credited with popularizing the JIT system. Toyota's implementation of JIT has been widely studied and emulated by other companies due to its success in reducing waste, improving efficiency, and enhancing overall operational performance.

Overall, the successful implementation of JIT inventory management can be observed across various industries, demonstrating its versatility and effectiveness in optimizing material and information management activities within the production context.

32. Amazon's Supply Chain Management Strategy

Sector: E-commerce and Retail

Description: Amazon has optimized its supply chain process by integrating various elements such as product storage, inventory management, pricing, and logistics. The company offers two fulfilment options – Amazon fulfilment and merchant fulfilment, and has incorporated high-tech equipment like the Kiva system to enhance efficiency. Additionally, Amazon uses FedEx and UPS for shipping, operates in multiple countries, and strategically integrates stocks in distribution centres and partner warehouses to save on storage costs. The company has also built Prime Now Delivery hubs and incorporated a drone system for faster and more convenient delivery.

Benefits: Amazon's supply chain strategy has led to reduced order processing time, faster delivery, cost savings on storage, and improved customer satisfaction. The integration of technology and strategic partnerships has allowed Amazon to efficiently manage its supply chain on a global scale.

Tool as best practice: The use of high-tech equipment like the Kiva system, strategic integration of stocks in distribution centres, and the incorporation of a drone system for delivery can be considered as best practices in supply chain management.

Amazon has implemented various strategies to optimize its supply chain process, including offering fulfilment options, utilizing advanced technology for picking and packaging, integrating stocks in partner warehouses, establishing Prime Now Delivery hubs and sorting centres, implementing drone delivery systems, and introducing motorcycle delivery services. These efforts have enabled Amazon to deliver products within 24 hours and offer one-day services in certain areas.

Amazon's supply chain management strategy is a prime example of how a company can optimize its processes to achieve efficiency and customer satisfaction. By offering fulfilment options such as Amazon fulfilment and merchant fulfilment, Amazon provides flexibility for sellers while maintaining control over the logistics process. This allows the company to manage inventory effectively and ensure timely delivery to customers.

The integration of high-tech equipment, such as the Kiva system, for picking and packaging has significantly improved the efficiency of Amazon's warehouses. This automation reduces the time and labour required for order processing, contributing to faster delivery times and cost savings. Additionally, the strategic integration of stocks in distribution centres and partner warehouses allows Amazon to optimize inventory management and minimize storage costs.

The introduction of Prime Now Delivery hubs and sorting centres further enhances Amazon's ability to provide fast and convenient delivery to customers. By strategically locating these facilities, Amazon can ensure that products are closer to the end consumer, reducing delivery times and improving overall customer satisfaction.

The incorporation of a drone system for delivery and motorcycle delivery services in some countries demonstrates Amazon's commitment to leveraging innovative technologies to enhance its supply chain operations. These initiatives not only contribute to faster delivery but also showcase Amazon's willingness to explore new and unconventional methods to meet customer demands.

Overall, Amazon's supply chain management strategy has resulted in reduced order processing times, faster delivery, cost savings on storage, and improved customer satisfaction. The company's ability to integrate technology, strategic partnerships, and innovative delivery methods on a global scale sets a high standard for best practices in supply chain management. This case study serves as a valuable example for other companies seeking to optimize their supply chain processes and enhance customer experience.

Amazon has optimized its supply chain process through various strategic initiatives. This includes offering two fulfilment options – Amazon fulfilment and merchant fulfilment, integrating high-tech equipment like the Kiva system for efficient picking and packaging of products, and strategically integrating stocks in distribution centres and partner warehouses to save on storage costs. Additionally, Amazon has built Prime Now Delivery hubs, incorporated a drone system for faster and more convenient delivery, and introduced motorcycle delivery services in some countries. These efforts have contributed to reduced order processing time, faster delivery, cost savings on storage, and improved customer satisfaction.

Amazon offers two fulfilment options: Amazon fulfilment and merchant fulfilment. Amazon fulfilment provides storage, packaging, and shipping services for sellers who prefer a hands-off approach, while merchant fulfilment allows sellers to personally take care of product storage, packaging, and shipping.

Amazon has incorporated advanced technology into its logistics operations in several ways. One notable example is the use of the Kiva system, which involves the deployment of robots for picking and packaging products. By the end of 2017, Amazon had installed around forty-five thousand Kiva robots, significantly reducing the time from order receipt to order shipment.

Additionally, Amazon has integrated a drone system for delivery in limited city areas, with the proposal of Amazon Prime Air promising to deliver products within thirty minutes using drones. This not only reduces delivery costs but also makes delivery faster and more convenient.

Furthermore, Amazon has developed a network of Prime Now Delivery hubs in various cities, enhancing its ability to provide fast and convenient delivery to customers. These initiatives showcase Amazon's commitment to leveraging innovative technologies to enhance its logistics operations.

33. High Technology Systems

Sector: Electronic manufacturing services

Description: The case study focuses on High Technology Systems (HTS), an electronic manufacturing services company established in 2001 in Cittanova, Italy. HTS specializes in providing high-quality and complex electronic assembly services, with a focus on niche market products. The company has developed expertise in verifying electronic components' characteristics and behaviour, particularly under induced humidity stress, in compliance with international standards such as Jedec. Additionally, HTS has a strong focus on new product introduction and has developed a Fast Track procedure to prioritize urgent customer production needs. The company has also diversified its services to include a wide range of assembly offerings and has established a strong presence in the EMS sector.

Benefits: HTS has gained a competitive edge through its niche market focus, high-quality assembly services, and expertise in electronic component verification. The company's Fast Track procedure has been well-received by customers, contributing to customer satisfaction and loyalty. Additionally, HTS's diversification into the EMS sector has allowed it to offer a broader range of assembly services, further enhancing its market position.

Tool as best practice: The SWOT analysis has been utilized as a best practice tool to assess HTS. This analysis has helped the company identify areas for improvement, capitalize on its strengths, and navigate potential challenges in the industry.

The case study discusses various aspects of the product lifecycle, technological progress, industrial production process, and the electronic manufacturing services (EMS) market. It also provides details about the EMS sector's lifecycle and the phases of the PCB assembly process. The company's SWOT analysis and its strategies to address risks and challenges are also included. Additionally, the paper mentions the company's focus on supplier loyalty, investment in enabling technologies, and its plans to invest in advanced equipment. The EMS market is divided into four phases, and the company aims to maintain competitive barriers by leveraging its expertise and capital-intensive investments. The company is specialized in electronic component treatment and verification, has diversified into EMS, and obtained certification to enter the automotive market. It is investing in qualified personnel and equipment, implementing the "INDUSTRIA 4.0" plan, and prioritizing technological innovation to improve product quality and resilience. The company has received tax credits and is considering utilizing them for loan repayment. It also plans to invest in innovative equipment to enhance the production process and increase revenue.

The case study focuses on High Technology Systems (HTS), an EMS company based in Cittanova, Italy. HTS specializes in providing high-quality and complex electronic assembly services, with a focus on niche market products. The company has developed expertise in verifying electronic components' characteristics and behaviour, particularly under induced humidity stress, in compliance with international standards such as Jedec. Additionally, HTS has a strong focus on new product introduction and has developed a Fast Track procedure to prioritize urgent customer production needs. The company has also diversified its services to include a wide range of assembly offerings and has established a strong presence in the EMS sector.

The case study provides detailed information on the product life cycle, technological progress, industrial production process, and the EMS market. It also includes an analysis of the reference market for HTS and details on the EMS sector's life cycle. The study outlines the various phases of the SMT production process, including the insertion of printed circuits, solder paste application, electronic component placement, soldering, and electronic board collection. HTS is committed to following environmental regulations and uses lead-free tin-silver alloys. The company also offers PTH assembly services and has conducted a SWOT analysis to evaluate its strengths, weaknesses, opportunities, and threats. The study also highlights the main risks identified by the company, such as high competitiveness, reduced margins, competition from Chinese manufacturing, and high labor and raw material costs. To address these risks, the company focuses on innovation, automation, and quality.

HTS has diversified its services in the EMS sector and obtained IATF 16949:2016 certification to enter the automotive market. The company is investing in qualified personnel and equipment to be competitive in the electric vehicle market. Additionally, HTS is implementing the national industrial plan "INDUSTRIA 4.0" to improve digitalization, integrated automation, and predictive maintenance. Technological innovation is crucial for the company, which aims to improve the quality and resilience of its products and services. The study emphasizes the importance of understanding the product life cycle to comprehend technology evolution and its innovative process.

The EMS market is divided into four phases: introduction, growth, maturity, and decline. EMS companies are categorized into three groups: TOP CLASS, GLOBAL CLASS, and TERZI. The European EMS market is dominated by three major companies. HTS aims to maintain competitive barriers in the market, such as corporate know-how and intensive capital investment. The company has evolved from a focus on niche market products and electronic component verification to diversifying its services in the EMS sector and obtaining IATF 16949:2016 certification to enter the automotive market. Additionally, HTS is investing in qualified personnel and equipment to be competitive in the electric vehicle market and is implementing the national industrial plan "INDUSTRIA 4.0" to improve digitalization, integrated automation, and predictive maintenance.

High Technology Systems (HTS) addresses the risks and challenges in the EMS market through a strategic approach. The company has identified several key risks, including high labour costs, reduced profitability, competition from Chinese manufacturing, and the threat of shortages in electronic components. To overcome these challenges, HTS focuses on innovation and technological advancement as a driving force for success in the EMS sector. The company emphasizes the importance of automation and quality investments to reduce the reliance on manual labour and mitigate the impact of high labour costs. Additionally, HTS aims to compete successfully in the EMS market by leveraging its expertise in technological innovation and flexible production methods, such as Just in Time manufacturing and economical batch production.

Furthermore, HTS employs strategic barriers to secure its position in the European EMS market, which is dominated by major players such as Foxcon, Flex, and Jabil. The company establishes barriers to make it extremely difficult for competitors to enter its established market space. These barriers include a focus on high-value, high-margin production, positioning HTS among companies with promising growth potential and reducing the risk of decline in the short term.

In summary, HTS addresses the risks and challenges in the EMS market by prioritizing innovation, technological advancement, and strategic positioning to maintain a competitive edge and secure its position in the industry.

The EMS market is composed of different phases, including High Volume Low Mix (HVLV) and High Mix Low Volume (HMLV). High Technology Systems (HTS) positions itself within the HMLV segment, characterized by high value-added activities and potential for promising growth. This strategic positioning allows HTS to differentiate itself from competitors and mitigate the risk of decline in the short term.

Additionally, HTS focuses on high-value, high-margin production, which aligns with the characteristics of the HMLV segment. By emphasizing innovation and technological advancement, HTS aims to maintain a competitive edge within the EMS market and capitalize on the potential for growth in the HMLV segment.

In summary, HTS strategically positions itself within the HMLV segment of the EMS market, leveraging its expertise in high-value production and technological innovation to differentiate itself and drive growth.

HTS prioritizes technological innovation in the EMS market. The company places a strong emphasis on improving the quality and resilience of its products and services through innovation and technological advancement. HTS has taken several steps to achieve this, including investing in digitalization, integrated automation, and comprehensive predictive maintenance programs.

Furthermore, HTS has recognized the importance of product innovation in driving improvements in quality and resilience. The company is committed to implementing changes in its products to achieve higher quality and greater efficiency, leveraging technological progress to optimize resource utilization and enhance the technical and economic efficiency of its production system.

In summary, HTS's strategic initiatives demonstrate the company's commitment to leveraging technology to drive continuous improvement and maintain a competitive edge in the EMS market.

References

Abdulmalek, F.A. and Rajgopal, J., 2007. Analyzing the benefits of lean manufacturing and value stream mapping via simulation: A process sector case study. *International Journal of Production Economics*, 107(1), pp.223–236.

Accenture, 2021. Technology vision 2021: *Leaders wanted experts at change at a moment of truth*. Accenture.

Accerboni, F. and Sartor, M., 2019. ISO/IEC 27001. In: *Quality management: Tools, methods, and standards*. doi:10.1108/978-1-78769-801-720191015.

Adebanjo, D., 2001. TQM and business excellence: is there really a conflict?. *Measuring Business Excellence*, 5(3), pp.37–40.

Adel, A., 2022. Future of industry 5.0 in society: human-centric solutions, challenges and prospective research areas. *Journal of Cloud Computing*, 11(40). https://doi.org/10.1186/s13677-022-00314-5

Al-Azzam, N. and Shatnawi, I., 2021. Comparing supervised and semi-supervised machine learning models on diagnosing breast cancer. *Annals of Medicine and Surgery*, 62, pp.53–64.

Anderson, J.C., Rungtusanatham, M. and Schroeder, R.G., 1994. A theory of quality management underlying the Deming management method. *Academy of Management Review*, 19(3), pp.472–509.

Antony, J., Sunder M.V., Laux, C. and Cudney, E., 2019. *The ten commandments of lean Six Sigma*. Emerald Publishing Limited, pp.117–128. https://doi.org/10.1108/978-1-78973-687-820191014

Anvari, A., Ismail, Y. and Hojjati, S.M.H., 2011. A study on total quality management and lean manufacturing: through lean thinking approach. *World Applied Sciences Journal*, 12(9), pp.1585–1596.

Aven, T. and Renn, O., 2009. The role of quantitative risk assessments for characterizing risk and uncertainty and delineating appropriate risk management options, with special emphasis on terrorism risk. *Risk Analysis*, 29(4), pp.587–600.

Baines, T.S., et al., 2007. State-of-the-art in product-service systems. Proceedings of the Institution of Mechanical Engineers, Part B: *Journal of Engineering Manufacture*, 221(10):1543–1552. doi:10.1243/09544054JEM858

Bakri, A. and Januddi, M.A.F.M.S., 2020. *Systematic industrial maintenance to boost the quality management programs*. Springer Nature.

Baldwin, R.E., 2006. Multilateralising regionalism: Spaghetti bowls as building blocs on the path to global free trade. *The World Economy*, 29(11), pp.1451–1518.

Barlotti, M., 2013. Archimede risponde. *Archimede*, 65, pp.71–76.

Bartels, R., Dudink, J., Haitjema, S., Oberski, D. and van 't Veen, A., 2022. A perspective on a quality management system for AI/ML-based clinical decision support in hospital care. *Frontiers in Digital Health*, 4. doi:10.3389/fdgth.2022.942588.

Basu, R., 2009. *Implementing Six Sigma and Lean: A practical guide to tools and techniques*. Amsterdam: Elsevier.

Besterfield, D.H., Besterfield-Michna, C., Besterfield, G.H., Besterfield-Sacre, M., Urdhwareshe, H. and Urdhwareshe, R., 2019. *Total quality management*. London: Pearson.

Blackburn, R. and Rosen, B., 1993. Total quality and human resources management: lessons learned from Baldrige Award-winning companies. *Academy of Management Perspectives*, 7(3), pp.49–66.

Brawner, K., Wang, N. and Nye, B., 2023. Teaching artificial intelligence (AI) with AI for AI applications. In: *The International FLAIRS Conference Proceedings*, 36.

Bresciani, S., Rehman, S., Giovando, G. and Alam, G., 2022. The role of environmental management accounting and environmental knowledge management practices influence on environmental performance: mediated-moderated model. *Journal of Knowledge Management*. doi:10.1108/jkm-12-2021-0953.

Breyfogle III, F.W., 2003. *Implementing six sigma: smarter solutions using statistical methods*. John Wiley & Sons.

Browning, T.R. and Heath, R.D., 2009. Reconceptualizing the effects of lean on production costs with evidence from the F-22 program. *Journal of Operations Management*, 27(1), pp.23–44.

Campbell, S.D., 2005. *A review of backtesting and backtesting procedures*. Finance and Economics Discussion Series. Divisions of Research & Statistics and Monetary Affairs, Federal Reserve Board.

Cao, G., Duan, Y., Edwards, J. and Dwivedi, Y., 2021. Understanding managers' attitudes and behavioral intentions towards using artificial intelligence for organizational decision-making. *Technovation*. doi:10.1016/J.TECHNOVATION.2021.102312.

Charantimath, P.M., 2017. *Total quality management*. London: Pearson.

Chiarini, A., 2012. *From Total quality control to lean Six Sigma: Evolution of the most important management systems for the excellence*. Berlin: Springer.

Chiarini, A., Found, P. and Rich, N., 2016. *Understanding the lean enterprise*. Berlin: Springer.

Christopher, M. and Towill, D., 2001. An integrated model for the design of agile supply chains. *International Journal of Physical Distribution & Logistics Management*, 31(4), pp.235–246.

Crosby, P.B., 1979. *Quality is free: The art of making quality certain*. New American Library.

Crouch, E.A. and Wilson, R., 1982. *Risk/benefit analysis*. Cambridge, MA: Ballinger Publishing Company.

Crute, V., Ward, Y., Brown, S. and Graves, A., 2003. Implementing Lean in aerospace: Challenging the assumptions and understanding the challenges. *Technovation*, 23(12), pp.917–928.

Cudney, E., 2009. Improvement over waste. *Industrial Engineer*, 41(4), p.58.

Cunha, B., Madureira, A., Fonseca, B. and Matos, J., 2021. Intelligent scheduling with reinforcement learning. *Applied Sciences*, 11(8), p.3710. doi:10.3390/APP11083710.

Cusumano, M.A., 1988. Manufacturing innovation: lessons from the Japanese auto industry. *MIT Sloan Management Review*, 30(1), pp.29–39.

Dahlgaard, J., Kristensen, K. and Kanji, G.K., 1995. *The quality journey: A journey without an end*. Madras: Productivity Press (India).

De Treville, S. and Antonakis, J., 2006. Could lean production job design be intrinsically motivating? Contextual, configurational, and levels-of-analysis issues. *Journal of Operations Management*, 24(2), pp.99–123.

De Treville, S., Antonakis, J. and Edelson, N.M., 2005. Can standard operating procedures be motivating? Reconciling process variability issues and behavioural outcomes. *Total Quality Management & Business Excellence*, 16(2), pp.231–241.

Demeter, K. and Matyusz, Z., 2011. The impact of lean practices on inventory turnover. *International Journal of Production Economics*, 133(1), pp.154–163.

Deming, W.E., 1982. *Quality, productivity and competitive position*. Cambridge, MA: MIT.

Deming, W.E., 2018. *Out of the crisis*. Cambridge, MA: MIT Press.

Department of Defense, 2000. *MIL-STD-882D: Standard practice for system safety*. Washington, DC: U.S. Department of Defense.

Deutsche Bank Research, 2019. *Artificial intelligence in banking: A lever for profitability with limited implementation to date*. EU Monitor Global financial markets. Deutsche Bank Research.

Dey, D., Habibovic, A., Löcken, A., Wintersberger, P., Pfleging, B., Riener, A., . . . and Terken, J., 2020. Taming the eHMI jungle: A classification taxonomy to guide, compare, and assess the design principles of automated vehicles' external human-machine interfaces. *Transportation Research Interdisciplinary Perspectives*, 7, p.100174.

Dieniezhnikov, S.S. and Pshenychna, L.V., 2021. *Peculiarities of formation of value orientations of modern student youth*. Baltija Publishing.

Duncker, K., 1945. *On problem solving*. Psychological Monographs.

Emiliani, B., 2007. *Real lean: understanding the lean management system*. CLBM.

European Commission, Directorate-General for Research and Innovation, 2021. *Industry 5.0: Towards a sustainable, human-centric and resilient European industry*. Publications Office of the European Union. https://data.europa.eu/doi/10.2777/308407

Feigenbaum, A.V., 1960. *Total quality control*. New York: McGraw-Hill.

Feld, C.K., Sousa, J.P., Da Silva, P.M. and Dawson, T.P., 2010. Indicators for biodiversity and ecosystem services: towards an improved framework for ecosystems assessment. *Biodiversity and Conservation*, 19, pp.2895–2919.

Frick, J. and Grudowski, P., 2023. Quality 5.0: A paradigm shift towards proactive quality control in industry 5.0. *Asia-Pacific Journal of Business Administration*, 14, pp.51–56. doi:10.5430/ijba.v14n2p51.

Friedman, B. and Hendry, D.G., 2019. *Value sensitive design: Shaping technology with moral imagination*. MIT Press.

Friedman, J., Hastie, T. and Tibshirani, R., 2009. *The elements of statistical learning*. New York: Springer.

Fullerton, R.R. and Wempe, W.F., 2009. Lean manufacturing, non-financial performance measures, and financial performance. *International Journal of Operations & Production Management*, 29(3), pp.214–240.

Fullerton, R.R., McWatters, C.S. and Fawson, C., 2003. An examination of the relationships between JIT and financial performance. *Journal of Operations Management*, 21(4), pp.383–404.

Galgano, A., 2008. *Qualità totale. Il metodo scientifico nella gestione aziendale*. Milan: Guerini e Associati.

Garvin, D.A., 1984. Product quality: An important strategic weapon. *Business Horizons*, 27(3), pp.40–43.

Gattiker, T. and Carter, C., 2010. Understanding project champions' ability to gain intra-organizational commitment for environmental projects. *Journal of Operations Management*, 28, pp.72–85. doi:10.1016/J.JOM.2009.09.001.

Goldratt, E.M., 1988. Computerized shop floor scheduling. *The International Journal of Production Research*, 26(3), pp.443–455.

Golhar, D.Y. and Stamm, C.L., 1991. The just-in-time philosophy: a literature review. *The International Journal of Production Research*, 29(4), pp.657–676.

Golovianko, M., Terziyan, V., Branytskyi, V. and Malyk, D., 2023. Industry 4.0 vs. Industry 5.0: Co-existence, transition, or a hybrid. *Procedia Computer Science*, 217, pp.102–113. https://doi.org/10.1016/j.procs.2022.12.206

Gómez, C. and Heredero, C., 2020. Artificial intelligence as an enabling tool for the development of dynamic capabilities in the banking industry. *International Journal of Enterprise Information Systems*, 16, pp.20–33. doi:10.4018/ijeis.2020070102.

Gorbacheva, A. and Smirnov, S., 2017. Converging technologies and a modern man: emergence of a new type of thinking. *AI & Society*, 32, pp.465–473.

Grasso, L.P., 2005. Are ABC and RCA accounting systems compatible with lean management?. *Management Accounting Quarterly*, 7(1), p.12.

Gygi, C., DeCarlo, N. and Williams, B., 2002. *Six Sigma for dummies*. Hoboken, NJ: Wiley.

Harms-Ringdahl, L., 2001. *Safety analysis: Principles and practice in occupational safety*. 2nd ed. London: CRC Press. https://doi.org/10.4324/9780203302736

Harry, M. and Schroeder, R., 2000. *Six Sigma: The breakthrough management strategy revolutionizing the world's top corporations*. New York: Doubleday.

Harry, M.J., Mann, P.S., De Hodgins, O.C., Hulbert, R.L. and Lacke, C.J., 2010. *Practitioner's guide to statistics and Lean Six Sigma for process improvements*. Hoboken, NJ: Wiley.

Health and Safety Authority, 2005. Safety, Health and Welfare at Work Act 2005. Retrieved from www.hsa.ie/eng/topics/managing_health_and_safety/safety,_health_and_welfare_at_work_act_2005 (accessed 24 August 2024).

Hines, P., Holweg, M. and Rich, N., 2004. Learning to evolve: a review of contemporary lean thinking. *International Journal of Operations & Production Management*, 24(10), pp.994–1011.

Hines, P., Rich, N., Bicheno, J., Brunt, D., Taylor, D., Butterworth, C. and Sullivan, J., 1998. Value stream management. *International Journal of Logistics Management*, 4(2), pp.235–246.

Hoerl, R.W. and Snee, R., 2010. Statistical thinking and methods in quality improvement: A look to the future. *Quality Engineering*, 22(3), pp.119–129. doi:10.1080/08982112.2010.481485.

Hofer, C., Eroglu, C. and Hofer, A.R., 2012. The effect of lean production on financial performance: The mediating role of inventory leanness. *International Journal of Production Economics*, 138(2), pp.242–253.

Holweg, M., 2007. The genealogy of lean production. *Journal of Operations Management*, 25(2), pp.420–437.

Imai, M., 1986. *Kaizen: The key to Japan's competitive success*. London: Kaizen Institute.

Imai, M., 2007. Gemba Kaizen. A commonsense, low-cost approach to management. In: *Kaizen* (pp. 49–66). Springer, Berlin, Heidelberg.

International Organization for Standardization (ISO), 2002. *Risk management vocabulary*. ISO/IEC Guide 73:2002. Geneva: ISO. [New version available: ISO Guide 73:2009]

IRGC. 2005. Risk governance: Towards an integrative approach. Retrieved from www.irgc.org (accessed 24 August 2024).

Jackson, T.L., 2019. *Hoshin Kanri for the lean enterprise: developing competitive capabilities and managing profit*. Productivity press.

Jadhav, J.S. and Deshmukh, J., 2022. A review study of the blockchain-based healthcare supply chain. *Social Sciences & Humanities Open*, 6(1), p.100328.

Jaiswal, D., Kaushal, V., Kant, R. and Singh, P.K., 2021. Consumer adoption intention for electric vehicles: Insights and evidence from Indian sustainable transportation. *Technological Forecasting and Social Change*, 173, p.121089.

Jay, A., 2008. *Lean Six Sigma demystified*. New York: McGraw Hill Professional.

Kaizen-coach.com, n.d. Dizionario Lean. Retrieved from www.kaizen-coach.com/it/formazione-lean/dizionario-lean (accessed 24 August 2024).

Kaplan, R.S., 1996. Using the balanced scorecard as a strategic management system. *Harvard Business Review*, 74(1), pp.75–85.

Kaplinsky, R., 2004. Spreading the gains from globalization: What can be learned from value-chain analysis?. *Problems of Economic Transition*, 47(2), pp.74–115.

Keding, C. and Meissner, P., 2021. Managerial overreliance on AI-augmented decision-making processes: How the use of AI-based advisory systems shapes choice behavior in R&D investment decisions. *Technological Forecasting and Social Change*, 171, p.120970. doi:10.1016/J.TECHFORE.2021.120970.

Kennedy, F.A. and Widener, S.K., 2008. A control framework: Insights from evidence on lean accounting. *Management Accounting Research*, 19(4), pp.301–323.

Khurana, M.P.S. and Singh, P., 2012. Waste water use in crop production: A review. *Resources and Environment*, 2(4), pp.116–131.

Kiangala, K. and Wang, Z., 2020. An effective predictive maintenance framework for conveyor motors using dual time-series imaging and convolutional neural network in an industry 4.0 environment. *IEEE Access*, 8, pp.121033–121049. doi:10.1109/ACCESS.2020.3006788.

Kinney, M.R. and Wempe, W.F., 2002. Further evidence on the extent and origins of JIT's profitability effects. *The Accounting Review*, 77(1), pp.203–225.

Klefsjö, B., Wiklund, H. and Edgeman, R.L., 2001. Six sigma seen as a methodology for total quality management. *Measuring Business Excellence*, 5(1), pp.31–35.

Krafcik, J.F., 1988. Triumph of the lean production system. *Sloan Management Review*, 30(1), pp.41–52.

Kumamoto, H. and Henley, E.J., 1996. *Probabilistic risk assessment and management for engineers and scientists*. 2nd ed. New York: IEEE Press.

Laraia, A.C., Moody, P.E. and Hall, R.W., 1999. *The kaizen blitz: accelerating breakthroughs in productivity and performance*. John Wiley & Sons.

Lee, D. and Yoon, S., 2021. Application of artificial intelligence-based technologies in the healthcare industry: Opportunities and challenges. *International Journal of Environmental Research and Public Health*, 18. doi:10.3390/ijerph18010271.

Lee, K.J., Kwon, J.W., Min, S. and Yoon, J., 2021. Deploying an artificial intelligence-based defect finder for manufacturing quality management. *AI Magazine*, 42(2), pp.5–18. doi:10.1609/aimag.v42i2.15094.

Lee, S., Lee, D. and Kim, Y., 2019. The quality management ecosystem for predictive maintenance in the Industry 4.0 era. *International Journal of Quality Innovation*, 5, pp.1–11. doi:10.1186/S40887-019-0029-5.

Leordeanu, M., Sukthankar, R. and Hebert, M., 2009. Unsupervised learning for graph matching. *International Journal of Computer Vision*, 96, pp.28–45. doi:10.1007/s11263-011-0442-2.

Li, S., Rao, S.S., Ragu-Nathan, T.S. and Ragu-Nathan, B., 2005. Development and validation of a measurement instrument for studying supply chain management practices. *Journal of Operations Management*, 23(6), pp.618–641.

Linderman, K., Schroeder, R.G. and Choo, A.S., 2006. Six Sigma: The role of goals in improvement teams. *Journal of Operations Management*, 24(6), pp.779–790.

Lopez-Fresno, P., Savolainen, T. and Miranda, S., 2021. Trust building for integrative trade negotiations: Challenges posed by Covid-19. In: *22nd European Conference on Knowledge Management*, pp.501–508.

Luthra, S., Garg, D., Agarwal, A. and Mangla, S.K., 2021. *Total quality management (TQM): Principles, methods and applications*. Boca Raton: Taylor & Francis.

MacBryde, J., Radnor, Z., Taj, S. and Berro, L., 2006. Application of constrained management and lean manufacturing in developing best practices for productivity improvement in an auto-assembly plant. *International Journal of Productivity and Performance Management*, 55(3/4), pp.332–345.

Maharshi, S., 2019. *Lean problem solving and QC tools for industrial engineers*. Boca Raton: Taylor & Francis.

Maskell, B.H. and Kennedy, F.A., 2007. Why do we need lean accounting and how does it work?. *Journal of Corporate Accounting & Finance*, 18(3), pp.59–73.

Mayer, R.E., 1990. Problem solving. In: M.W. Eysenk, ed. *The Blackwell dictionary of cognitive psychology*. Oxford: Basil Blackwell, pp.284–288.

McAuley, E., Morris, K.S., Motl, R.W., Hu, L., Konopack, J.F. and Elavsky, S., 2007. Long-term follow-up of physical activity behaviour in older adults. *Health Psychology*, 26(3), p.375.

McBride, D., 2003. The 7 manufacturing wastes. Retrieved from www.emsstrategies.com/dm090203article2.html (accessed 24 August 2024).

McKone, K.E., Schroeder, R.G. and Cua, K.O., 1999. Total productive maintenance: a contextual view. *Journal of Operations Management*, 17(2), pp.123–144.

McLachlin, R., 1997. Management initiatives and just-in-time manufacturing. *Journal of Operations Management*, 15(4), pp.271–292.

McVay, G., Kennedy, F.A. and Fullerton, R.R., 2013. *Accounting in the lean enterprise: Providing simple, practical & decision-relevant information*. New York: CRC Press.

Melton, T., 2005. The benefits of lean manufacturing: What lean thinking has to offer the process industries. *Chemical Engineering Research and Design*, 83(6), pp.662–673.

Monden, Y., 1998. *Toyota production system: An integrated approach to just-in-time*. 3rd ed. Norcross, Georgia: Engineering & Management Press.

Moraes, A., Carvalho, A.M. and Sampaio, P., 2023. Lean and Industry 4.0: A review of the relationship, its limitations, and the path ahead with Industry 5.0. *Machines*, 11(4), p.443. doi:10.3390/machines11040443.

Morandini, S., Fraboni, F., De Angelis, M., Puzzo, G., Giusino, D. and Pietrantoni, L., 2023. The impact of artificial intelligence on workers' skills: Upskilling and reskilling in organisations. *Informing Science*, 26, pp.39–68.

Nahavandi, S., 2019. Industry 5.0: A human-centric solution. *Sustainability*, 11(16), p.4371. https://doi.org/10.3390/su11164371

Nahmias, S., 2001. *Production and operations analysis*. 4th ed. New York: McGraw Hill.

Newbold, P., Carlson, W. and Thorne, B., 2022. *Statistics for business and economics*, global edition. London: Pearson.

Nicholas, J., 2018. *Lean production for competitive advantage: A comprehensive guide to lean methods and management practices*. New York: Routledge.

Nosayba Al-Azzam and Ibrahem Shatnawi, 2021. Comparing supervised and semi-supervised machine learning models on diagnosing breast cancer. *Annals of Medicine and Surgery*, 62, pp.53–64. doi:10.1016/j.amsu.2020.12.043.

Ohno, T., 1978. *The Toyota production system: Beyond large-scale production*. Portland: Productivity Press.

Ortiz, C., 2010. Kaizen vs. lean: Distinct but related. *Metal Finishing*, 108(1), pp.50–51.

Paliukas, V. and Savanevičienė, A., 2018. Harmonization of rational and creative decisions in quality management using AI technologies. *Economics and Business*, 32(1), pp.195–208.

Pampanelli, A.B., Found, P. and Bernardes, A.M., 2013. A lean & green model for a production cell. Retrieved from www.sciencedirect.com/science/article/abs/pii/S0959652613004265 (accessed 24 August 2024).

Pandey, H., Garg, D. and Luthra, S., 2018. Identification and ranking of enablers of green lean Six Sigma implementation using AHP. *International Journal of Productivity and Quality Management*, 23(2), pp.187–217.

Park, S.H., 2003. *Six Sigma for quality and productivity promotion*. Tokyo: Asian Productivity Organization.

Pavanato, R., 2020. *The lean book*. Milan: Hoepli.

Pearce, A., Pons, D. and Neitzert, T., 2023. Understanding lean: Statistical analysis of perceptions and self-deception regarding lean management. *Operations Research Forum*, 4, p.28. doi:10.1007/s43069-023-00198-4.

Piccarozzi, M., Stefanoni, A., Silvestri, C. and Ioppolo, G., 2023. Industry 4.0 technologies as a lever for sustainability in the communication of large companies to stakeholders. *European Journal of Innovation Management*. doi:10.1108/EJIM-09-2022-0485.

Poornima, M.C., 2017. *Total quality management*. London: Pearson.

Porter, M.E. and van der Linde, C., 1995. Green and competitive: ending the stalemate. *Harvard Business Review*, 73(5), pp.120–134.

Prentice, C. and Nguyen, M., 2020. Engaging and retaining customers with AI and employee service. *Journal of Retailing and Consumer Services*, 56, p.102186. doi:10.1016/j.jretconser.2020.102186.

Przekop, P., 2003. *Six Sigma for business excellence: A manager's guide to supervising Six Sigma projects and teams*. McGraw-Hill.

Pyzdek, T., 2009. *The Six Sigma handbook: A complete guide for green belts, black belts, and managers at all levels*. McGraw-Hill.

Pyzdek, T. and Keller, P., 2013. *The handbook for quality management: A complete guide to operational excellence*. New York: McGraw-Hill.

Radzwill, N.M., 2018. Quality 4.0: Let's get digital: The many ways the fourth industrial revolution is reshaping the way we think about quality. *arXiv* preprint. arXiv:1810.07829.

Raines, S., 2002. Implementing ISO 14001: An international survey assessing the benefits of certification. *Corporate Environmental Strategy*, 9, pp.418–426. doi:10.1016/S1066-7938(02)00009-X.

Raju, N.V.S., 2014. *Total quality management*. New Delhi: Cengage Learning.

Rao, P., Kumar, S., Chavan, M. and Lim, W.M., 2023. A systematic literature review on SME financing: Trends and future directions. *Journal of Small Business Management*, 61(3), pp.1247–1277.

Robertson, I.S., 2001. Problem solving. Boca Raton: Taylor & Francis.

Rosa, E.A., 1998. Metatheoretical foundations for post-normal risk. *Journal of Risk Research*, 1(1), pp.15–44. https://doi.org/10.1080/136698798377303

Roser, C., 2021. *All about pull production: Designing, implementing, and maintaining Kanban, CONWIP, and other pull systems in lean production*. AllAboutLean.com.

Rother, M. and Shook, J., 1999. *Learning to see: Value stream mapping to add value and eliminate muda*. Brookline, MA: The Lean Enterprise Institute.

Sader, S., Husti, I. and Daroczi, M., 2021. A review of quality 4.0: definitions, features, technologies, applications, and challenges. *Total Quality Management & Business Excellence*, 33(9–10), pp.1164–1182. doi:10.1080/14783363.2021.1944082.

Salinas Navarro, D.E., Garay-Rondero, C. and Arana-Solares, I., 2023. Digitally enabled experiential learning spaces for engineering education 4.0. *Education Sciences*, 13(1), p.63. doi:10.3390/educsci13010063.

Sambasivan, M. and Fei, N., 2008. Evaluation of critical success factors of implementation of ISO 14001 using analytic hierarchy process (AHP): a case study from Malaysia. *Journal of Cleaner Production*, 16, pp.1424–1433. doi:10.1016/J.JCLEPRO.2007.08.003.

Sang, L., Yu, M., Lin, H., Zhang, Z. and Jin, R., 2021. Big data, technology capability and construction project quality: a cross-level investigation. *Engineering, Construction and Architectural Management*, 28(3), pp.706–727.

Senapati, N.R., 2004. Six Sigma: Myths and realities. *International Journal of Quality & Reliability Management*, 21(6), pp.683–690.

Senoner, J., Netland, T. and Feuerriegel, S., 2020. Using explainable artificial intelligence to improve process quality: Evidence from semiconductor manufacturing. *Management Science*, 68(8), pp.5704–5723. https://doi.org/10.1287/mnsc.2021.4190

Serrat, O., 2017. The five why technique. In: *Knowledge solutions*. Singapore: Springer.

Seth, D. and Gupta, V., 2005. Application of value stream mapping for lean operations and cycle time reduction: an Indian case study. *Production Planning & Control*, 16(1), pp.44–59.

Shah, R. and Ward, P.T., 2003. Lean manufacturing: context, practice bundles, and performance. *Journal of Operations Management*, 21(2), pp.129–149.

Shah, R. and Ward, P.T., 2007. Defining and developing measures of lean production. *Journal of Operations Management*, 25(4), pp.785–805.

Shewhart, W.A., 1939. Application of statistical method in mass production. *Journal of the Franklin Institute*, 227(6), pp.831–832.

Shiba, S. and Walden, D., 2006. *Breakthrough management*. New Delhi: Confederation of Indian Industry.

Shin, W.S., Dahlgaard, J.J., Dahlgaard-Park, S.M. and Kim, M.G., 2018. A quality scorecard for the era of Industry 4.0. *Total Quality Management & Business Excellence*, 29(9–10), pp.959–976.

Shingo Institute, 2010. *The Shingo model for operational excellence*. Logan, UT: Shingo Institute.

Shingo, S., 1986. *New Japanese manufacturing philosophy*. Cambridge, MA: Productivity Press.

Simpson, D.F. and Power, D.J., 2005. Use the supply relationship to develop lean and green suppliers. *Supply Chain Management: An International Journal*, 10(1), pp.60–68.

Slack, N., Chambers, S. and Johnston, R., 2010. *Operations management*. London: Pearson.

Solomon, J.M. and Fullerton, R., 2007. *Accounting for world class operations*. Fort Wayne, IN: WCM Associates.

Souza, R., Ferenhof, H. and Forcellini, F., 2022. Industry 4.0 and Industry 5.0 from the lean perspective. *International Journal of Management, Knowledge and Learning*. doi:10.53615/2232-5697.11.145-155.

Spear, S.J. and Bowen, H.K., 1999. Decoding the DNA of the Toyota production system. *Harvard Business Review*, 77(5), pp.96–106.

Sugiura, T. and Yamada, Y., 1995. *The QC storyline: A guide to solving problems and communicating the results*. Tokyo: Asian Productivity Organization.

Sun, D., Wang, L., Song, X. and Yang, F., 2010. RFID based automatic worker identification for lean production. In: *2010 IEEE International Conference on Automation and Logistics*. Hong Kong, China: IEEE, pp.292–297.

Suri, R., 2004. *Quick response manufacturing*. Cambridge, MA: Productivity Press.

Swink, M., Narasimhan, R. and Kim, S.W., 2005. Manufacturing practices and strategy integration: Effects on cost efficiency, flexibility, and market-based performance. *Decision Sciences*, 36(3), pp.427–457.

Szwedzka, K. and Kaczmarek, J., 2018. One point lesson as a tool for work standardization and optimization: Case study. In: R. Goossens, ed. *Advances in social & occupational ergonomics*. AHFE 2017. Advances in Intelligent Systems and Computing, vol 605. Cham: Springer.

Tadić, D., 2022. Phases of quality development: Concluding with the concept of quality 5.0. Tehnika. doi:10.5937/tehnika2205643t.

Tague, N.R., 2005. Plan–do–study–act cycle. In: *The quality toolbox*, 2nd ed. Milwaukee: ASQ Quality Press.

Taj, S., 2008. Lean manufacturing performance in China: assessment of 65 manufacturing plants. *Journal of Manufacturing Technology Management*, 19(2), pp.217–234.

Tozawa, B., 1995. *The improvement engine: creativity & innovation through employee involvement: the Kaizen teian system*. Japan Human Relations Association, Productivity Press.

Tu, Q., Vonderembse, M.A., Ragu-Nathan, T.S. and Sharkey, T.W., 2006. Absorptive capacity: enhancing the assimilation of time-based manufacturing practices. *Journal of Operations Management*, 24(5), pp.692–710.

Upadhyay, A., Mukhuty, S., Kumar, V. and Kazancoglu, Y., 2021. Blockchain technology and the circular economy: Implications for sustainability and social responsibility. *Journal of Cleaner Production*, 293, p.126130.

Vela-Valido, J., 2021. *Translation quality management in the AI age: New technologies to perform translation quality management operations*. Revista Tradumàtica.

Wada, K., 2004. *Kiichiro Toyoda and the birth of the Japanese automobile industry: Reconsideration of the Toyoda-Platts agreement*. Working paper CIRJE-F-288. University of Tokyo.

Wamba-Taguimdje, S., Wamba, S., Kamdjoug, J. and Wanko, C., 2020. Influence of artificial intelligence (AI) on firm performance: the business value of AI-based transformation projects. *Business Process Management Journal*, 26, pp.1893–1924. doi:10.1108/bpmj-10-2019-0411.

Ward, P. and Zhou, H., 2006. Impact of information technology integration and lean/just-in-time practices on lead-time performance. *Decision Sciences*, 37(2), pp.177–203.

Warnecke, H.J. and Huser, M., 1995. Lean production. *International Journal of Production Economics*, 41(1–3), pp.37–43.

WBCSD. 2010. Vision 2050: The new agenda for business. Retrieved from www.wbcsd.org (accessed 24 August 2024).

Wei, R. and Pardo, C., 2022. Artificial intelligence and SMEs: How can B2B SMEs leverage AI platforms to integrate AI technologies?. *Industrial Marketing Management*, 107, pp.466–483.

Womack, J., Jones, D. and Roos, D., 1990. *The machine that changed the world*. New York: Rawson Associates.

Wonigeit, J., 1994. *Total quality management: Basics and efficiency analysis*. Wiesbaden: Deutscher Universitätsverlag.

World Economic Forum, 2016. *World Economic Forum annual meeting 2016: Mastering the fourth industrial revolution*. World Economic Forum.

Xue, Y., Zhang, R., Deng, Y., Chen, K. and Jiang, T., 2017. A preliminary examination of the diagnostic value of deep learning in hip osteoarthritis. *PLoS ONE*, 12. doi:10.1371/journal.pone.0178992.

Yang, M.G., Hong, P. and Modi, S.B., 2010. Impact of lean manufacturing and environmental management on business performance: An empirical study of manufacturing firms. *International Journal of Production Economics*, 129(2), pp.251–261.

Yuan, X., Pan, Y., Yang, J., Wang, W. and Huang, Z., 2021. Study on the application of reinforcement learning in the operation optimization of HVAC system. *Building Simulation*, 14, pp.75–87. doi:10.1007/s12273-020-0602-9.

Zeng, N., Ye, X., Liu, Y. and König, M., 2024. BIM-enabled Kanban system in construction logistics for real-time demand reporting and pull replenishment. *Engineering, Construction and Architectural Management*, 31(8), pp.3069–3096.

Zhang, K., Li, Y., Qi, Y. and Shao, S., 2021. Can green credit policy improve environmental quality? Evidence from China. *Journal of Environmental Management*, 298, p.113445.

Zirar, A., Ali, S.I. and Islam, N., 2023. Worker and workplace artificial intelligence (AI) coexistence: Emerging themes and research agenda. *Technovation*, 124, p.102747.

Index